Organic S

Organic Spectroscopy

William Kemp

Senior Lecturer in Organic Chemistry
Heriot-Watt University, Edinburgh

THIRD EDITION

MACMILLAN

First edition 1975
Reprinted six times
Second edition 1987
Reprinted once
Third edition 1991

Published by
MACMILLAN PRESS LTD
Houndmills, Basingstoke, Hampshire RG21 6XS
and London
Companies and representatives
throughout the world

ISBN 0–333–51953–1 hardcover
ISBN 0–333–51954–X paperback

A catalogue record for this book is available
from the British Library.

12   11   10   9
03   02   01   00

Printed in China

To Hannah, Lorna and Lauren

Contents

Preface
to the First Edition

This book is an introduction to the application of spectroscopic techniques in organic chemistry. As an introduction it presupposes very little foreknowledge in the reader and begins at a level suitable for the early student. Each chapter is largely self-contained, beginning with a basic presentation of the technique and developing later to a more rigorous treatment. A supplement to each of the principal chapters covers recent and recondite areas of the main fields, so that the book will also serve to refresh and update the postgraduate student's knowledge. Sufficient correlation data are given to satisfy the average industrial or academic user of organic spectroscopic techniques, and these tables and charts constitute a useful reference source for such material.

SI units are used throughout, including such temporarily unfamiliar expressions as *relative atomic mass* and *relative molecular mass*. A major break with the British conventions in organic nomenclature has also been made in favor of the American system (thus, 1-butanol rather than butan-1-ol). This step is taken both in recognition of the vast amount of chemical literature that follows American rules (including the UKCIS computer printouts) and in the expectation that these conventions will in due course be adopted for use by more and more British journals and books.

Chapter 1 takes a perspective look at the electromagnetic spectrum, and introduces the unifying relationship between energy and the main absorption techniques.

The next four chapters deal with the four mainstream spectroscopic methods—methods which together have completely altered the face of organic chemistry in little over a decade and a half. Most students will use infrared spectroscopy first (chapter 2), and nuclear magnetic resonance spectroscopy will follow (chapter 3). Electronic spectroscopy is more limited in scope (Chapter 4), and mass spectroscopy (chapter 5) is the most recent and, in general, the most expensive. Chapter 6 provides both worked and problem examples in the application of these techniques, both singly and conjointly.

The emphasis throughout has been unashamedly 'organic', but interpretative theory has been included even where controversy exists: the theory of nuclear magnetic resonance is particularly satisfying and logical when treated semiempirically, but infrared theory is often conflicting in its predictions, mass spectroscopy theory is often speculative and electronic theory can be very mathematical. These strengths and weaknesses are emphasized throughout.

Students of chemistry, biochemistry or pharmacy at university or college will hopefully find the book easy to read and understand: the examples chosen for illustration are all simple organic compounds, and chapter 6 includes problems at an equally introductory level (so that students can *succeed* in problem-solving!). It is likely that the chapter supplements will be studied by students at honours chemistry degree or postgraduate level, and on the whole this is reflected in degree of complexity.

The author is reluctant to admit it, but he graduated at a time when no spectroscopic technique (other than X-ray crystallography) was taught in the undergraduate curriculum. He hopes that his own need to learn has given him a sympathetic insight into those dark areas that students find difficult to understand, and that the treatment accorded them in this book reflects their travail. His own colleagues have been of immense support, and did not laugh when he sat down to play. He thanks them all for it.

Heriot-Watt University, Edinburgh, January 1975   WILLIAM KEMP

Preface
to the Second Edition

When first produced, this book tapped into a floodstream of progress in the application of spectroscopy to organic chemistry, and the unabated flow of developments has made a second edition necessary and timely.

The developments have not been entirely spectroscopic *per se*, but have been associated with the considerable reduction in the cost of computers, so that new spectroscopic information can be elicited, and the data then manipulated in new ways, too. This is equally true in the parallel working of spectroscopy with chromatography, and a new section is devoted to outlining this successful marriage.

A major area of expansion is in the coverage of nuclear magnetic resonance, where carbon-13 NMR is now given equal prominence with proton NMR, although the theory is developed around a protocentric *Weltanschauung*. New tables of data, new worked examples and more problem examples (with answers) give a student-oriented coverage of all interpretative applications of NMR.

Pulsed Fourier Transform methods make it possible to observe NMR from even the most unfavorable of magnetic nuclei, and multinuclear spectrometers are now less expensive and esoteric than before—a brief look at nitrogen-15 and oxygen-17 NMR is included in deference to their importance to organic chemists. Time and money are still needed to record NMR from these nuclei, but the interpretation of the spectra is no more difficult than for carbon-13 (and usually much simpler than for the proton); the book emphasizes these simplicities.

Infrared spectroscopy has undergone a renaissance with the advent of extremely sensitive Fourier Transform instruments, but although the spectra can be obtained from small (and very unusual) samples, the structural application of the method is not much different from before.

New techniques and devices have appeared in ultraviolet spectroscopy and in mass spectroscopy and these are introduced as part of the updating of the text.

Two changes in units have been agreed internationally, and in accordance with new recommendations chemical shift in NMR is now quoted as, for example, δ 7.3 (and not 7.3 δ), and mass-to-charge ratios in MS are quoted in units of m/z (and not m/e).

Throughout these alterations and expansions the character of the book has survived, particularly in the use of simple examples to illustrate sophisticated science. Hopefully they will enhance its usefulness not only for reference but also in learning.

Heriot-Watt University, Edinburgh, 1986   W. K.

Preface
to the Third Edition

Since the publication of the second edition, the rate of change in these various fields of spectroscopy has maintained its pace. Some of the developments have been in instrumentation rather than in the exploitation of new spectroscopic phenomena, but this has led to the ready availability of spectra which were regarded, only a few years ago, as in the exotic class. As a consequence of these advances, we are witnessing changes in the emphases which the organic spectroscopist places on particular techniques: his time will be spent more with the carbon-13 and proton NMR spectra than with the infrared and ultraviolet.

The publication of a third edition has been centered around three main themes. The first change acknowledges a need to minimize discussion of obsolete instruments or techniques, and many more details of spectrometer operation have been added; Fourier Transforms and computers are no longer optional extras in the spectroscopy laboratory.

The second change is in the introduction of almost one hundred new student exercises throughout the book, in the form of both worked 'examples' (showing the working of a model problem, with a model answer to the question) and problem 'exercises' for the student to practice, having seen the method demonstrated in the model; answers to all of these self-assessment exercises are given at the back of the book. Several more difficult problems have also been added.

The third and major change is to the chapter on nuclear magnetic resonance, which has been considerably extended to take cognizance of its position as the preeminent method for structural determination in organic chemistry. This has been done mainly through the use of the Supplements, so that the beginning student can still come to grips with the simpler ideas of NMR and thereafter, at his or her own developing pace, tackle the conceptual complexities of rotating frames, pulse angles, and the like.

For students coming fresh to spectroscopy, it is difficult to anticipate how each of the methods can help in deducing the structure of an organic compound, and so an extended introduction in chapter 1 sets out a comparison among them; even then, much of this will need to be reread *post hoc* before any perspective can be gained. Progress in the development of one technique may also downvalue a particular strength of another; a clear example of this is in the way that NMR has stolen IR thunder in the analysis of substitution patterns in benzene rings, identification of alkyl groups (methyl, ethyl, isopropyl, *tert*-butyl), differentiation of aldehydes from ketones from esters, and so on.

In infrared spectroscopy, relatively cheap Fourier Transform infrared (FTIR) spectrometers have become more readily available and it is quite probable that all new instruments designed by the major manufacturers will be based on FTIR, although the dispersive instruments at present in use throughout the world will expect to live on for a time yet. Because of this important imminent change, the section on infrared instrumentation has been completely rewritten, with FTIR spectroscopy brought out from the Supplement to its proper place in equal prominence with dispersive instrumentation. In addition to the principal advantages of FTIR instruments (speed and sensitivity), the spectral data are digitized, allowing many manipulations such as spectral subtraction: an example of this has been included. To minimize the chore of having to skip back and forth from text to spectrum, most of the infrared spectra are now annotated with assignments for the bands.

Only one manufacturer is still producing a low-cost continuous wave nuclear magnetic resonance spectrometer, all other instruments on the market being FTNMR machines. Superconducting magnets are now able to reach 14.1 T, corresponding to 600 MHz in proton frequency, and it is projected that 700 MHz will be achievable—with a stable magnet—in a few years' time. Unlike infrared spectra, where the spectrum will often look the same whether it has been recorded on a grating or on an FTIR instrument, proton NMR spectra from CW instruments exhibit 'ringing' and therefore do not look the same as those from FTNMR machines (even if the field strength is unchanged). The older literature, and most of the spectra catalogs, contain only CW spectra, whereas new spectra are virtually always from the FTNMR mode, and one consequence of this for the student is the necessity to recognize these differences: the interpretation of a spectrum from either mode (all other things being equal) poses the same challenges. Thus, in this third edition, a few of the simple first-order proton spectra shown in earlier editions at 60 MHz CW have been retained, but most of the CW spectra have been replaced by FTNMR spectra, from 80 MHz up to 600 MHz: the additional abilities of FT instruments to record DEPT spectra and 2-D spectra are also exemplified.

It has been decided to follow IUPAC recommendations and to eliminate almost completely the use of the terms 'high field', 'low field', 'upfield' and 'downfield' from the book. The fact that almost all new NMR instruments are FTNMR machines (in which the field strength is constant) means that it is in the interests of new students to refer exclusively to relative chemical shift positions in the context of 'lower frequency' and 'higher frequency', respectively, even though these terms may initially be unfamiliar to former users. Discussion of techniques with outdated usefulness has been eliminated (spin tickling) or severely curtailed (INDOR).

There have been few changes in the science of ultraviolet and visible spectroscopy since the second edition, and in mass spectrometry the major change of interest to the compass of this book has been the ubiquitous availability of computerized data handling. This has been reflected in the discussion of the subject: the importance of library searching as a means of compound identification has also been given increased prominence. An additional section deals with laser ionization, and its importance, particularly in surface analysis.

The development of separation science (mainly chromatography) has continued steadily in parallel with the development of spectroscopy, but especially dynamic growth is taking place in those joint techniques in which the separation of mixtures is coupled with spectroscopic analysis of the separated constituents, as in supercritical fluid chromatography–mass spectrometry (SFC–MS) or gas chromatography–Fourier Transform infrared spectroscopy (GC–FTIR). The discussion of these so-called *hyphenated techniques* has been extended in recognition of their importance and novelty, and their derived acronyms have been included in Appendix 2, as a guide through the maze.

As must ever be the case, thanks are due to the many people who have helped in this revision by supplying information or argument, but especial thanks must go to my colleague Dr Alan Boyd for the amount of work involved in rerunning so many of the NMR spectra (and in carrying out the musical Fourier Transforms in chapter 1).

Heriot-Watt University, Edinburgh, 1990   W.K.

Acknowledgments

The author wishes to place on record his grateful thanks to the many people who supplied material, information, spectra and a share of their valuable time; comments critical and encouraging were received from many colleagues, and changes in this edition reflect these indications.

The book, in its third edition, is being jointly published for the first time by W. H. Freeman, New York, and much advice has been accepted from chemists in the USA to try and meet the needs of students there. Some early help came from Dr Donna Wetzel, Rohm and Haas, Bristol, Pennsylvania; Dr Daniel F. Church, Louisiana State University; Dr William Closson, State University of New York; Dr John Gratzner, Purdue University; and Dr Neil Schore, University of California. The entire manuscript was read by Prof. George B. Clemans, of Bowling Green State University, and Prof. Harold M. Bell, of Virginia Tech.: their many suggestions for improvement have been incorporated wherever possible. Gary Carlson, of W. H. Freeman, was responsible for inviting the whole project to the USA, and for arranging its review by faculty on that side of the Atlantic.

CHAPTER 1. The portraits of Isaac Newton and Joseph Fourier at the chapter head are reproduced with the permission, respectively, of The Royal Society of Chemistry, London, England, and Librairie Larousse, Paris, France. The music featured in figure 1.6 had Ailsa Boyd on clarinet, Iona Boyd on violin and the author on bagpipes: the recording and subsequent Fourier Transformation were by Dr Alan Boyd, of Heriot-Watt University.

CHAPTER 2. The infrared spectra reproduced in this chapter were recorded on a Perkin-Elmer Model 700 infrared spectrophotometer, with the exceptions of figures 2.10 and 2.11, which were recorded, respectively, on a Perkin-Elmer Model 1720-X FTIR spectrometer (by Dr Iain McEwan, of Heriot-Watt University) and a Pye-Unicam Model SP3-100 instrument. The photographs at the chapter head were supplied by Perkin-Elmer, and show examples of the Models 1600 and 1700 series. The correlation charts on pages 60–71 are reproduced with permission from *Qualitative Organic Analysis*, by W. Kemp, McGraw-Hill, Maidenhead (2nd edn, 1986).

CHAPTER 3. All of the NMR spectra and all of the spectra simulations reproduced in this chapter were recorded by Dr Alan Boyd, with the exceptions of the following. Figure 3.4 is reproduced from *High Resolution NMR Spectra Catalog*, with the permission of the publishers, Varian Associates, Palo Alto, California. The 360 MHz and 600 MHz proton NMR spectra for menthol in figure 3.30 were recorded by Dr Ian Sadler, of Edinburgh University. Figures 3.32 and 3.40 are reproduced from *NMR Quarterly*, with permission of Perkin-Elmer, the publishers. The photographs at the chapter head were kindly supplied by Japanese Electronic and Optical Laboratories, JEOL UK, London. Figures 3.36, 3.37, 3.38, 3.41, 3.42, 3.43, 3.44 and 3.45 are reproduced from *NMR in Chemistry: A Multinuclear Introduction*, by W. Kemp, Macmillan, London (1986), with permission. The MRI brain scan (Figure 3.49) was furnished by Bruker Spectrospin, Karlsruhe, Germany.

CHAPTER 4. The photographs at the chapter head were supplied by Perkin-Elmer, Beaconsfield, England. The chromascan in figure 4.3 is reproduced with permission of Pye-Unicam, Cambridge. Tables 4.5 and 4.6 are reproduced with permission from *Qualitative Organic Analysis*, by W. Kemp, McGraw-Hill, Maidenhead (2nd edn, 1986).

CHAPTER 5. Figure 5.5 is reproduced with permission from Beynon, J. H., *Mass Spectrometry and its Application to Organic Chemistry*, Elsevier, Amsterdam (1960). The photographs at the chapter head were supplied by VG Analytical Ltd, Manchester, England, and figure 5.10 is reproduced with permission from Bruker Spectrospin, Coventry.

CHAPTER 6. The proton NMR spectra reproduced in this chapter are from *High Resolution NMR Spectra Catalog*, with permission of the publishers, Varian Associates, Palo Alto, California (except figure 6.9(a), recorded on a Bruker WM200 spectrometer). The infrared spectra were recorded on a Perkin-Elmer Model 700 spectrometer, except figure 6.11 (recorded on a Pye-Unicam Model SP3-100 instrument).

Energy and the Electromagnetic Spectrum

Isaac Newton
1642–1720

'In the year 1666 I procured me a triangular glass prism, to try therewith the celebrated phenomena of colours.' His classic experiments constitute the first scientific study of spectroscopy.

Jean-Baptiste Joseph Fourier
1768–1830

His mathematical analysis of periodic systems led to Fourier Transforms. Although he lived through the French Revolution, he was not able to witness the spectroscopic revolution associated with FT.

When the sun's rays are scattered by raindrops to produce a rainbow, or in a triangular glass prism (as in the famous early experiments of Sir Isaac Newton in 1666), the white light is separated into its constituent parts—the visible spectrum of primary colors. This rainbow spectrum is a minute part of a much larger continuum, called the *electromagnetic spectrum*: why 'electromagnetic'?

Visible light is a form of energy, which can be described by two complementary theories: the wave theory and the corpuscular theory. Neither of these theories alone can completely account for all the properties of light: some properties are best explained by the wave theory, and others by the corpuscular theory. The wave theory most concerns us here, and we shall see that the propagation of light by light waves involves both electric and magnetic forces, which give rise to their common class name *electromagnetic radiation*.

1.1 UNITS

Units are best named following the stipulations of the SI system (Système Internationale d'Unités). These are:

wavelength in meters, m
frequency in reciprocal seconds, s^{-1}, or hertz, Hz ($1\ s^{-1} = 1$ Hz)
wavenumber in reciprocal meters, m^{-1}
energy in joules, J

Multiplying prefixes are used as convenient—for example, $1000\ m = 1$ km, $10^{-6}\ m = 1$ micrometer or $1\ \mu m$, $10^{-9}\ m = 1$ nanometer or 1 nm, $1\ 000\ 000$ Hz $= 1$ megahertz or 1 MHz, etc.

We can represent a light wave travelling through space by a sinusoidal trace as in figure 1.1. In this diagram λ is the *wavelength* of the light; different colors of light have different values for their wavelengths, so that, for example, red light has wavelength $\approx$ 800 nm, while violet light has wavelength $\approx$ 400 nm.

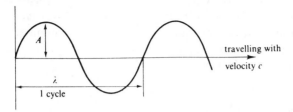

Figure 1.1 *Wave-like propagation of light (λ = wavelength, A = amplitude, c = 2.998 $\times$ 10^8 m s^{-1}, ca. 3 $\times$ 10^8 m s^{-1}).*

If we know the wavelength λ, we can calculate the inverse of this, $1/\lambda$, which is the number of waves per unit of length. This is most frequently used as the number of waves per cm, and is called the *wavenumber* $\bar{\nu}$, in reciprocal centimeters (cm^{-1}).

Also, provided that we know the *velocity* with which light travels through space ($c = 2.998 \times 10^8$ m s^{-1}), we can calculate the number of waves per second as the *frequency* of the light, $\nu = c/\lambda$ (s^{-1}).

In summary, we can describe light of any given 'color' by quoting either its wavelength, λ, or its wavenumber, $\bar{\nu}$, or its frequency, ν.

The following are the relationships among the four quantities wavelength, wavenumber, frequency and velocity:

Quantity	Relationship	Units
wavelength	$\lambda = \dfrac{1}{\bar{\nu}} = \dfrac{c}{\nu}$	m, μm, nm
wavenumber	$\bar{\nu} = \dfrac{1}{\lambda} = \dfrac{\nu}{c}$	m^{-1}, cm^{-1}
frequency	$\nu = \dfrac{c}{\lambda} = c\bar{\nu}$	s^{-1} (Hz)
velocity	$c = \nu\lambda = \dfrac{\nu}{\bar{\nu}}$	m s^{-1}

Example 1.1

Question. Calculate the frequency and wavenumber of infrared light of wavelength $\lambda = 10$ μm (micrometers, formerly called microns) = 10×10^{-6} m or 1.0×10^{-5} m.

Model answer. Since frequency $\nu = c/\lambda$, then $\nu = (3 \times 10^8 \text{ m s}^{-1})/(1.0 \times 10^{-5} \text{ m}) = 3 \times 10^{13} \text{ s}^{-1}$ or (as frequency in s^{-1} is usually quoted in hertz, Hz) 3×10^{13} Hz. Wavenumber is simply the reciprocal of wavelength, so if $\lambda = 1.0 \times 10^{-5} \text{ m}$, then $\bar{\nu} = 1.0/(1.0 \times 10^{-5} \text{ m}) = 1.0 \times 10^5 \text{ m}^{-1}$: it is conventional to express wavenumber in reciprocal centimeters rather than reciprocal meters, so this value corresponds to $1.0 \times 10^3 \text{ cm}^{-1}$, or 1000 cm^{-1}

Example 1.2
Question. A local radio station transmits (a) at approximately 95 MHz on its VHF (very high frequency) transmitter and (b) at 810 kHz on medium wave. Calculate the wavelengths of these transmissions.

Model answer. In each case, the broadcast frequency is given: $c = 3 \times 10^8 \text{ m s}^{-1}$, and $\lambda = c/\nu$. Thus, the wavelength corresponding to (a) 95 MHz, or $95 \times 10^6 \text{ s}^{-1}$, is found from $(3 \times 10^8 \text{ m s}^{-1})/(95 \times 10^6 \text{ s}^{-1}) = 3.158$ m. The wavelength for (b) is $(3 \times 10^8 \text{ m s}^{-1})/(810 \times 10^3 \text{ Hz}) = 370$ m.

Unfortunately, users of the different spectroscopic techniques we shall meet in this book do not all use the same units, although it would be possible for them to do so. In some techniques the common unit is wavelength, in other techniques most workers use wavenumber, while in others we find that frequency is the unit of choice. This is merely a question of custom and usage, but it makes comparison among the techniques a little less clear than it might be.

Exercise 1.1   Calculate the frequencies and wavelengths of infrared light of wavenumber (a) 2200 cm^{-1} and (b) 3000 cm^{-1}.
    Express frequency in s^{-1}, Hz, and wavelength in micrometers, μm.

Exercise 1.2   Calculate the wavelengths of the following useful radio broadcast frequencies: (a) CB radio at 27 MHz; (b) International Distress broadcasts at 2182 kHz long wave and 156.8 MHz (Channel 16 on VHF); (c) Emergency Position-Indicating Radio Beacons (EPIRBs) at 121.5 MHz for civil aircraft and 243 MHz for military aircraft.

1.2   THE ELECTROMAGNETIC SPECTRUM

The sensitivity limits of the human eye extend from violet light ($\lambda = 400$ nm, 4×10^{-7} m) through the rainbow colours to red light ($\lambda = 800$ nm, 8×10^{-7} m). Wavelengths shorter than 400 nm and longer than 800 nm exist, but they cannot be detected by the human eye. *Ultraviolet* light ($\lambda < 400$ nm) can be detected on photographic film or in a photoelectric cell, and *infrared* light ($\lambda > 800$ nm) can be detected either photographically or using a heat detector such as a thermopile.

Beyond these limits lies a continuum of radiation, which is shown in figure 1.2. Although all of the different divisions have certain properties in common (all possess units of λ, ν, $\bar{\nu}$, etc.), they are sufficiently different to require different handling techniques. Thus, visible light (together with ultraviolet and infrared) can be transmitted in the form of 'beams', which can be bent by reflection or by diffraction in a prism; X-rays can pass through glass and muscle tissue and can be deflected by collision with nuclei; microwaves are similar to visible light, and are conducted through tubes or 'waveguides'; radiowaves can travel easily through air, but can also be conducted along a metal wire; alternating current travels only with difficulty through air, but easily along a metal conductor.

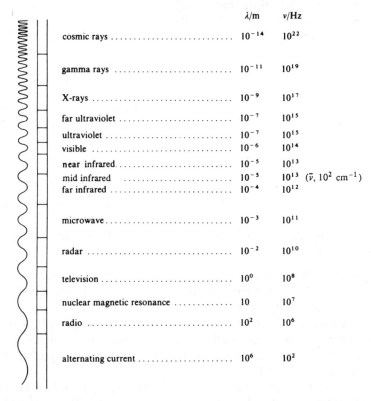

	λ/m	ν/Hz
cosmic rays	10^{-14}	10^{22}
gamma rays	10^{-11}	10^{19}
X-rays	10^{-9}	10^{17}
far ultraviolet	10^{-7}	10^{15}
ultraviolet	10^{-7}	10^{15}
visible	10^{-6}	10^{14}
near infrared	10^{-5}	10^{13}
mid infrared	10^{-5}	10^{13} ($\bar{\nu}$, 10^2 cm^{-1})
far infrared	10^{-4}	10^{12}
microwave	10^{-3}	10^{11}
radar	10^{-2}	10^{10}
television	10^{0}	10^{8}
nuclear magnetic resonance	10	10^{7}
radio	10^{2}	10^{6}
alternating current	10^{6}	10^{2}

Figure 1.2 *The electromagnetic spectrum, with wavelengths λ and frequency ν shown.*

Alternating current is familiar as a wave-like electrical phenomenon, setting up fluctuating electrical fields as it travels through space or along a conductor. Associated with these *electric vectors* (and at 90° to them) are *magnetic vectors*. This relates to the simple experiment of placing a compass needle near a current-carrying conductor; the fluctuating electrical forces generate magnetic forces which deflect the compass needle. The relationship between these two quite different forms of energy is shown in figure 1.3. Alternating current is an electromagnetic phenomenon; all other parts of the spectrum in figure 1.2 possess electric and magnetic vectors and the name *electromagnetic radiation* is given to all the energy forms of this genre.

The energy associated with regions of the electromagnetic spectrum is related to wavelengths and frequency by the equations

$$E = h\nu = hc/\lambda$$

where E = energy of the radiation in joules/J,
    h = Planck's constant/6.626×10^{-34} J s,
    ν = frequency of the radiation/Hz,
    c = velocity of light/2.998×10^8 m s^{-1},
    λ = wavelength/m.

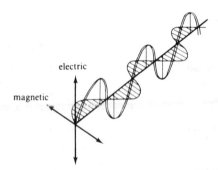

Figure 1.3   *Propagation of alternating electric forces and the related magnetic fields.*

The higher the frequency, the higher the energy; the longer the wavelength, the lower the energy. Cosmic radiation is of very high energy; ultraviolet light is of higher energy than infrared light; etc.

Most references to energy will be expressed in joules (J), but the electron volt will be mentioned in mass spectrometry: 1 electron volt, 1 eV $\approx 1.6022 \times 10^{-19}$ J, so that ultraviolet light of wavelength 100 nm has an energy of about 12 eV.

To express energy in terms of J mol^{-1}, the expressions $E = h\nu$, etc., must be multiplied by the Avogadro constant, N_A (= 6.02×10^{23} mol^{-1}). A numerical example will help to make this clear.

Example 1.3
Question. For ultraviolet light of wavelength 200 nm, calculate (a) the frequency of this light, (b) the amount of energy absorbed by one molecule

when it interacts with this light, and (c) the corresponding amount of energy absorbed by one mole of substance.

Model answer. (a) Frequency is given by $c/\lambda = (3 \times 10^8 \text{ m s}^{-1})/(200 \times 10^{-9} \text{ m}) = 1.5 \times 10^{15} \text{ s}^{-1}$ (Hz). (b) The energy associated with this is given by $E = h\nu = (6.6 \times 10^{-34} \text{ J s}) \times (1.5 \times 10^{-15} \text{ s}^{-1}) = 9.9 \times 10^{-19} \text{ J}$; this would be the energy absorbed by one molecule interacting with the ultraviolet light. (c) To find the amount of energy absorbed by one mole of substance, we must multiply by the Avogadro constant N_A ($= 6.02 \times 10^{23} \text{ mol}^{-1}$), giving approximately 6×10^5 J mol^{-1} or 600 kJ mol^{-1}.

Exercise 1.3   Typical bond dissociation energies in organic molecules are around 400 kJ mol^{-1}. Calculate (a) the frequency and (b) the wavelength of the electromagnetic radiation which corresponds to this dissociation energy. In which part of the spectrum does this radiation lie?

Exercise 1.4   Gamma-irradiation of food uses cobalt-60 as a source of electromagnetic radiation with frequency around 10^{19} Hz. Calculate the energy of this radiation (a) in joules and then (b) in kJ mol^{-1}.

Exercise 1.5   Routine mass spectra of organic compounds are recorded by bombardment of the molecule with an electron beam of energy 70 eV. Calculate the corresponding energy (a) in joules and then (b) in kJ mol^{-1}.

1.3   ABSORPTION OF ELECTROMAGNETIC RADIATION BY ORGANIC MOLECULES

If we pass light from an ultraviolet lamp through a sample of an organic molecule such as benzene, some of the light is absorbed. In particular, some of the wavelengths (frequencies) are absorbed and others are virtually unaffected.

We can plot the changes in absorption against wavelength as in figure 1.4 and produce an *absorption spectrum*. The spectrum presented in figure 1.4 shows *absorption bands* at several wavelengths—for example, 255 nm.

The organic molecule is absorbing light of $\lambda = 255$ nm, which corresponds to energy absorption of 470 kJ mol^{-1}. Energy of this magnitude is associated with changes in the electronic structure of the molecule, and when a molecule absorbs this wavelength, electrons are promoted to higher-energy orbitals, as represented in figure 1.5. The energy transition $E_1 \rightarrow E_2$ corresponds to the absorption of energy *exactly* equivalent to the energy of the wavelength absorbed:

$$\Delta E = (E_2 - E_1) = hc/\lambda = h\nu$$

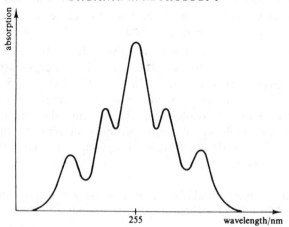

Figure 1.4 *An absorption spectrum.*

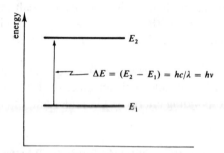

Figure 1.5 *Energy transition for the absorption of light or other electromagnetic radiation.*

While this example refers specifically to ultraviolet light, the same principle holds for the absorption of energy from any part of the electro-magnetic spectrum.

A molecule can only absorb a particular frequency, if there exists within the molecule an energy transition of magnitude $\Delta E = h\nu$.

Although almost all parts of the electromagnetic spectrum are used for studying matter, in organic chemistry we are mainly concerned with energy absorption from three or four regions—ultraviolet and visible, infrared, microwave and radiofrequency absorption.

Table 1.1 shows the kind of information that can be deduced from studying the absorption of these radiations.

The last of the spectroscopic techniques summarized in table 1.1 is different from the others. In *mass spectrometry* we bombard the molecule with high-energy electrons (≈ 70 eV, or 6000 kJ mol^{-1}), and cause the molecule first to ionize and then to disperse into an array (or spectrum) of fragment ions of different masses. This *mass spectrum* presents us with a jigsaw pattern of fragments from which we have to reconstruct a picture of the whole molecule.

Table 1.1 Summary of spectroscopic techniques in organic chemistry and the information obtainable from each

Radiation absorbed	Effect on the molecule (and information deduced)
ultraviolet–visible λ, 190–400 nm and 400–800 nm	changes in electronic energy levels within the molecule (extent of π-electron systems, presence of conjugated unsaturation, and conjugation with nonbonding electrons)
infrared (mid infrared) λ, 2.5–25 μm $\bar{\nu}$, 400–4000 cm^{-1}	changes in the vibrational and rotational movements of the molecule (detection of functional groups, which have specific vibration frequencies—for example, $C=O$, NH_2, OH, etc.)
microwave ν, 9.5×10^9 Hz	electron spin resonance or electron paramagnetic resonance; induces changes in the magnetic properties of unpaired electrons (detection of free radicals and the interaction of the electron with, for example, nearby protons)
radiofrequency ν, 60–600 MHz	nuclear magnetic resonance; induces changes in the magnetic properties of certain atomic nuclei, notably that of hydrogen and the ^{13}C isotope of carbon (hydrogen and carbon atoms in different environments can be detected and counted, etc.)
electron-beam impact 70 eV, 6000 kJ mol^{-1}	ionization and fragmentation of the molecule into a spectrum of fragment ions (determination of relative molecular mass (molecular weight) and deduction of molecular structures from the fragments produced)

The spectrometers used to record the IR, NMR or UV/VIS spectra vary enormously, and yet have certain essential features in common:

1. a *source* of radiation of the appropriate frequency range (a UV or IR lamp, a radiofrequency transmitter, etc.);
2. a *sample holder* to permit efficient irradiation of the sample;
3. a *frequency analyzer* which separates out all of the individual frequencies generated by the source (the most familiar being the triangular glass prism as used by Isaac Newton for visible light);

4. a *detector* for measuring the intensity of radiation at each frequency, allowing the measurement of how much energy has been absorbed at each of these frequencies by the sample; and

5. a *recorder*—either a pen recorder or a computerized data station, with a VDU for initial viewing of the spectrum, with the possibility of manipulation in scale, etc.

Mass spectrometers do not measure the interaction of molecules with electromagnetic radiation, but apart from having a mass analyzer in place of a frequency analyzer, they have all of the other spectrometer features listed above.

No one spectroscopic technique supplies enough information to deduce the structure of an organic molecule of any complexity, but, *depending on the molecule*, some methods are more helpful and amenable to analysis than others; part of the enjoyable challenge of organic spectroscopic problems is the interplay among the various spectra, and the iterative process that leads to a solution. The costs of instruments vary enormously, as does the ease of operation, even between a simple teaching instrument in one branch and a state-of-the-art research spectrometer in the same branch.

Table 1.2 is an approximate league table, full of challengeable generalities, setting out a comparison among the main spectroscopic methods; the award of three stars is good; one star is not.

Table 1.2 Some comparisons among the principal spectroscopic methods: good features score three stars

Feature \ Method	^{13}C NMR	^{1}H NMR	IR	MS	UV/VIS
identification of functional groups	**	**	***	**	*
measurement of molecular complexity	**	**	*	***	*
sensitivity (sample size needed)	*	*	***	***	***
quantitative information	*	**	**	*	***
interpretability of all of the data	***	***	**	*	*
theory needed to interpret spectra	*	*	***	**	**
ease of instrument operation	**	**	***	*	***
instrument cost, running costs	**	**	***	*	***

SUPPLEMENT 1

1S.1 SPECTROSCOPY AND COMPUTERS

An absorption spectrum such as that in figure 1.4 may be recorded directly on a pen recorder; if the same information is digitized and stored in a microcomputer interfaced to the spectrometer instrument, then many additional manipulations of the data become possible, provided that the appropriate hardware and software are available.

Expansions and contractions of the spectrum, on either the ordinate or the abscissa, can be selected on a video display unit before printing out the hard copy.

Difference spectra. The spectra of mixtures (including the common circumstance of compounds in solution, or containing impurities) can be simplified by computer subtraction of one spectrum from another (for example, removing the spectrum of the solvent or impurity). See figure 2.10.

Spectra summation. Weak or noisy spectra can be improved by summing several spectra together. In these summations the transient signals of random noise average to a minimum, leaving the true spectrum peaks with enhanced intensity; the increase in intensity when n spectra are summed in this way is a factor of $n^{1/2}$, so that 100 summations increase the ratio of signal to noise by a factor of 10. The technique is also called *signal averaging* or *computer averaging of transients* (CAT).

Smoothing routines can further improve the appearance of spectra by removal of unwanted noise.

Baseline corrections can eliminate skew.

Integration of the areas under peaks can be used for quantitative studies.

Library searches can be carried out to compare the spectrum of an unknown compound with standard spectra held in the computer memory or disk store. This search may use the entire digitized spectrum for comparison, or else important salient features alone (such as the carrying out of a search for spectra with an absorption band at a specified frequency).

Graphics displays. Having stored several spectra in the computer, the information can be presented in several different graphics formats, including various styles of stack plot which make the visual comparison of several spectra more easy than on separate paper sheets. An example is shown in figure 4.3.

Derivative spectroscopy. It may be important in analyzing certain spectra to decide where minor inflections arise on the main curve of a spectral band; a common case is in the analysis of mixtures, where broad absorption bands may arise from the near-overlap of absorptions from the several constituents of the mixture. Minor inflections on a curve can be highlighted by calculating the first derivative of the slope (so that, instead of plotting *intensity* versus frequency, a plot of *rate of change of intensity* versus frequency is obtained). An example of the use of derivative curves is given in the discussion in section 3S.7 on electron spin resonance (ESR) spectroscopy. It is also possible to calculate and plot the second derivative of the curve, which further sharpens the peaks and effectively separates the overlapping peaks to a greater extent.

1S.2 FOURIER TRANSFORMS—FREQUENCY AND TIME

If the G string of a violin is bowed and a microphone used to transmit the signal to an oscilloscope, the complex (but regular) interference pattern shown on the screen signifies that in a violin string the pure G frequency is mixed with other frequencies (harmonics). A clarinet sounding the same note produces a different pattern of harmonics, and, hence, a different interference pattern on the oscilloscope. The Scottish great highland bagpipe is different again.

These instrument sounds can be compared in two ways (other than aurally). See Figure 1.6.

a. The oscilloscope shows the interference patterns (interferograms) as *changes in intensity* versus *time*; the interferograms are visually complex, because several frequencies are involved and these interact with one another to give beats.

b. The instruments can also be compared by plotting the frequencies emitted, showing the *relative intensity of each frequency*: this is much more easily comprehended and compared than the three interferograms.

Note that in the former presentation the plot is of intensity against time (in units of seconds), while the latter presentation is a plot of intensity against frequency (in units of reciprocal time, s^{-1}, or Hz); both contain the same information, and one form is the mathematical reciprocal of the other.

We refer to the former as being in the *time domain* and to the latter as being in the *frequency domain*.

There is a relationship between a complex interferogram and the set of sine and cosine frequencies which go to produce it. This relationship was studied and defined by Jean-Baptiste Joseph Fourier (1768–1830) and the conversion of one to the other is called a Fourier Transform (FT).

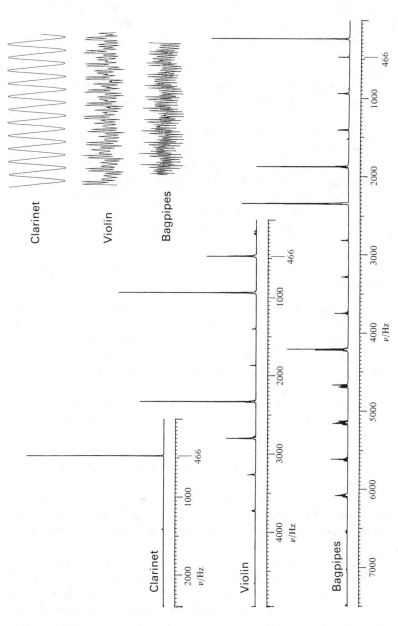

Figure 1.6 *The clarinet, the violin and the highland bagpipe playing the same note (B flat above middle C, 466 Hz); the interferograms (above) were subsequently Fourier Transformed to show the individual frequencies emitted (below). The clarinet note is almost pure, the violin shows several higher frequency overtones, and the bagpipe shows both lower- (from the bass drone) and higher-frequency harmonics.*

Several spectroscopic methods produce interferograms which are difficult to interpret unless they are first transformed (by FT) into a plot of individual frequencies. The computer program for executing the FT is complex: spectrometers provided with this facility first digitize the interferogram, perform the FT (in a few seconds) and then plot the absorption spectrum on a hard-copy printer or pen recorder.

Examples are given of the FT method in all four of the techniques discussed in this book (infrared, ultraviolet–visible, nuclear magnetic resonance and mass spectrometry). See sections 2.4.5, 3.3.3, 4.3.1 and 5S.5.

There are several advantages to be gained in using FT methods to record spectra, the most striking of which is the saving in time: an infrared spectrum can be recorded easily in one-hundredth of the time required by nonFT spectrometers, and indeed for routine use the limitation in sample throughput is the speed of the final printer/ plotter. In mass spectrometry speed is increased a thousandfold, and in carbon-13 NMR spectroscopy this factor is of the order of 5000.

1S.3 SPECTROSCOPY AND CHROMATOGRAPHY—HYPHENATED TECHNIQUES

Separation of mixtures by chromatographic processes is a central part of analytical and preparative chemistry. The methods include (a) liquid chromatography, LC (in columns, or on thin layers, TLC, or in high-performance liquid chromatography, HPLC); (b) gas chromatography, GC (in packed or capillary columns); and (c) the more specialized ion exchange chromatography, gel permeation chromatography, GPC, size exclusion chromatography (SEC) and supercritical fluid chromatography (SFC). The direct conjunction of these techniques with spectroscopic examination of the separated fractions constitutes several powerful analytical partnerships.

The quantities of materials separated out are inevitably small and the concentrations low, so this presents special problems to the spectroscopist, but notwithstanding the practical difficulties, it is possible to pair virtually all of the chromatographic methods with all of the spectroscopic methods; equipment cost is the only countering penalty.

With a dedicated microcomputer, spectrometer instruments are capable of performing elaborate procedures repetitively and at high speed, not only in the accumulation, manipulation and presentation of data, but also in the control of the spectroscopic experiment being performed. In chromatographic processes where components elute from the column over a period of a few seconds or less, the absorption spectrum of the solute can only be recorded in real time ('on-the-fly') if the spectrometer is fast, sensitive and capable of storing sufficient data in digital form to allow accurate reconstruction of the spectrum. Manufacturers of spectroscopic instruments have invested enormous effort in producing instruments capable of doing just this.

1S.3.1   Gas chromatography and spectroscopy
Gas chromatography can only be used to separate compounds with a substantial vapor pressure at the column temperature.

The concentration of a compound in the gas phase in GC is commonly of the order of a few nanograms per milliliter. The targeted component in the eluent from the chromatograph can be condensed out, but this is time-consuming and of low efficiency. (It is, however, the only practical method usable in coupling GC with NMR spectroscopy.)

The carrier gas can be removed by various diffusion devices which utilize the higher diffusion rates of gases, especially hydrogen and helium, compared with the higher molecular weight component being eluted. This is a technique commonly used to couple gas chromatography with mass spectrometry (GC–MS); see section 5S.2. Some designs of mass spectrometer do not require that the carrier gas be removed, and the gas chromatograph eluent is led directly into the mass spectrometer to give truly on-the-fly GC–MS; see section 5S.1.1.

In gas chromatography coupled to infrared spectroscopy (GC–IR) the eluent from the GC can be led through a light-pipe fitted with transparent windows at each end, so that the infrared beam can be directed from the source, along the light-pipe, and thence to the detector. This on-the-fly technique can record the IR spectrum of each eluted component, but it demands a fast-scanning and sensitive spectrometer to record an entire spectrum during the short period of elution of each peak; see section 2.4 on FTIR.

Gas chromatography coupled to ultraviolet spectroscopy (GC–UV) is possible but it is rarely used.

1S.3.2   Liquid chromatography and spectroscopy
Liquid chromatography can be used to separate volatile and nonvolatile solutes.

Large-scale column chromatography at atmospheric pressure gives eluent fractions which may be quite rich in solute, and consequently this poses no special problems in the search for good spectra. For high-performance liquid chromatography, HPLC (formerly called high-pressure liquid chromatography) small volumes and low concentrations arise.

Ultraviolet spectroscopy detection at fixed wavelengths has long been used as one of the mainstays of HPLC detectors, but with fast-scanning UV detectors using diode arrays, the entire UV spectrum of each eluting component can be measured on-the-fly; see section 4.3 and figure 4.3.

Such stack plots present the *change in absorption spectrum* versus *elution time*; they are often called *chromascans* or *spectrochromatograms*. Fluorescence spectra are complementary to the UV absorption spectra in these applications; see section 4S.2.

Infrared detection of eluting components can also be carried out at fixed wavelength (e.g. near 3000 cm^{-1}) to monitor the presence of compounds with C—H groups, but, as in UV spectroscopy, the entire infrared spectrum of each component can be recorded on-the-fly, using fast-scanning and sensitive Fourier Transform infrared instruments; see section 2.4 on FTIR.

To couple HPLC with mass spectrometry, it is possible to remove solvent manually, but direct interfacing can be done by passing the eluent from the column through a very narrow orifice (*ca* 10 μm) to generate a fine cloud of vapor (i.e. to 'nebulize' the eluent). In another variant the components condense out when the eluent impinges on a cooled band of metal. See also section 5S.2.

Because of its inherent low sensitivity, NMR cannot be coupled directly to HPLC except after total or partial solvent removal.

The components of a mixture separated by TLC are static (in contrast to the flow system of HPLC) and they can be removed from the plates for spectroscopic examination by any of the methods. It is also possible to record the infrared spectra directly by reflectance, using the high sensitivity of FTIR instruments; see sections 2.4 and 2S.2.

In supercritical fluid chromatography (SFC) a gas, commonly carbon dioxide or ammonia, is used as the mobile phase, but at a pressure above its critical pressure (hence *supercritical*). For certain polar compounds of high molecular weight the method gives superior separations to those given by HPLC, but this is not a generality. Enrichment of the eluent by selective removal of the gaseous mobile phase is often easier than in HPLC, but the development of the thermospray (see section 5S.2) for combined HPLC–MS has limited the acceptance of SFC–MS to specialized analyses.

FURTHER READING

Willard, H. H., Merritt, L. L., Dean, J. A. and Settle, F. A. Jr, *Instrumental Methods of Analysis*, Wadsworth, New York (7th edn, 1989).

Wayne, R. P., Fourier transformed, *Chemistry in Britain*, **23**, 440 (1987).

Mills, I. (Ed.), *Quantities, Units and Symbols in Physical Chemistry*, Blackwell, Oxford (1987).

Homann, K. H. (Ed.), *The Abbreviated List of Quantities, Units and Symbols in Physical Chemistry*, Blackwell, Oxford (1987); indispensable, inexpensive.

Banwell, C. N., *Fundamentals of Molecular Spectroscopy*, McGraw-Hill, New York (3rd edn, 1983).

Barrow, G. M., *Introduction to Molecular Spectroscopy*, McGraw-Hill, New York (1962).

(These last two books dwell more on physical interpretations and introduce the mathematics of spectroscopy.)

Davis, R. and Wells, C. H. J., *Spectral Problems in Organic Chemistry*, Blackie, Glasgow (1984).

SUPPLEMENTARY TEXTS

Carrick, A., *Computers and Instrumentation*, Heyden, London (1979).

Hollas, J. M., *Modern Spectroscopy*, Wiley, Chichester (1986).

MacRae, M. (Ed.), *Spectroscopy International*, Aster Publishing Corp., Eugene, Oregon. Bimonthly magazine for spectroscopists and analytical chemists, discussing all of the spectroscopic techniques in this book, and several others; free to bona fide practitioners in Europe (including the UK).

Instrument and accessory manufacturers publish information updates on their own products, and these useful documents are available free of charge.

Infrared Spectroscopy

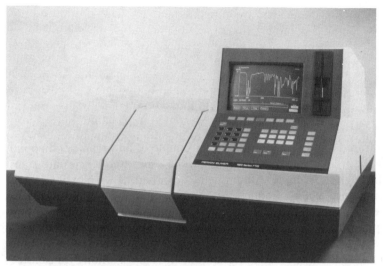

Low cost Fourier Transform infrared spectrometer.

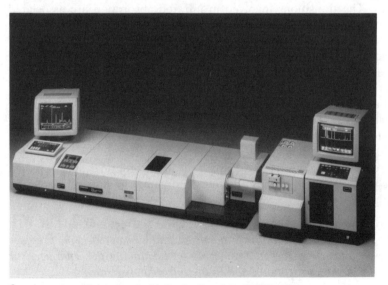

Gas chromatograph interfaced with Fourier Transform IR (GC–FTIR).

When infrared light is passed through a sample of an organic compound, some of the frequencies are absorbed, while other frequencies are transmitted through the sample without being absorbed. If we plot absorbance or transmittance against frequency, the result is an infrared spectrum.

Figure 2.1 is the infrared spectrum of a mixture of long-chain alkanes (liquid paraffin, or Nujol), showing that *absorption bands* appear in the regions around 3000 cm^{-1} and 1400 cm^{-1}; other frequencies do not interact with the sample and are consequently almost wholly transmitted.

The alkane molecules will only absorb infrared light of a particular frequency if there is an energy transition within the molecule such that $\Delta E = h\nu$. The transitions involved in infrared absorption are associated with *vibrational* changes within the molecule; for example, the band near 3000 cm^{-1} (that is, corresponding to 9.3×10^{13} Hz) has exactly the same frequency as a C—H bond undergoing *stretching vibrations*.

The absorption band near 3000 cm^{-1} is therefore called the C—H (*stretch*) *absorption*, usually represented as C—H str.

The bands around 1400 cm^{-1} correspond to the frequency of the *bending vibrations* of C—H bonds, and are called the C—H (*bend*) *absorptions*. Alternatively, bending vibrations are referred to as *deformations*, so that C—H deformation bands can be labelled as C—H *def*.

Infrared spectroscopy is therefore basically vibrational spectroscopy, and the principal value of the technique to organic chemists relates to the following observation.

Different bonds (C—C, C=C, C≡C, C—O, C=O, O—H, N—H, etc.) have different vibrational frequencies, and we can detect the presence of these bonds in an organic molecule by identifying this characteristic frequency as an absorption band in the infrared spectrum.

For example, the spectrum in figure 2.3 is that of a carbonyl compound, and we know that the strong band at 1700 cm^{-1} is associated with the

21

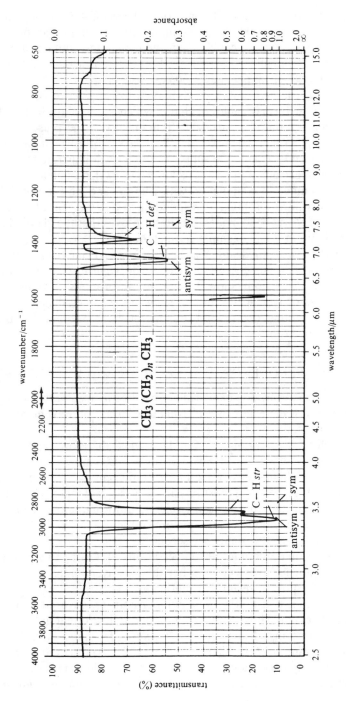

Figure 2.1 *Infrared spectrum of mixed long-chain alkanes (liquid paraffin, Nujol). Liquid film.*

stretching vibration of the C=O bond (thus, we say C=O *str* appears at 1700 cm^{-1}, or $\bar{\nu}_{C=O}$ = 1700 cm^{-1}). Similarly, by examining the spectrum in figure 2.18 (page 85), we can say that the compound contains a nitrile group; the strong band at 2250 cm^{-1} is the C≡N *str* absorption. (Alternatively, we say C≡N *str* appears at 2250 cm^{-1}, or $\bar{\nu}_{C≡N}$ = 2250 cm^{-1}.)

By examining a large number of compounds of a given class, we can draw up data in the form of tables or charts, which allow us to correlate the presence of absorption at a particular frequency with the presence of a functional group within the molecule. Such a set of *correlation charts* appears on pages 60–71, and two simple examples will illustrate their use.

The infrared spectrum in figure 2.2 is known to be that of an alkyne: the spectrum is examined in conjunction with chart 1, and we can see that it shows characteristic absorptions corresponding to C≡C *str* (at 2150 cm^{-1}) *and* to ≡C—H *str* (at 3320 cm^{-1}); this latter proves that the alkyne has a terminal triple bond. (The example is 1-octyne.)

Using chart 2, we can distinguish acetophenone from benzaldehyde, because although the spectra of benzaldehyde (figure 2.3) and acetophenone (figure 2.4) both show C=O *str* absorptions (around 1700 cm^{-1}), the benzaldehyde spectrum also shows the characteristic aldehyde C—H *str* absorptions (near 2800 cm^{-1}), which are absent from the acetophenone spectrum.

2.1 UNITS OF FREQUENCY, WAVELENGTH AND WAVENUMBER

The position of an absorption band can be specified in units of frequency, ν (s^{-1}, or Hz), or wavelength, λ (micrometers, μm), or wavenumber, $\bar{\nu}$ (reciprocal centimeters, cm^{-1}).

The stretching absorption of C—H bonds therefore appears at $\approx 9.3 \times 10^{13}$ s^{-1} ≡ 9.3×10^{13} Hz ≡ 3.3 μm ≡ 3000 cm^{-1}.

The majority of chemists use wavenumber units (cm^{-1}), although a very small minority, together with most physicists, use wavelength (μm). True frequencies are virtually never used (which is unfortunate, since only by quoting frequency can one obtain a picture of the C—H bond stretching and contracting 9.3×10^{13} times per second).

We shall occasionally commit the universal sin of referring to a vibration as having a 'frequency' of x cm^{-1} (rather than a wavenumber of x cm^{-1}). The unit cm^{-1} is spoken 'reciprocal centimeter'.

In the discussions that follow, wavenumber units are favored, although spectra and chart data show both wavenumber and wavelength units.

Organic applications of infrared spectroscopy are almost entirely concerned with the range of 650–4000 cm^{-1} (15.4–2.5 μm) (the mid infrared). The region of frequencies lower than 650 cm^{-1} is called the *far infrared*, and that of frequencies higher than 4000 cm^{-1} is called the *near infrared*.

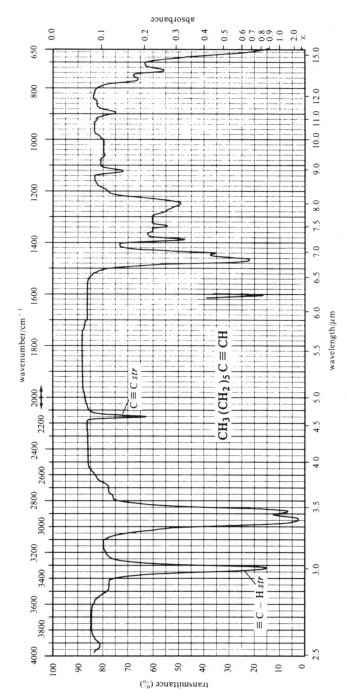

Figure 2.2 *Infrared spectrum of 1-octyne* ($CH_3(CH_2)_5C \equiv CH$). *Liquid film.*

24

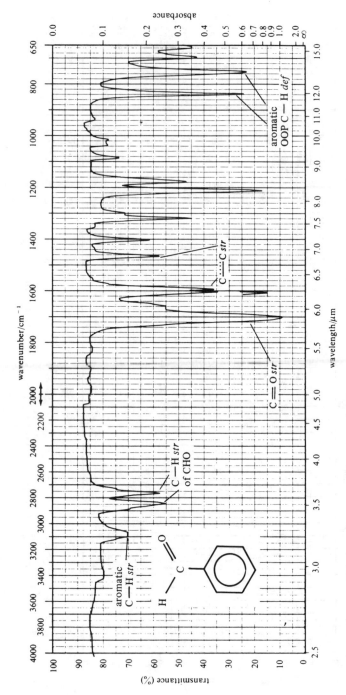

Figure 2.3 *Infrared spectrum of benzaldehyde (PhCHO). Liquid film.*

25

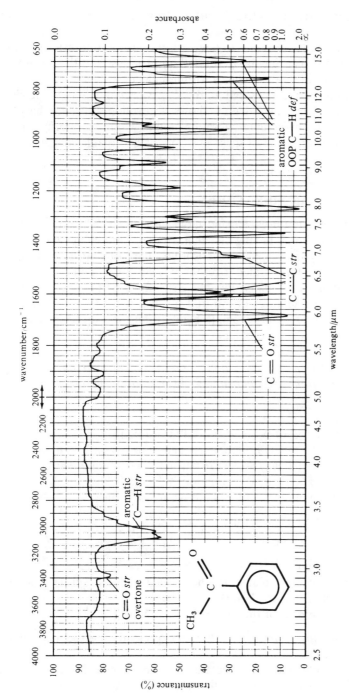

Figure 2.4 *Infrared spectrum of acetophenone (PhCOCH₃). Liquid film.*

These regions are, respectively, farther from and nearer to the visible spectrum.

The far infrared contains a few absorptions of interest to organic chemists, notably carbon–halogen bond absorptions, and the absorptions associated with rotational changes within molecules.

The near infrared reaches right to the long wavelength limit of the visible spectrum (0.75 μm), and mainly shows absorptions that are harmonic overtones of the fundamental vibrations found within the 'normal' range. Little use has been made of these absorptions by organic chemists.

2.2 MOLECULAR VIBRATIONS

At ordinary temperatures organic molecules are in a constant state of vibration, each bond having its characteristic stretching and bending frequency, and being capable of absorbing light of that frequency. The vibrations of two atoms joined together by a chemical bond can be likened to the vibrations of two balls joined by a spring: using this analogy, we can rationalize several features of infrared spectra. For example, to stretch a spring requires more energy than to bend it; thus, the stretching energy of a bond is greater than the bending energy.

Stretching absorptions of a bond appear at higher frequencies in the infrared spectrum than the bending absorptions of the same bond.

2.2.1 CALCULATION OF VIBRATIONAL FREQUENCIES

We can calculate the vibrational frequency of a bond with reasonable accuracy, in the same way as we can calculate the vibrational frequency of a ball and spring system; the equation is Hooke's law, which correlates frequency with bond strength and atomic masses, since

$$\nu \propto \sqrt{\frac{\text{bond strength}}{\text{mass}}}; \text{ thus, } \nu = \frac{1}{2\pi} \left(\frac{k}{m_1 m_2/(m_1+m_2)} \right)^{1/2}$$

where ν = frequency,
  k = a constant related to the strength of the spring (the *force constant* of the bond),
m_1, m_2 = the masses of the two balls (or atoms).

The quantity $m_1 m_2/(m_1 + m_2)$ is often expressed as μ, the *reduced mass* of the system.

As an example, we can calculate the approximate frequency of the C—H stretching vibration from the following data:

$k = 500 \text{ N m}^{-1} = 5.0 \times 10^5 \text{ g s}^{-2}$ (since 1 newton $= 10^3 \text{ g m s}^{-2}$)

m_C = mass of the carbon atom = 20×10^{-24} g

m_H = mass of the hydrogen atom = 1.6×10^{-24} g

$$\nu = \frac{7}{2 \times 22} \left(\frac{5.0 \times 10^5 \text{ g s}^{-2}}{(20 \times 10^{-24} \text{ g})(1.6 \times 10^{-24} \text{ g})/(20 + 1.6)10^{-24} \text{ g}} \right)^{1/2}$$

$$= 9.3 \times 10^{13} \text{ s}^{-1}$$

To express this in wavenumbers ($\bar{\nu}$), we use the relationship shown in section 1.2 (where c is the velocity of light = 3.0×10^8 m s^{-1}):

$$\bar{\nu} = \frac{\nu}{c} = \frac{9.3 \times 10^{13} \text{ s}^{-1}}{3.0 \times 10^8 \text{ m s}^{-1}}$$

$$= 3.1 \times 10^5 \text{ m}^{-1}$$

$$= 3100 \text{ cm}^{-1}$$

It is important qualitatively to restate the principles embodied in these calculations.

The vibrational frequency of a bond is expected to increase when the bond strength increases, and also when the reduced mass of the system decreases.

We can then predict that C=C and C=O *str* will have higher frequencies than C—C and C—O *str*, respectively: we also expect to find C—H and O—H *str* absorptions at higher frequencies than C—C and C—O *str*. Similarly, we would predict O—H *str* to be of higher frequency than O—D *str*.

Without accurate data on force constants, however, some caution should be exercised in predicting exact trends other than in this general way.

For example, on the basis of mass, we expect X—H *str* frequencies to fall along the series C—H, N—H, O—H, F—H; in fact they rise, mainly owing to increasing electronegativity. There are also extreme circumstances in which O—H *str* has lower frequency than O—D *str*. Section 2.3 deals with other factors influencing vibrational frequencies.

2.2.2 MODES OF VIBRATION

Molecules with large assemblages of atoms possess very many vibrational frequencies; for a nonlinear molecule with n atoms, the number of vibrational modes is ($3n - 6$), so that methane theoretically possesses 9 and ethane has 18. Does this lead to methane having 9 (and not more than 9) absorption bands in the infrared?

Figure 2.5 shows that for a single methylene group several vibrational modes are available, and any atom joined to two other atoms will undergo comparable vibrations (for example, any AX$_2$ system such as NH$_2$, NO$_2$).

STRETCHING MODES FOR CH₂

symmetric   antisymmetric

BENDING OR DEFORMATION MODES

In-plane Deformations   Out-of-plane Deformations

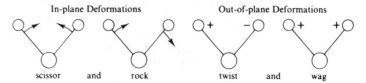

scissor   and   rock   twist   and   wag

Figure 2.5   *Vibration modes in methylene groups. Similar* AX_2 *groups* (—NH₂, —NO₂, *etc.*) *and methyl groups behave analogously.*

Each of the different vibration modes may (and frequently does) give rise to a different absorption band, so that CH₂ groups give rise to two C—H *str* bands at ν_{sym} and ν_{anti}. Other vibrations may not give rise to absorption, since some may have frequencies outside the normal infrared region being examined. Some of the vibrations may have the same frequency (that is, they are degenerate) and their absorption bands will overlap. It is also necessary, in order to 'see' an absorption band, that the particular vibration should produce a fluctuating dipole (and thus a fluctuating electric field), otherwise it cannot interact with the fluctuating electric fields of the infrared light. Thus, the stretching of a symmetrically substituted bond (for example, C≡C in acetylene) produces no change in the dipole of the system, and therefore this vibration cannot interact with infrared light. (See Raman spectroscopy, section 2S.3.)

In addition to these *fundamental vibrations* thus far discussed, other frequencies can be generated by modulation, etc., of the fundamentals. *Overtone bands* (harmonics) appear at integer multiples of fundamental vibrations, so that strong absorptions at, say, 800 cm⁻¹ and 1750 cm⁻¹ will also give rise to weaker absorptions at 1600 cm⁻¹ and 3500 cm⁻¹, respectively. Two frequencies may interact to give *beats* which are *combination* or *difference* frequencies; thus, absorptions at x cm⁻¹ and y cm⁻¹ interact to produce two weaker beat frequencies at $(x \pm y)$ cm⁻¹.

Modulation by addition or subtraction of two frequencies has important consequences in Raman spectroscopy (see section 2S.3).

2.2.3   QUANTUM RESTRICTIONS

When a vibrating ball and spring system gains energy, the *frequency* of vibration does not alter, but the *amplitude* increases. Thus, in the stretching of a spring, increased energy leads to greater degrees of extension and contraction of the spring. While this is also true of the vibrations of interatomic bonds, quantum theory applies to chemical bonds and imposes certain additional restrictions. Vibrational energy can only increase by quantum jumps, so that the energy difference between successive vibrational levels is the familiar $\Delta E = h\nu$ or $\Delta E = N_A h\nu$ (see section 1.3).

Absorption of infrared light around 3000 cm^{-1} (corresponding to C—H str) involves an increase in the energy of the molecule of $N_A h\nu$, or about 37 kJ mol^{-1}. Once in the vibrationally excited state, the molecule can give up this extra energy by rotational, collision or translational (kinetic) processes, etc.

2.3 FACTORS INFLUENCING VIBRATIONAL FREQUENCIES

Many factors influence the precise frequency of a molecular vibration, and it is usually impossible to isolate one effect from another. For example, the C=O str frequency in the ketone $RCOCH_3$ is lower than in RCOCl; is the change in frequency of the C=O str due to the difference in *mass* between CH_3 and Cl, or is it associated with the *inductive* or *mesomeric* influence of Cl on the C=O bond; perhaps there is some *coupling* interaction between the C=O and C—Cl bonds, or is there some steric effect which alters the *bond angles*?

We shall discuss here frequency shifts, which are brought about by structural changes in the molecule, or by interaction between functional groups. Due emphasis will be placed on those features that are most valuable in explaining the characteristic appearance and positions of the group frequencies.

Primary mass effects (for example, the mass effect of changing C—H to C—Cl) have been mentioned in section 2.2; secondary mass effects (for example, the effect on C=O str of changing $CO—CH_3$ to CO—Cl) are very difficult to study, because of the unavoidable intrusion of electronic effects. Frequency shifts also take place on moving from condensed phases to dilute solutions, as mentioned in the section on sampling techniques (see section 2.5).

2.3.1 VIBRATIONAL COUPLING

An isolated C—H bond has only one stretching frequency, but the stretching vibrations of C—H bonds in CH_2 groups combine together to produce two *coupled vibrations* of different frequencies—the antisymmetric, $\bar{\nu}_{anti}$, and symmetric, $\bar{\nu}_{sym}$, combinations discussed earlier and illustrated in figure 2.5. The C—H bonds in CH_3 groups also give rise to symmetric and antisymmetric vibrations. These are of different frequencies from those of CH_2 groups, and all four vibrations can be seen in high-resolution spectra of compounds containing both CH_2 and CH_3 groups.

Vibrational coupling takes place between two bonds vibrating with similar frequency, provided that the bonds are reasonably close in the molecule; the coupling vibrations *may both be fundamentals* (as in the coupled stretching vibrations of AX_2 groups) or a *fundamental vibration may couple with the overtone* of some other vibration. This latter coupling is frequently called Fermi resonance, after Enrico Fermi, who first described it.

Vibrational coupling is a feature of other AX_2 groups, so that the functions listed in table 2.1 exhibit not one, but two, stretching bands—antisymmetric and symmetric A—X *str* (antisymmetric usually being of higher frequency).

Carboxylic acid anhydrides. These give rise to two C=O *str* absorptions, $\bar{v}_{anti}$ and $\bar{v}_{sym}$ (around 1800–1900 cm^{-1}, with a separation of about 65 cm^{-1}); coupling occurs between the two carbonyl groups, which are *indirectly* linked through —O—: the interaction is presumably encouraged because of the slight double-bond character in the carbonyl–oxygen bonds brought about by resonance, since this will keep the system coplanar. The high-frequency band in this case is the symmetric C=O *str*.

Table 2.1 Typical antisymmetric and symmetric stretching frequencies for common AX_2 groups

Group	Antisymmetric, $\bar{v}_{anti}$/cm^{-1}	Symmetric, $\bar{v}_{sym}$/cm^{-1}
—CH$_2$—	3000	2900
—NH$_2$	3400	3300
—NO$_2$ $\left(-N\begin{smallmatrix}O\\\\O\end{smallmatrix}\right)$	1550	1400
—SO$_2$— $\left(\begin{smallmatrix}O\\\|\\-S-\\\|\\O\end{smallmatrix}\ \text{or}\ \begin{smallmatrix}O\\\|\\-S-O-\\\|\\O\end{smallmatrix}\right)$	1350	1150
$-C\begin{smallmatrix}O\\\\O\end{smallmatrix}$ (salts of $-C\begin{smallmatrix}O\\\\OH\end{smallmatrix}$)	1600	1400

anhydrides—resonance forms

Amides. These show two absorption bands around 1600–1700 cm^{-1} corresponding mainly to C=O *str* and N—H *def*, but because of vibrational coupling, the original characters of the vibrations are modified. The two bands are not pure C=O *str* and N—H *def*, and are usually referred to as the amide I and amide II bands. Amide I may be as high as 80 per cent C=O *str* in character, but amide II is a strongly coupled interaction between N—H *def* and C—N *str*. (See also section 2.9.)

In *aldehydes* the C—H *str* absorption usually appears as a doublet because of interaction between the C—H *str* fundamental and the overtone of C—H *def*.

2.3.2   HYDROGEN BONDING

Hydrogen bonding, especially in O—H and N—H compounds, gives rise to a number of effects in infrared spectra, and its importance here can scarcely be overemphasized. While most routine organic work will involve relatively nonassociating solvents (CCl$_4$, CS$_2$, CHCl$_3$), more polar solvents such as acetone or benzene will certainly influence O—H and N—H absorptions. Carbonyl groups or aromatic rings *in the same molecule* as the O—H or N—H group may cause similar shifts by intramolecular action.

Alcohols and phenols. Figure 2.6 shows the infrared spectrum of an alcohol (1-butanol) recorded as a liquid film; the dotted line insert around 3500 cm^{-1} was recorded in dilute solution (about 1 per cent in CCl$_4$). At low concentrations a sharp band appears at 3650 cm^{-1} in addition to the broad band at 3350 cm^{-1}.

The sharp band is O—H *str* in *free* alcohol molecules; the broad band is O—H *str* in hydrogen-bonded alcohol molecules.

Alcohols in phenols in condensed phases (bulk liquid or KBr disks, etc.) are strongly hydrogen-bonded, usually in the form of a dynamic polymeric association; dimers, trimers and tetramers also exist, and this leads to a wide envelope of absorptions and, hence, to broadening of the absorption band. In dilute solution in inert solvents (or in the vapor phase) the proportion of free molecules increases and these give rise to the 3650 cm^{-1} band.

Is it reasonable that bonded O—H *str* should appear at lower frequency than free O—H *str*?

The hydrogen bond can be regarded as a resonance hybrid of I and II (approximating overall to III), so that hydrogen bonding involves a lengthening of the original O—H bond. This bond is consequently

32

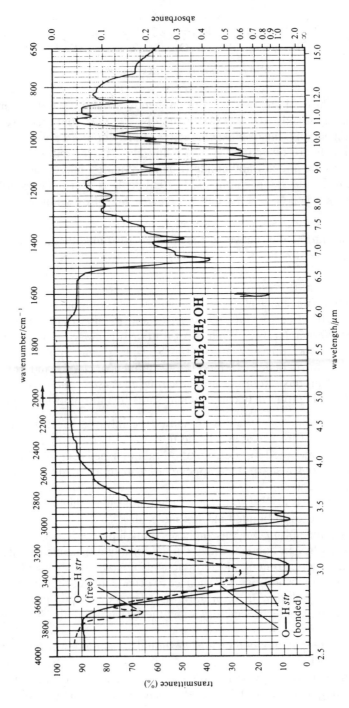

Figure 2.6 *Infrared spectrum of 1-butanol (CH₃CH₂CH₂CH₂OH). Complete spectrum, liquid film; dotted line insert near 3500 cm⁻¹, dilute solution in carbon tetrachloride.*

weakened (that is, its force constant is reduced), so the stretching
frequency is lowered.

polymeric association of O—H compounds

lengthening of O—H bond in hydrogen bonding

Enols and chelates. Hydrogen bonding in enols and chelates is particularly strong, and the observed O—H *str* frequencies may be very low
(down to 2800 cm^{-1}). Since these bonds are not easily broken on dilution
by an inert solvent, free O—H *str* may not be seen at low concentrations.

enol   chelate   dimer
(CH$_3$COCH$_2$COCH$_3$)   (methyl salicylate)   (of benzoic acid)

polymer
(of benzoic acid)

Carbonyl compounds. In enols and in chelates such as methyl salicylate hydrogen bonding will influence not only the O—H vibration frequency but also the C=O vibration to which it hydrogen-bonds. The key factor here is the basicity of the C=O group: the more basic it is, the stronger will be the hydrogen bond that it can form. The extreme case of

hydrogen bonding   protonation

protonation shows that the C=O bond has increased single-bond cha-racter and longer length: the same tendency occurs in hydrogen bonding, leading to a lowering of the vibration frequency.

Carboxylic acids. Figure 2.7 shows the infrared spectrum of benzoic acid, and the exceedingly broad band reaching from 2500 cm^{-1} to 3500 cm^{-1} is hydrogen-bonded O—H *str*. We are seeing here the O—H *str* band for the carboxylic acid *dimer* structure: in condensed phases, all carboxylic acids exist in this stable *dimeric association* in which the hydrogen bonds are particularly strong. (The fine structure on the O—H *str* peak is usually attributed to vibrational coupling with overtones of lower frequencies.) In very dilute solution in hexane it is just possible to distinguish free O—H *str*, but this is extreme dilution. Even in CCl$_4$ some degree of hydrogen bonding to solvent arises and in extreme dilution in CCl$_4$ the free O—H *str* absorption is seen at lower frequency than in hexane. Polymeric association is also known to occur in carboxylic acids, although dimeric association is the norm; the proportion of monomer to dimer increases in solvents such as benzene, and in dioxan there is no dimer formed, since the acid hydrogen-bonds preferentially to the solvent.

π-Cloud interactions. Since alkene and aromatic π bonds can behave as Lewis bases, it is not surprising that they can form hydrogen bonds to acidic hydrogens; the frequency of O—H *str* in phenols can be lowered by 40–100 cm^{-1} when the spectrum is recorded in benzene solution, com-pared with carbon tetrachloride solution.

Amines. In condensed phase spectra amines show bonded N—H *str* around 3300 cm^{-1} and in dilute solution a new band near 3600 cm^{-1} corresponds to free N—H *str*. Since nitrogen is less electronegative than oxygen, hydrogen bonds in amines are weaker than in alcohols, and the shifts in frequency are also correspondingly less dramatic than in alcohols.

35

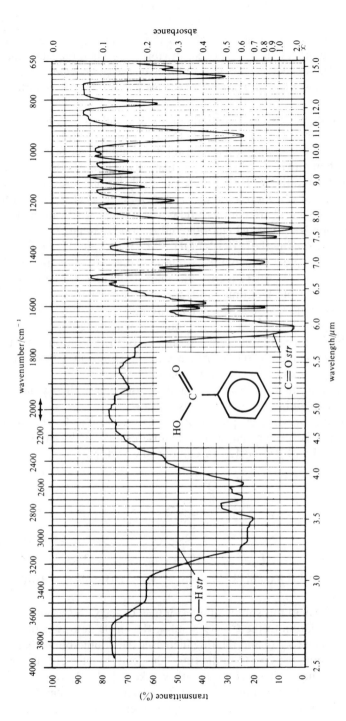

Figure 2.7 *Infrared spectrum of benzoic acid* (PhCO$_2$H). *KBr disc.*

2.3.3 Electronic Effects

One can use the theoretical principles of the organic chemist to explain many of the frequency shifts that occur in vibrations when the substituents are altered. The expected *inductive* and *mesomeric* (or *resonance*) effects are seen to be at work, together with an occasional through-space influence (or *field effect*).

Many unresolved problems remain, however, and we cannot concentrate on successes and ignore the many instances where simple theory fails to offer a reasonable explanation. Vibrational coupling (see section 2.3.1) often means that an observed absorption band is *not* purely associated with one bond alone and this will complicate our explanations; most C—H *def* modes are coupled vibrations, and we have seen that C=O *str* is a coupled vibration in amides and anhydrides, and is also a coupled vibration in such simple compounds as benzoyl chloride and cyclopentanone.

Again, if we examine the series MeOH, PhOH, MeCOOH, we find that the O—H *str* frequency decreases, while in the series $MeNH_2$, $PhNH_2$, $MeCONH_2$ the N—H *str* frequency inexplicably increases. We must also consider the effect of electronic influences on the strengths of the bonds *adjacent* to the bond whose frequency we are measuring: thus, pictorially, if we stiffen up the bonds to a C=O group, we will make it more difficult for the carbonyl carbon atom to move, and all of the vibration amplitude will have to be taken up by the oxygen atom, with almost inevitable shift in the C=O *str* frequency (see section 2.3.4).

With caution in mind, we can now look at cases where theory has been successful in explaining frequency shifts.

Conjugation lowers the frequency of C=O *str* and C=C *str*, whether the conjugation is brought about by αβ unsaturation or by an aromatic ring. Compare I with II and III; or compare IV with V.

$\bar{v}_{C=O} \approx 1720\ cm^{-1}$ $1700\ cm^{-1}$ $1700\ cm^{-1}$ $\bar{v}_{C=C} \approx 1650\ cm^{-1}$ $1610\ cm^{-1}$

The explanation of this shift is similar for C=O and C=O, but we shall illustrate it in relation to the C=O bond in III. In III delocalization of π electrons between C=O and the ring increases the double-bond character of the bond joining the C=O to the ring. This leads to a lower bond order in the C=O bond, which is consequently weakened; the decrease in force constant lowers the stretching vibration frequency by 20–30 cm^{-1}.

One can also attribute such C=O frequency shifts to the mesomeric (or resonance) effect: any substituent that enhances the mesomeric shift will decrease the bond order of the $C{=}O$ bond and lead to lower $C{=}O$ *str*

| VI | VII | VIII | IX |

frequency. Conjugation with phenyl (in VII) does so, and a +M group such as *p*-MeO in VIII will lead to even lower frequencies. A *p*-NO$_2$ group (−M) will oppose these trends and lead to higher frequencies (as in IX).

Inductive effects are difficult to consider in isolation from mesomeric effects: in some molecules I is more important than M, while in others the reverse is true.

In amides, XI, the +M effect produces a lengthening (weakening) of the $C{=}O$ bond, leading to lower frequency than in the corresponding ketone, X: the −I effect of nitrogen is here being dominated by +M. In contrast, the −I effect of chlorine in acyl chlorides, XII, is more influential than +M, and here an opposite shift (to higher frequency) occurs.

| X | XI | XII | XIII | XIV |

Esters represent another example of the conflict between I and M effects. In alkyl esters, XIII, the nonbonding electrons on oxygen increase the +M conjugation, tending to lower the $C{=}O$ frequency. The electronegativity of oxygen, −I, operates in the opposite sense, but +M is apparently dominant. In phenyl esters, XIV, however, the nonbonding electrons are partly drawn into the ring, and their conjugation with $C{=}O$ is consequently diminished. When this happens, the −I effect of oxygen becomes dominant, and $C{=}O$ moves to higher frequency.

In examples such as these it is easier to rationalize the shifts than it is to predict them, and caution should be exercised in applying the rules to new situations. The importance of vibrational coupling requires constant restatement.

2.3.4 BOND ANGLES

In *ketones*, the correlation charts show that highest $C{=}O$ frequencies arise in the strained cyclobutanones, and we can explain this in terms of bond-angular strain: the $C{-}CO{-}C$ bond angle is reduced below the normal 120°, leading to increased s character in the $C{=}O$ bond. The

$C=O$ bond is shortened and therefore strengthened and so $\nu_{C=O}$ increases. If the bond angle is pushed outwards above 120°, the opposite effect operates, and for this reason di-*tert*-butyl ketone has a very low $\bar{\nu}_{C=O}$ (1697 cm^{-1}).

An alternative view involves no change in the $C=O$ force constant, but merely an increased rigidity in the $C—CO—C$ bond system as ring size decreases: $C=O$ stretching must in these circumstances couple more effectively with $C—C$ stretching, leading to higher $C=O$ *str* frequencies.

Cycloalkenes also show such an effect, but a less simple relationship holds. Thus, in cycloalkenes, XV, $\bar{\nu}_{C=C}$ falls with increasing strain, but reaches a minimum in cyclobutene. In cyclobutene, XVI, stretching of $C=C$ involves only *bending* of the attached $C—C$ bonds: in all the others (where the internal angles are not 90°) $C=C$ stretching must involve some stretching of the adjacent $C—C$ bonds, which involves increasing the energy (frequency) of $C=C$ *str*.

$$\left(\overset{=}{\underset{(CH_2)_n}{C}} \right)$$

$n = 1$ to 6	except	cyclobutene
XV		XVI
$\bar{\nu}_{C=C} \approx 1610\text{–}1650$ cm^{-1}		1566 cm^{-1}

$C—H$ *stretching* vibrations move to higher frequency in the sequence alkane–alkene–alkyne. As hybridization goes from sp^3 to sp^2 to sp, the s character of the $C—H$ bond increases; bond lengths become shorter, and frequencies rise. Cyclopropanes have high $C—H$ *str* frequencies for the same reason (typical values being 3040–3070 cm^{-1}): the $C—\hat{C}—C$ bond angle is substantially contracted below the normal 109.5°, leading to increased s character in the $C—H$ bonds, and thus to higher frequencies.

2.3.5 FIELD EFFECTS

Two groups often influence each other's vibrational frequencies by a through-space interaction, which may be electrostatic and/or steric in nature. The best examples of this *field effect* are interactions between carbonyl groups and halogen atoms; for example, in the α-chloroketone derivatives of steroids, XVII, $C=O$ *str* frequency is higher when Cl is equatorial than when it is axial. Presumably the nonbonding electrons of oxygen and chlorine undergo repulsion when they are close together in the molecule; this results in a change in the hybridization state of oxygen, and therefore a shift in $C=O$ *str* frequency.

In *o*-chlorobenzoic acid esters this field effect shifts the $C=O$ frequency in the rotational isomer XVIII, and not in the isomer XIX; both isomers are normally present, so that *two* $C=O$ *str* absorptions are observed in the spectrum of this compound.

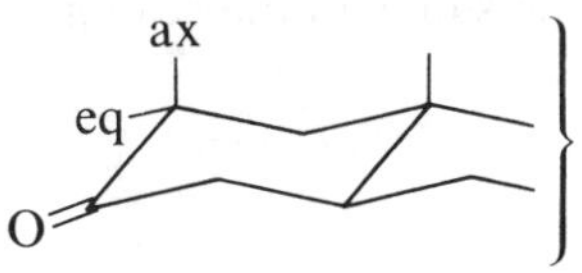

XVII 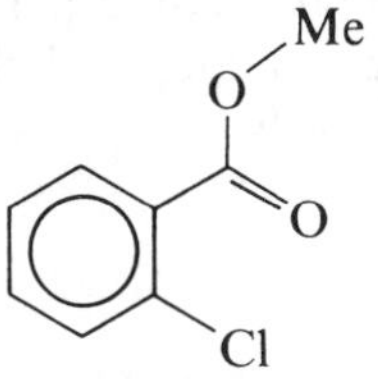   XVIII 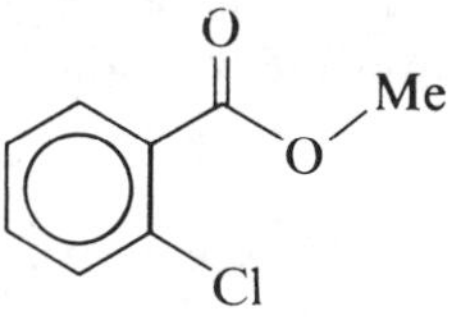   XIX

2.4   INSTRUMENTATION—THE INFRARED SPECTROMETER—DISPERSIVE AND INTERFEROMETRIC INSTRUMENTS

There are two basic types of infrared spectrophotometer, characterized by the manner in which the infrared frequencies are handled: the first type has had a long history, and in it the infrared light is separated into its individual frequencies by dispersion, using a grating monochromator, whereas in the second type the infrared frequencies are allowed to interact to produce an interference pattern, and this pattern is then analyzed mathematically, using Fourier Transforms, to determine the individual frequencies and their intensities.

Dispersive instruments are widely in use at present, but the modern Fourier Transform infrared instruments (FTIR machines—or interferometric spectrometers) are becoming more commonplace as their costs fall.

Both types of instrument require, in addition, a suitable source of infrared light and a detector to measure light intensity.

The high-precision optical system in all of these spectrophotometers represents decades of development, with physics and engineering involved jointly in their manufacture: more details on instrument construction are to be found in the books listed in 'Further Reading', at the end of this chapter, and also in the text by H. H. Willard *et al.*, listed at the end of chapter 1.

2.4.1   INFRARED SOURCES
Common sources are rods of the following, electrically heated to near 1800 °C:

'Nernst glower' or 'Nernst filament' (sintered mixtures of the oxides of Zr, Th, Ce, Y, Er, etc.)
  'Globar' (silicon carbide)
  Various other ceramic materials

The infrared output from these sources varies in intensity over the required frequency range, being at a maximum around 5000–7000 cm^{-1}, dependent on the source, and reducing by a factor of several hundreds near

600 cm^{-1}. To accommodate this, in dispersive instruments a compensating variable slit is programmed to open and close in unison with the scanning over the individual frequencies.

2.4.2 MONOCHROMATORS

The original infrared instruments used prisms to diffract the light, and this simple principle is illustrated in figure 2.8, but all modern dispersive instruments use diffraction gratings. In *ruled gratings* a series of micro-scopically close parallel triangular grooves is engraved onto a reflective surface (spaced anything from 20 grooves per mm in far infrared gratings to nearly 4000 grooves per mm in ultraviolet gratings). Light reflected from the grating is diffracted, interference arises at certain angles, and so specific wavelengths appear with constructive interference at specific angles of reflection. *Holographic gratings* are made by depositing a photoresist material onto optically flat glass, and then irradiating this with two beams of laser light, thus producing interference fringes in the photosensitive material of the resist. The irradiated parts of the resist are dissolved away, leaving a groove pattern which is coated with a reflective material; the result is similar to a ruled grating, but more than 5000 grooves per mm can be produced, and the wavelength 'purity' is higher in the holographic grating.

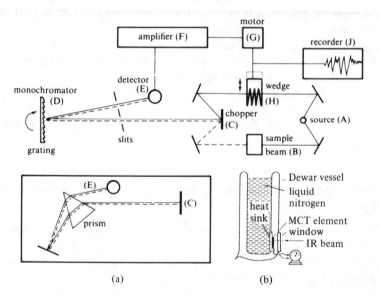

(a)   (b)

Figure 2.8   *Schematic layout of a dispersive spectrometer. Main diagram—grating optics; insert (a) prism monochromator, which is simple but now obsolete; insert (b) MCT detector.*

2.4.3 DETECTORS

Most dispersive instruments use thermopile detectors: these consist of several thermocouples connected in series so that their outputs are added together for greater sensitivity. A thermocouple works on the principle that if wires of two dissimilar metals (such as bismuth and antimony) are joined head to tail, then a difference in temperature between head and tail causes a current to flow in the wires. In the infrared spectrometer this current will be proportional to the intensity of radiation falling on the thermopile.

In FTIR instruments increased sensitivity and speed in recording the spectrum must be matched in the speed and photometric accuracy of the detector; this is achieved by thermal detectors based on pyroelectric materials or on solid state semiconductor devices based on photovoltaic or photoconductive principles.

An extremely sensitive *photovoltaic* detector uses a diode-type device based on doped silicon or indium antimonide (InSb) or indium–gallium arsenide (InGaAs). There are three main ways in which these devices can be incorporated into detector circuits. (1) Infrared irradiation excites electrons to the conduction band, and the generation of the resulting electron–hole pair leads to a corresponding increase in the voltage across the p–n junction, varying in a logarithmic way with intensity of irradiation. (2) Alternatively, we can take account of the existing voltage difference across the p–n junction; if the p and n zones are shorted together, a small current flows through the short-circuit, the current increasing proportionately if the device is irradiated by infrared photons. (3) Lastly, a reverse bias voltage can be applied, acting in opposition to the normal potential difference across the p–n junction; irradiation by infrared photons of sufficient energy generates holes in the p-type layer, with increased conduction (increased signal output). This arrangement gives very fast response time, but it is more noisy than the former—and most widely used—alternative (2).

A very common high-sensitivity detector is that based on the *photoconductive* properties of mercury cadmium telluride (MCT). This is a homogeneous (undoped) semiconductor, operating at the temperature of liquid nitrogen, whose resistivity changes with temperature—for example, during infrared irradiation. The resistance is measured by applying voltage and measuring current, or vice versa.

A diagram of the detector construction is shown in figure 2.8. It consists in the main of a vacuum Dewar flask, approximately 25×8 cm, holding sufficient liquid nitrogen for about eight hours of operation: set into the outer wall is an infrared transparent window and in line with this, in the inner wall, is the MCT. The sensitive element measures approximately 1 mm square, and it is mounted on a heat sink to assist cooling after irradiation, for fast recovery.

Less sensitive, by a factor of about 10, are the *pyroelectric* substances such as deuteriated triglycine sulphate (DTGS) or lithium tantalate (LiTaO$_3$), which are used at room temperature. Crystals of these highly polar materials exhibit an electric polarization along certain axes, and if they are heated in one area (by, for example, incident infrared radiation) the changes in the crystal lattice in this area will lead to an imbalance of polarization charge with respect to the rest of the crystal. This imbalance (which is proportional to the infrared intensity) can be measured by connecting electrodes to the different faces of the crystal—for example, by forming a sandwich of the DTGS between two plate electrodes, one of which is infrared-transparent.

This detector is also used in the more expensive 'ratio recording' dispersive instruments discussed below.

2.4.4 Mode of Operation—Dispersive Instruments—Optical Null and Ratio Recording

The following is a simplified description of the operation of a typical infrared machine, and figure 2.8 is a schematic representation of such a machine.

Light from the source (A) is split into two equal beams, one of which (B) passes through the sample (the *sample beam*), the other behaving as a *reference beam*; the function of such a *double-beam operation* is to measure the difference in intensities between the two beams at each wavelength.

The two beams are now reflected to a *chopper* (C), which consists of a rotating segmented mirror; as the chopper rotates ($\approx$ 10 times per second) it causes the sample beam and the reference beam to be reflected *alternately* to the monochromator grating (D). As the grating slowly rotates, it sends individual frequencies to the detector thermopile (E), which converts the infrared (thermal) energy to electrical energy.

When a sample has absorbed light of a particular frequency, the detector will be receiving *alternately* from the chopper an intense beam (the reference beam) and a weak beam (the sample beam). This will in effect lead to a pulsating or *alternating* current flowing from the detector to the amplifier (F). (If the sample had not absorbed any light, the sample beam and the reference beam would have been of equal intensity, and the signal from the detector would have been *direct* current. The amplifier is designed only to amplify alternating current.)

The amplifier, which is now receiving this out-of-balance signal, is coupled to a small servo-motor (G), which drives an optical wedge (H) into the reference beam until eventually the detector receives light of equal intensity from the sample and reference beams. This movement of the wedge (or attenuator) is in turn coupled to a pen recorder (J), so that movement of the wedge in and out of the reference beam shows as absorption bands on the printed spectrum.

Since this instrument balances out by optical means the differential between the two beams, it is a double-beam *optical null* recording spectrometer.

In more expensive instruments the intensity of the sample and reference beams can both be measured and ratioed (*ratio recording*). This circumvents a major deficiency in optical null methods, in which the comb attenuator blocks off much of the light beam; not only does this instrumental feature lead to reduced sensitivity, but also, since the comb is never either perfectly logarithmic in its geometry or infinitely smooth in its movement, poor quantification and slow scan speeds are inevitable. See also section 4.3.1.

2.4.5 MODE OF OPERATION—INTERFEROMETRIC INSTRUMENTS—FOURIER TRANSFORM INFRARED SPECTROSCOPY

Dispersive infrared spectrometers suffer from several disadvantages in sensitivity, speed and wavelength accuracy. Most of the light from the source does not in fact pass through the sample to the detector, but is lost in the narrowness of the focussing slits; only poor sensitivity results. Since the spectrum takes minutes to record, the method cannot be applied to fast processes, such as recording the infrared spectra of peaks being eluted from a chromatography column (see Section 1S.3).

Dispersive infrared spectrometers scan over the wavelength range and disperse the light by use of a grating, as shown in figure 2.8. Consequently, these spectrometers suffer from wavelength inaccuracies associated with the backlash in the mechanical movements, such as in the rotation of mirrors and gratings.

An entirely different principle is involved in *Fourier Transform infrared spectroscopy*, which centres on a Michelson interferometer, so that the method can also be called *interferometric infrared spectroscopy*.

The Michelson interferometer

In 1887 Albert Michelson (German-born American physicist) perfected this instrument and used it for several measurements in his study of light and relativity. The operation can be understood with reference to figure 2.9.

At (a) in the figure monochromatic light from the source strikes a half-silvered mirror B, which is designed to split the beam A exactly in half. One half is *transmitted* as beam C to a plane mirror D and reflected there back to B; the other half of beam A is *reflected* as beam E to another plane mirror F, where it is reflected back to B. Both of these beams are recombined as beam G (by being reflected and transmitted, respectively) and pass to a detector; for visible wavelengths this can be an eye. Provided that the path lengths from B to the mirrors D and F differ by an integer number of wavelengths (including the situation where the path lengths are

44

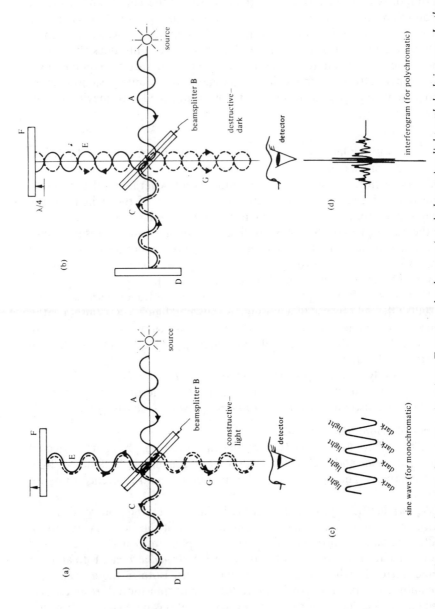

Figure 2.9 *The Michelson interferometer. As mirror F moves, the detector 'sees' alternating light and dark images. In the FTIR spectrometer, the sample is placed between the source and the beamsplitter; the instrument operates in single-beam mode.*

equal), the two emergent beams (G) will recombine with *constructive interference*, and high-intensity radiation will reach the detector.

At (b) in the figure the mirror F has moved a distance of $\lambda/4$, so that the total path lengths differ by exactly one-half of a wavelength, and thus the two emergent beams (G) are recombined with *destructive interference*, and low-intensity radiation will reach the detector.

(For visible light the beam splitter is made of glass, and coated with a very thin film of silver insufficiently thick to reflect the beam completely; for infrared light it is often made of KBr or CsI coated with germanium.)

The moving mirror F is the key to the interferometer, because alternate light and dark images will reach the detector if this mirror is slowly moved either away from or toward B. The signal from the detector will then be of the form shown at (c).

If the source supplies two different frequencies of monochromatic light, the signal from the detector will exhibit a different periodicity, which will be the result of combining the two separate sine-wave signals for the two different wavelengths.

For more complicated polychromatic light sources, a complex interferogram (see (d) in the figure) arises. The individual frequencies of light which were the genesis of this interferogram can be calculated by the use of a Fourier Transform (see section 1S.2).

Strictly, the Fourier Transform being computed here is the transform of distance (for example, in meters or centimeters) into its reciprocal (in reciprocal meters or reciprocal centimeters).

The Fourier Transform infrared spectrometer
Infrared light from a suitable source (see section 2.4.1) passes through a scanning Michelson interferometer, and Fourier Transformation gives a plot of intensity versus frequency. When a sample compound is placed in the beam (either before or after the interferometer), it absorbs particular frequencies, so that their intensities are reduced in the interferogram and the ensuing Fourier Transform is the infrared absorption spectrum of the sample.

The scan time for the moving mirror dictates the speed with which the infrared spectrum can be recorded; digitization of the data and calculation of the Fourier Transform take a few seconds more, but the information which constitutes the spectrum can be acquired in exceedingly short times, even in a few milliseconds. Slower scans (lasting several seconds) allow the accumulation of more intense signals, and signal-to-noise ratios of 100 000 : 1 can be reached. Unlike dispersive instruments, no slits are required for monochromating and so all of the energy of the source is utilized. Signal averaging gives additional improvement.

In the physics of Fourier Transform spectroscopy the advantage of collecting the spectroscopic data nearly simultaneously is referred to as the Fellgett advantage: the advantage gained by using all of the energy of the source, and passing through circular holes rather than slits, is called the Jacquinot advantage.

Since the only moving part in the spectrometer is the moving mirror, wavelength accuracy rests in the precision with which this is carried out, so that this is an expensive piece of high-technology engineering; resolution of less than 0.1 cm^{-1} is obtainable. In addition, the instrument will have a He-Ne laser, the red light from which is used for continuous spectrum calibration; the wavelength is 632.8 nm, corresponding to $15\,804$ cm^{-1}.

The spectrometer is not double-beam in its working, but elimination of unwanted absorptions, principally those of water and carbon dioxide in the atmosphere, can be carried out by summing a few spectra with the sample in position, and summing the same number without; computer subtraction of the two is then carried out. Alternatively, the sample may be moved (automatically) in and out of the sample beam, and alternate scans subtracted as before.

A Fourier Transform infrared spectrometer can do everything that can be done by a simple IR spectrometer—but at greater cost. Its *forte* therefore lies in performing where others cannot—with respect to both speed and sensitivity.

Picogram quantities of sample can give good spectra, particularly if the beam is focused on to the sample by mirrors or a KBr lens. Beam diameters of a few micrometers are possible, and it is additionally advantageous if a microdetector is used in conjunction, since a large detector will merely produce noise from those areas which are not in the beam. Spectra have been obtained from such diverse samples as small paint chips and a single pollen grain.

One of the striking advantages of FTIR instruments is the ability to perform spectra summation or subtraction on the digitized spectra which they produce; figure 2.10 illustrates a typical application. At (a) is the IR spectrum of a polymer blend, known to contain poly(methyl vinyl ether), PMVE, and poly(α-methylstyrene), PAMS, with the possibility of other polymers also being present. The spectrum of PMVE at (b) is subtracted from (a), leaving the difference spectrum (c), which is identical with that of PAMS, (d). A second subtraction of (d) from (c) produced a flat baseline, establishing that only these two constituents were present in the blend.

Accurate spectra subtractions can only be achieved when the spectra have very high wavenumber accuracy, and in FTIR instruments this is assured by the frequency referencing to the internal He–Ne laser.

By coupling an optical microscope to a suitably adapted FTIR spectrometer (a technique named FTIR microscopy), sample sizes down to a few micrometers can be examined.

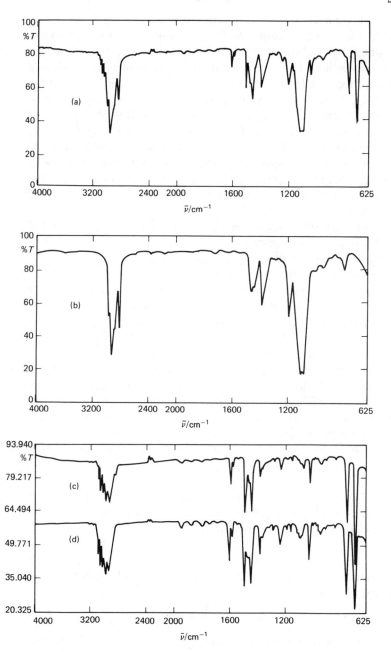

Figure 2.10 *Difference spectra in analysis of mixtures. Infrared spectra of* (a) *a blend of poly(methyl vinyl ether), PMVE, and poly (a-methylstyrene), PAMS;* (b) *PMVE alone;* (c) *difference spectrum by subtraction of* (b) *from* (a); *and* (d) *PAMS alone.*

2.4.6 CALIBRATION OF THE FREQUENCY SCALE

Since dispersive instruments use preprinted recorder chart paper, the scanning of the frequencies and the driving of the recorder must be carefully adjusted so that accurate frequencies are traced onto the charts. Calibration can be carried out using the spectrum of polystyrene (or of indene); these spectra show many sharp bands whose frequencies are accurately known (see figures 2.11 and 2.24). It is good practice to check the instrument calibration frequently, and even changes in humidity can cause inaccuracies because of the shrinking or stretching of the paper charts. All of the spectra in this book show the 1601 cm^{-1} peak of polystyrene, marked on as a check of frequency accuracy.

As we saw above, wavenumber accuracy in FTIR instruments is calibrated from the internal laser.

Most chemists prefer to use linear-wavenumber spectra, and all spectra in this book are so presented.

With the advent of cheap grating instruments, together with a linear-wavenumber presentation, full advantage can now be taken of the remarkable resolution these machines give at high frequencies. (Note, however, that in all of these there is a 2:1 gear change at 2000 cm^{-1}, which avoids overcrowding of the bands at low frequencies).

2.4.7 ABSORBANCE AND TRANSMITTANCE SCALES

The intensity of absorption bands in infrared spectra cannot easily be measured with the same accuracy as in ultraviolet spectra (see chapter 4). It is usually sufficient for an organic chemist to know that a band is of strong, medium, weak or variable intensity, indicated on charts, etc., as s, m, w or v, respectively.

The *absorbance* of a sample at a particular frequency is defined as

$$A = \log(I_0/I)$$

where I_0 and I are the intensities of the light before and after interaction with the sample, respectively. Absorbance is therefore a logarithmic ratio.

The *transmittance* of a sample is defined as

$$T = I/I_0$$

Transmittance therefore bears a reciprocal and logarithmic relationship to absorbance:

$$A = \log(1/T)$$

The ordinate scale of figure 2.11 shows how these might be presented on a typical spectrum. For reasons that are largely instrumental, most infrared spectra record the intensities of bands as a linear function of T, and this has important consequences in quantitative work (see section 2S.1).

49

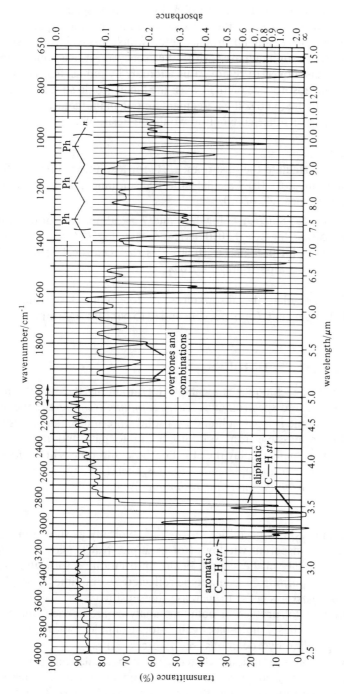

Figure 2.11 *Infrared spectrum of polystyrene. 0.05 mm film. Presented linear-in-wavenumber.*

Note also the common convention in infrared spectra of plotting increasing transmittance on the y axis, whereas all other spectroscopic techniques plot increasing absorbance, etc., which makes infrared spectra appear 'upside-down' compared with other spectra. At least one infrared spectra atlas has moved to remove this anomaly by publishing the spectra with increasing absorbance on the y axis: the facility to do so is built into all modern instruments, and some agreement among manufacturers and users to establish one or other convention would be desirable.

Example 2.1

Question. (a) Calculate the absorbance, A, for a solution showing 50 per cent transmittance, T. (b) Calculate the transmittance, T, for a solution showing absorbance, A, of 1.0.

Model Answer. (a) Since $A = \log(1/T)$ and $T = 50$ per cent or 0.50, then $A = \log(1/0.50) = \log 2 = 0.3010$. (b) Since $A = \log(I_0/I) = 1.0$, we must find the antilog of 1.0, which is 10.00. T is the reciprocal of this $= 0.1$, which expressed as a percentage $= 10\%$.

Exercise 2.1. Calculate the absorbances, A, for solutions showing (a) 90 per cent, (b) 95 per cent, (c) 10 per cent and (d) 5 per cent transmittance, T.

2.5 SAMPLING TECHNIQUES

A wide range of techniques is available for mounting the sample in the beam of the infrared spectrometer. These *sampling techniques* depend on whether the sample is a gas, a liquid or a solid. Intermolecular forces vary considerably in passing from solid to liquid to gas, and the infrared spectrum will normally display the effect of these differences in the form of frequency shifts or additional bands, etc. It is, therefore, most important *to record on a spectrum the sampling technique used*.

2.5.1 GASES

The gas sample is introduced into a *gas cell*, typically as shown in figure 2.12; infrared-transparent windows (for example, NaCl) allow the cell to be mounted directly in the sample beam. In a modified form the use of internal mirrors permits the beam to be reflected several times through the sample (*multi-pass* gas cells) to increase the sensitivity.

In environmental protection pollutant analyses are carried out on air samples, using these multi-pass cells—for example, to measure un-combusted gasoline hydrocarbons; see Exercise 2.11.

In the vapor phase rotational changes in the molecule can occur freely, and these very-low-frequency (low-energy) processes can modulate the higher-energy vibrational bands; the vibrational bands are split, often with the production of considerable fine structure.

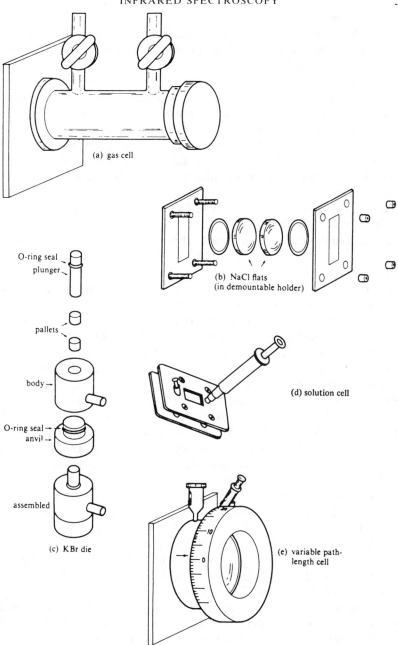

O-ring seal
plunger

pallets

body →

O-ring seal →
anvil →

assembled

(a) gas cell

(b) NaCl flats
(in demountable holder)

(c) KBr die

(d) solution cell

(e) variable path-
length cell

Figure 2.12 *Equipment used in infrared sampling techniques: (a) gas cell; (b) sodium chloride flats (rock salt flats) and demountable holder for liquid samples; (c) KBr die (exploded view and assembled view) for solid samples; (d) solution or liquid cell being filled; (e) variable path-length cell.*

2.5.2   LIQUIDS

The simplest infrared technique of all consists of sampling a liquid as a *thin film* squeezed between two infrared-transparent windows (for example, NaCl flats: see figure 2.12).

These *salt plates* or *rock salt flats* must be optically polished (using jewellers' rouge, for example) and must be cleaned immediately after use by rinsing in a suitable solvent such as toluene, chloroform, etc. They must be kept dry, and should be handled only by their edges.

The thickness of the film can be adjusted by varying the pressure used to squeeze the flats together; the film thickness is ≈ 0.01–0.1 mm. The assembled pair of flats and liquid film are mounted in the sample beam as shown in figure 2.12.

For spectra down to 250 cm^{-1}, CsI flats are used, and for samples that contain water, CaF$_2$ flats are used.

Liquid samples can also be examined in solution (see section 2.5.4).

2.5.3   SOLIDS

There are three common techniques for recording solid spectra: *KBr disks*, *mulls* and *deposited films*. Solids can also be examined in solution (see section 2.5.4) but solution spectra may have different appearances from those of solid spectra, since intermolecular forces will be altered.

Many organic compounds exist as polymorphic variations, and these different crystalline forms may also lead to different infrared absorptions. If polymorphism is suspected, the substance should be examined in solution, where all polymorphic forms lose their differences.

KBr disks are prepared by grinding the sample (0.1–2.0 per cent by weight) with KBr and compressing the whole into a transparent wafer or disk. The KBr must be dry, and it is an advantage to carry out the grinding under an infrared lamp to avoid condensation of atmospheric moisture, which gives rise to broad absorption at 3500 cm^{-1}. This can be alleviated by having a blank disk in the reference beam, but the best remedy is prevention. The grinding is usually done with an agate mortar and pestle, although commercial ball mills are available; considerable work is needed to achieve good dispersion, and poorly ground mixtures lead to disks that scatter more light than they transmit. The particle size that must be achieved to avoid scattering is less than the wavelength of the infared radiation—that is, less than 2 μm.

Compression to a cohesive disk requires high pressure, the commonest technique being to use a special die (see figure 2.12) from which air can be evacuated before hydraulic compression to about 10 tonnes load. Disks produced in this way are fairly easy to handle (by their edges!), and measure commonly 13 mm in diameter and 0.3 mm in thickness. Less expensive equipment can consist of the special die, used with a simple screw-jack; even a large steel nut with two bolts screwed in from opposite ends with the KBr between them can give satisfactory disks.

Mulls, or pastes, are prepared by grinding the sample with a drop of oil; the mull is then squeezed between transparent windows as for liquid samples. The mulling agent should ideally be infrared-transparent, but this is never true, and the spectrum produced always shows the absorptions of the mulling agent superimposed on that of the sample.

Liquid paraffin (Nujol), whose infrared spectrum is shown in figure 2.1, is transparent over a wide range, and *Nujol mulls* are by far the most widely utilized. The absorptions of the Nujol effectively blank out the regions of C—H *str* and C—H *def*, but these regions can be studied by preparing a mull with a complementary agent containing no C—H bonds: hexachlorobutadiene and chlorofluorocarbon oils are used for this purpose.

The regions of absorption of these three mulling agents are shown in figure 2.13.

Solid films can be deposited onto NaCl plates by allowing a solution in a volatile solvent to evaporate drop by drop on the surface of the flat. Polymers and various waxy or fatty materials often give excellent spectra in this way, but in many other cases the film is too sharply crystalline and therefore opaque.

The infrared spectra of solid surfaces, powders and pastes, etc., are best recorded by the reflectance techniques discussed in section 2S.2, using a FTIR spectrometer.

2.5.4 SOLUTIONS

The sample can be dissolved in a solvent such as carbon tetrachloride, carbon disulfide or chloroform, and the spectrum of this solution recorded. The solution (usually 1–5 per cent) is placed in a *solution cell* consisting of transparent windows with a spacer between them of known thickness; the spacer is commonly of lead or polytetrafluoroethylene, and its thickness determines the path length of the cell—usually 0.1–1.0 mm. A second cell containing pure solvent is placed in the reference beam, so that solvent absorptions are cancelled out and the spectrum recorded is that of solute alone. In FTIR instruments the spectra are subtracted digitally.

Although solvent cancellation is in principle possible throughout the entire range, in those regions where the solvent has strong absorption bands too little light passes to the detector (in either sample or reference beam) for the detector to respond sensitively. For this reason two spectra should be run, using solvents whose transparent regions are complementary. Figure 2.13 shows where the common solvents have strong absorption bands; note that carbon tetrachloride and carbon disulfide are almost perfectly complementary, whereas chloroform is less suitable and is usually only called upon because of its superior solvent powers.

For really careful work it should be appreciated that using two cells of the same path length will lead to overcompensation of solvent absorptions in the reference beam; as an example, with 5 per cent solute and 95 per

54

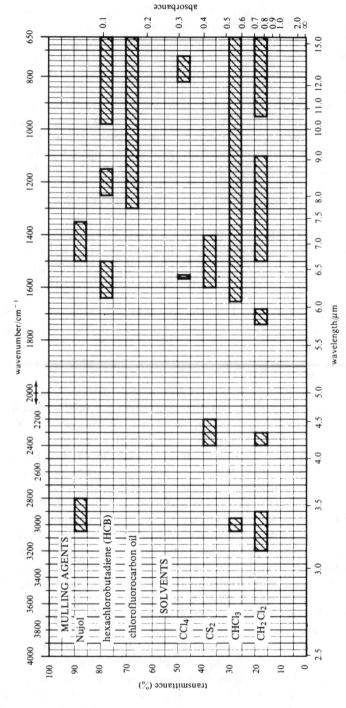

Figure 2.13 *Infrared absorption regions for common mulling agents and solvents. (Not all of these absorption regions contain strong bands.)*

cent solvent in the sample beam, exact compensation would be achieved when the reference cell had a path length only 95 per cent of that of the sample cell. When this is important (as in quantitative measurements; see section 2S.1) a *variable path-length cell* can be used whose path length can be accurately adjusted in deference to the concentration of the solution (see figure 2.12).

Solution spectra, with their absence of complicating features introduced by intermolecular forces and polymorphism, in general make the exact comparison of infrared data from different sources more valid. The solvent, however, must always be specified on the spectrum, since band frequencies do shift on changing to solvents of different polarity, etc. This is particularly true of $C=O$ bands and hydrogen-bonded systems (see section 2.3.2), and for this reason routine organic solution spectra are invariably recorded using carbon tetrachloride, carbon disulfide, chloroform, and occasionally dichloromethane, as solvents.

2.6   APPLICATIONS OF INFRARED SPECTROSCOPY—IDENTITY BY FINGERPRINTING

Infrared spectra contain many absorptions associated with the complex interacting vibrating systems in the molecule, and this pattern of vibrations, since it is uniquely characteristic of each molecule, gives rise to a uniquely characteristic set of absorption bands in the spectrum.

This band pattern serves as a *fingerprint* of the molecule; the region that contains a particularly large number of unassigned vibrations (and is most valuable in this respect) is roughly from 900 cm^{-1} to 1400 cm^{-1}, and this general area is often called the *fingerprint region*.

To identify an unknown compound, one need only compare its infrared spectrum with a set of standard spectra recorded under identical conditions.

Substances that give the same infrared spectra are identical.

This proof of identity is far more characteristic than the comparison of any other physical property.

Having stated the principle, a few cautionary words should be added. For two spectra to be really identical would involve recording the spectra on the same machine under identical conditions of sampling, scan speed, slit widths, etc. Where this condition does not apply, some discretion must be allowed, but, in general, the greater the number of peaks in the fingerprint region the more reliable the proof of identity.

Digitized infrared spectra lend themselves very well to automatic computer library searching (see section 1S.1), and large compilations of spectra are held by many industrial companies: Sadtler Research Laboratories publish collections of infrared spectra in digital form for use by individuals and institutions on their own computers. Even if big enough computers existed which were able to store the spectra from all known compounds (and there are about seven million, increasing by a quarter of a million each year), a library search could never determine the structure of a new unknown compound; at best, it could print out those stored spectra which have similar features to those of the unknown, but the chemist would thereafter have to interpret this information in conjunction with other data before a structure could be deduced.

Small changes in large molecules may produce very little change in the spectrum. For example, the infrared spectrum of a C_{20} straight-chain alkane is quite indistinguishable from that of its next higher straight-chain homolog. For a distinction of this magnitude, mass spectrometry would be the method of choice (see chapter 5).

2.7 APPLICATIONS OF INFRARED SPECTROSCOPY— IDENTIFICATION OF FUNCTIONAL GROUPS

By examining a large number of compounds known to contain a functional group, we can establish which infrared absorptions are associated with that functional group; we can also assess the range of frequencies within which each absorption should appear. Exactly this kind of information is set out in the *correlation charts* which follow (pp. 60–71).

Now, working in the converse, if we have an unknown compound whose functional groups we wish to identify, we can examine its infrared spectrum and use the correlation data to deduce which are the functional groups present.

It is impossible to arrive at a wholly systematic method for dealing with an infrared spectrum. All other evidence should be assessed simultaneously, be it chemical, physical or spectroscopic; even the known history of the compound can be revealing. It is *not* possible to identify a compound merely by interpreting its infrared spectrum from correlation data.

We shall discuss in detail the strengths and weaknesses of the method in the following pages, but it is useful now to make a few clear statements on the general principles involved.

(i) Most weight can be placed on the absorptions above 1400 cm^{-1} and below 900 cm^{-1}. (The fingerprint region, 900–1400 cm^{-1}, contains many unassigned absorptions.)

(ii) *Group frequencies* are more valuable than single absorption bands. In other words, a functional group that gives rise to *many* characteristic absorptions can usually be identified more definitely than a function that gives rise to only one characteristic absorption. (Thus, ketones (C=O *str*) are less easily identified than esters (C=O *str* and C—O *str*); esters are less easily identified than amides (C=O *str*, N—H *str*, N—H *def*); etc.)

(iii) The absence of a characteristic absorption may be more illuminating than its presence. (Consider the relative implications of the presence or absence of a C=O *str* absorption.)

(iv) Multifunctional compounds will show the separate absorptions of the individual functional groups, unless these interact. (Examples of *interacting functional types* are β-diketones, aliphatic amino acids, γ-hydroxy acids, etc.)

(v) The frequencies shown graphically on the correlation charts do not take account of any exceptional features in specific molecules; important circumstances, which might in this way lead to frequency shifts outside the quoted ranges, are discussed below.

(vi) *Graphically presented correlation charts are invariably sufficiently accurate for functional group identification*; coupled with this, frequencies of bands cannot easily be measured more accurately than ±5 cm^{-1} on low-cost routine instruments (with lesser accuracy at higher frequencies).

More accurate tabular data are to be found in the specialist texts of Bellamy, Cross and van der Maas, and such tables should be used for studying the restricted frequency ranges of more narrowly specific classes of organic molecule. The discussions of group frequencies that follow contain some more detailed frequency data (see sections 2.8–2.13).

A reasonable initial plan of attack on the infrared spectrum of a totally unknown compound would be to spend a few minutes searching out the most commonly successful correlations.

The *carbon skeleton* should be tackled first (see section 2.8): look for evidence of alkane, alkene, alkyne and aromatic residues (using C—H *str*, C—H *def* and the various carbon–carbon bond stretching frequencies). Evidence from the NMR spectrum is of great complementary value.

Look for C=O *str*; if present, it may be associated with C—H *str* in aldehydes, N—H *str* in amides, C—O *str* in esters; etc.

Look for O—H *str* or N—H *str*.

Look for C≡N *str*.

In sulfur compounds look for S—H *str*, S=O *str* and —SO$_2$— *str*.

In phosphorus compounds look for P=O *str*.

Chemical and/or mass spectrometric evidence for the presence of nitrogen, sulfur, halogens or phosphorus, etc., is essential: infrared evidence of these is self-sufficient *only* in exceptional cases.

The correlation charts are set out in the order indicated by the above approach to an infrared identification of functional groups. Thus, 'aromatic carbon skeletons' precede alkane, since the former are frequently easier to confirm. The charts themselves contain commentary, which will frequently suffice to make an assignment, but additional discussion of each class is given after the charts.

The frequency ranges shown on the charts apply to spectra recorded on liquid films or KBr disks, or Nujol mulls, etc., since most routine spectra will be recorded thus. Where dilute solution spectra produce substantial shifts, this is indicated on the charts and in the discussion following.

CORRELATION CHARTS

CHART 1 contains assignments that will identify the nature of *the carbon skeleton*, identifying (i) aromatic, (ii) alkane, (iii) alkene and (iv) alkyne residues. (See also section 2.8.)

CHART 2 is concerned with *carbonyl compounds*, including aldehydes, ketones, enols and all carboxylic acid derivatives. (See also section 2.9.)

CHART 3 deals with *hydroxy compounds and ethers*, but excluding carboxylic acids (see previous chart); alcohols, phenols, epoxides and carbohydrates appear here. (See also section 2.10.)

CHART 4 begins with the assignments of *nitrogen compounds* that are basic (amines, etc.); nitro compounds, nitriles and isonitriles also appear here. (See also section 2.11.)

CHART 5 shows the limited number of carbon–halogen bond assignments that routine infrared spectra afford in *halogen compounds*; acyl halides have the more valuable C=O *str* correlation, and appear in chart 2: sulfonyl halides have the more valuable S=O *str* and appear in chart 6. (See also section 2.12.)

CHART 6 covers *sulfur* and *phosphorus compounds*, which may contain other heteroatoms (nitrogen and halogen). (See also section 2.13.)

Abbreviations: s = strong, m = medium, w = weak and v = variable intensity, respectively: anti = antisymmetric, sym = symmetric.

2.8 THE CARBON SKELETON (CHART 1)

In deducing the nature of carbon residues in an organic molecule by infrared spectroscopy, it is useful to note that (a) aromatic groups are most easily detected from C=C *str* and C—H *def* absorptions; (b) alkene groups may be detected from C=C *str* absorptions, unless aromatic residues are also present; (c) alkane residues are detected from C—H *str* and C—H *def* absorptions (but NMR, in general, is more specific in detecting particular groupings such as Me, Et, Pr^n, Pr^i, Bu^t); and (d) terminal alkynes are very easily detected from C≡C *str* and C—H *str* absorptions, but nonterminal alkynes may cause extreme difficulty.

2.8.1 Aromatics (Chart 1(i))

Four regions of the spectrum are associated with the confidently assignable vibrations of aromatic residues (C—H *str*, C═C *str*, C—H *def* and a group of overtone-combination bands).

The positions of C—H *str* absorptions are shown on the charts: these weak absorptions may appear merely as a shoulder on the stronger alkane C—H *str* bands, and indeed may be swamped by them. The distinction can usually be clearly made that aromatic and alkene C—H *str* is just *above* 3000 cm^{-1}, while alkane C—H *str* is just below.

Most aromatic compounds show three of the four possible C═C *str* bands: the band at 1450 cm^{-1} is often absent, and the two bands near 1600 cm^{-1} occasionally coalesce, so that two, three or four bands must be expected for this assignment.

Out-of-plane C—H deformations are strongly coupled vibrations, and the pattern of absorption is reasonably characteristic of the number of hydrogen atoms on the ring (and, hence, also characteristic of the *substitution pattern* on the ring). The most reliable is the strong band above 800 cm^{-1} for *p*-substituted benzenes. The out-of-plane C—H *def* bands are strong and broad, being typically 30–50 cm^{-1} wide at half height, and are therefore easily identified on the spectrum.

Because of the ambiguity in OOP C — H def band positions, NMR evidence for substitution pattern should always accompany infrared evidence.

Chart 1 shows the OOP C—H *def* absorptions for benzene and naphthalene rings: corresponding data for heterocyclic rings, etc., are given in Bellamy (see 'Further Reading').

Bands corresponding to aromatic C—H *str*, C═C *str* and out-of-plane C—H *def* are clearly seen in figures 2.3, 2.4 and 2.11. The aromatic C—H *str* bands can *not* be assigned in figures 2.7 and 2.16.

The bands shown in figure 2.11 (polystyrene film) between 1650 cm^{-1} and 2000 cm^{-1} are overtone and combination bands mainly of the aromatic C—H *def* absorptions: the nature of such bands is discussed in section 2.2.2. They appear in all spectra of aromatic compounds, but are usually too weak to be clearly interpreted unless large sample sizes are used. Since they are related to the out-of-plane C—H *def* vibrations, they too show a dependence on ring substitution; the band positions do not move on changing substitution; rather, their relative intensities alter. They are now seldom used for structural deductions where NMR evidence is available.

2.8.2 Alkanes and Alkyl Groups (Chart 1(ii))

Both C—H *str* and C—H *def* absorptions in saturated aliphatic groups are strong or medium absorptions, and there is seldom difficulty in assigning these bands on the spectrum.

CHART 1. The Carbon Skeleton

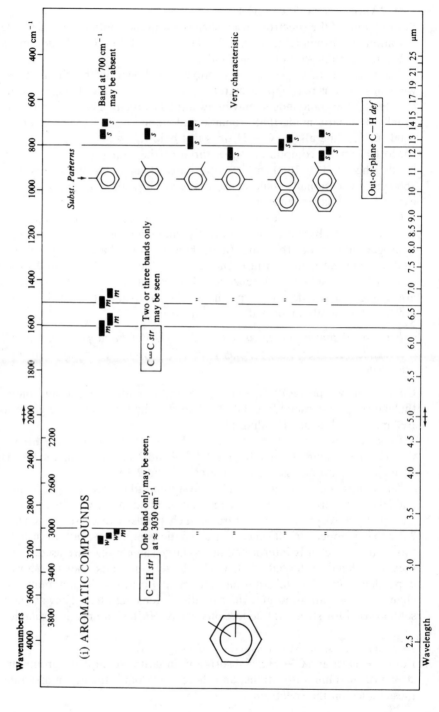

CHART 1. The Carbon Skeleton (*cont'd*)

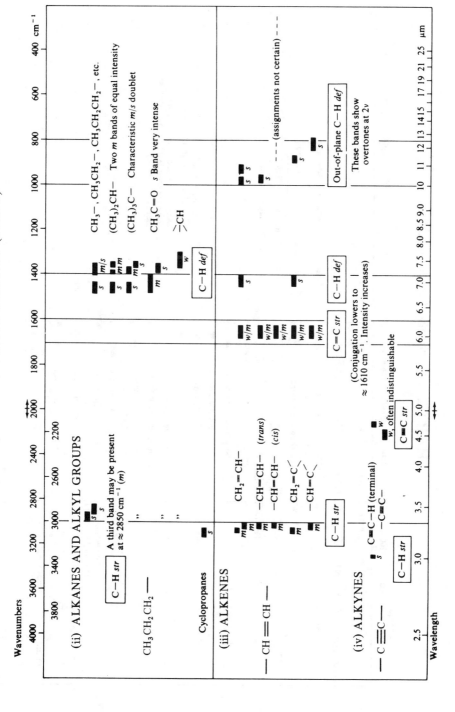

CHART 2. Carbonyl Compounds

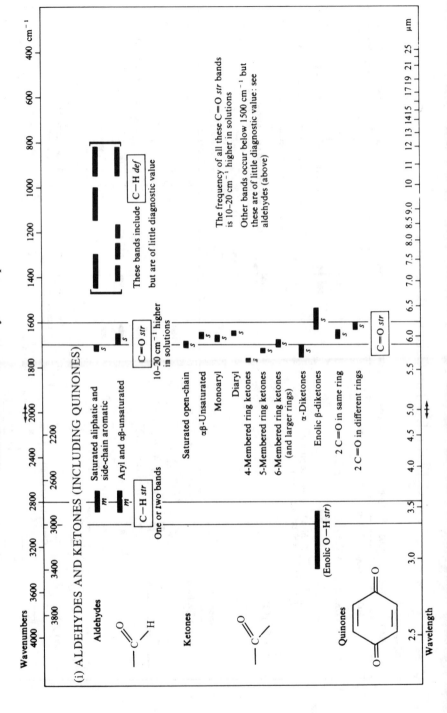

CHART 2. Carbonyl Compounds (*cont'd*)

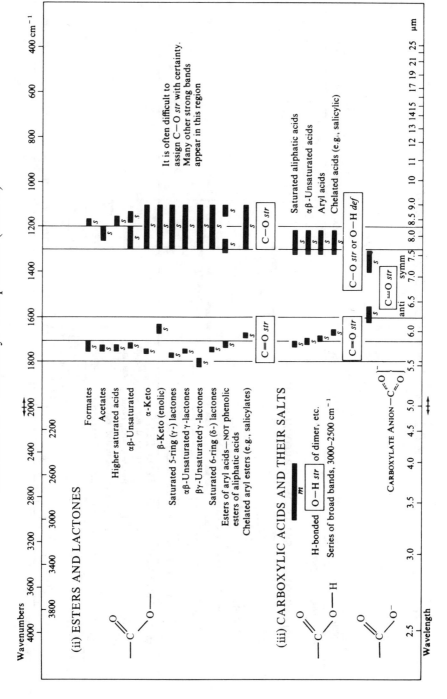

64

CHART 2. Carbonyl Compounds (*cont'd*)

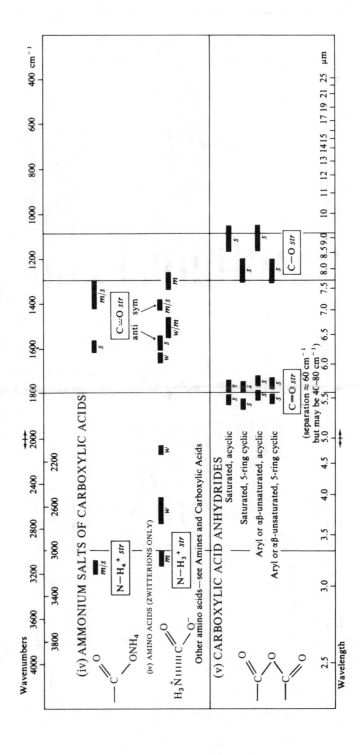

CHART 2. Carbonyl Compounds (*cont'd*)

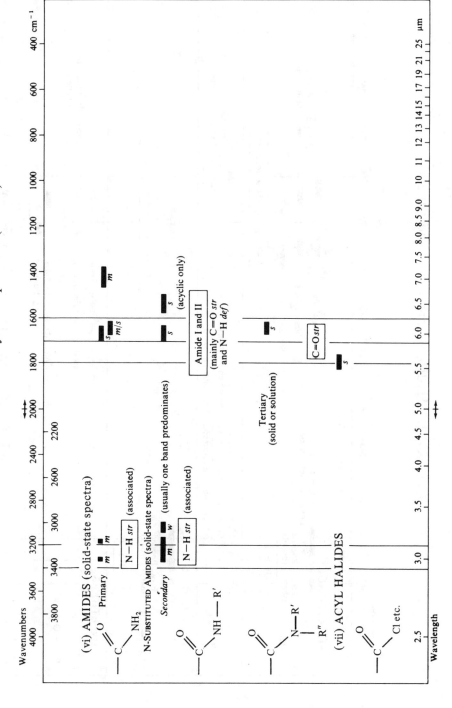

CHART 3. Hydroxy Compounds and Ethers

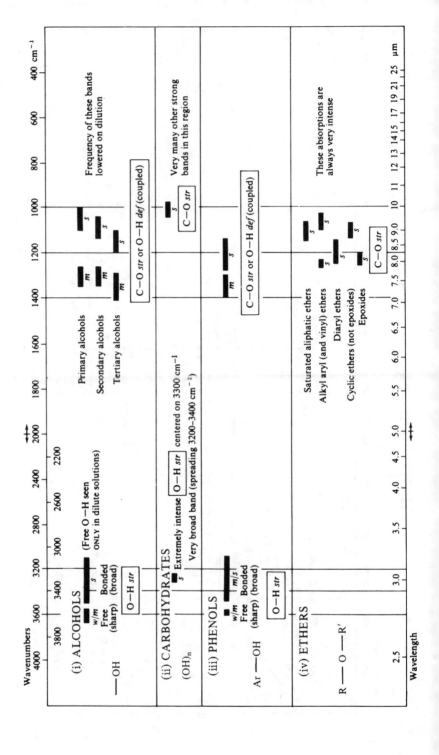

CHART 4. Nitrogen Compounds

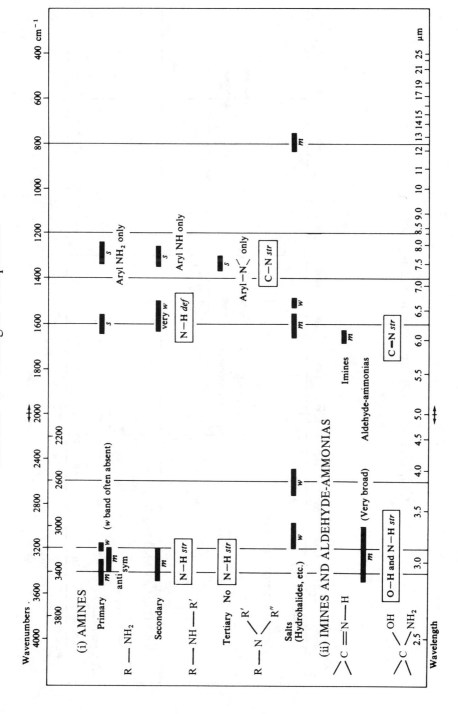

68

CHART 4. Nitrogen Compounds (*cont'd*)

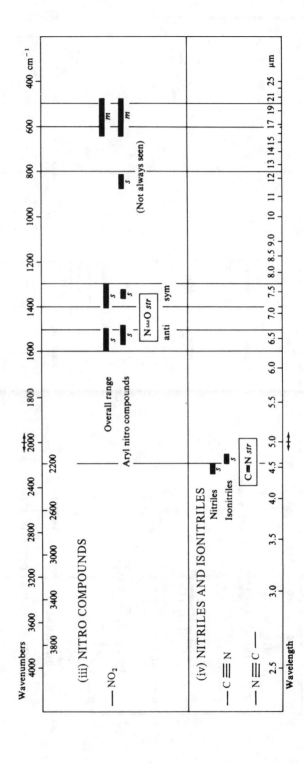

CHART 5.   Halogen Compounds

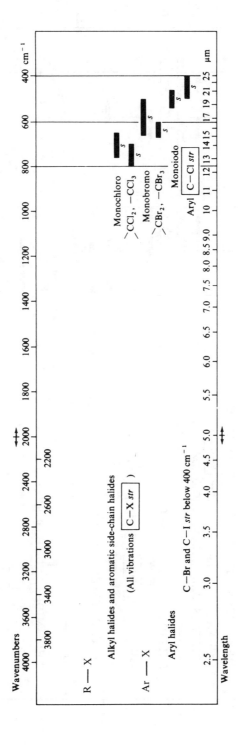

70

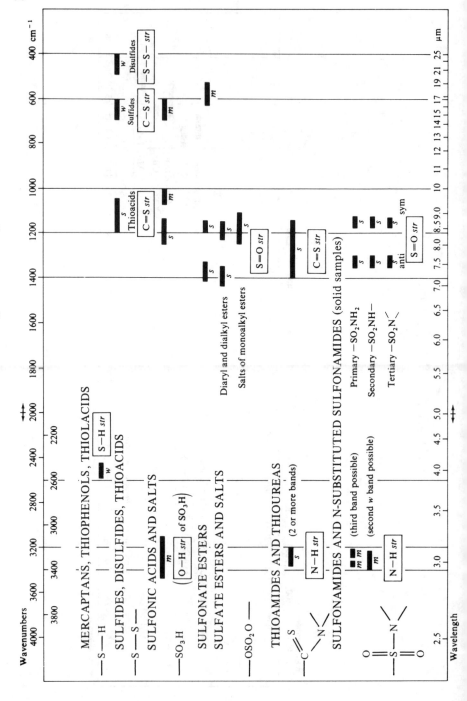

CHART 6.   Sulfur and Phosphorus Compounds

CHART 6. Sulfur and Phosphorus Compounds (*cont'd*)

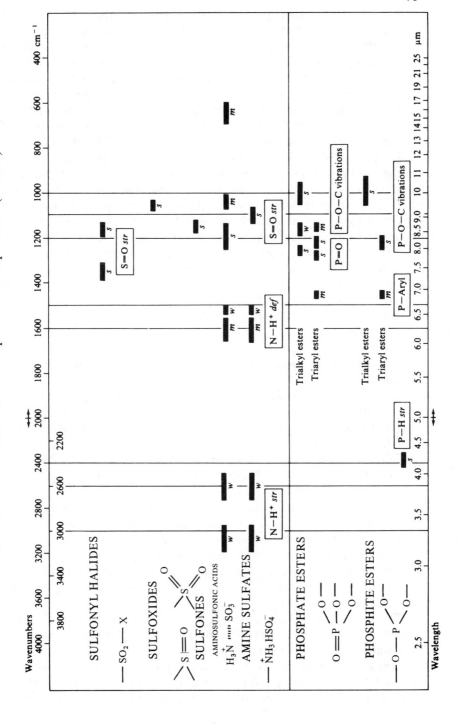

By far the commonest appearance of C—H *str* absorptions is two strong bands just below 3000 cm^{-1} (of which the higher frequency is antisymmetric stretch. See figures 2.1, 2.2, 2.6 and 2.10: this last figure illustrates nicely that aliphatic C—H *str* lies just below 3000 cm^{-1}, with aromatic C—H *str* just above. In spectra recorded on very-high-resolution instruments, simple alkanes show four C—H *str* bands corresponding to CH_3 and CH_2 antisymmetric C—H *str* (high-frequency pair) and the corresponding symmetric C—H *str* (pair at lower frequency): this degree of resolution is exceptional in compounds other than simple alkanes.

The pattern of bands corresponding to C—H *def* may be characteristic of the alkyl groups present, *provided that* no strong electrical influence is close by in the molecule. In particular, the antisymmetric CH_3 *def* around 1390 cm^{-1} is split in $(CH_3)_2C=$ and $(CH_3)_3C—$ groups, and such groups can be detected with reasonable certainty from the infrared spectrum (see figure 2.15): otherwise, the commonest appearance of the C—H *def* bands is as shown in figures 2.1 and 2.6, which gives no clue to the alkyl grouping present. In general, NMR evidence is preferred.

Other bands in the infrared spectra of alkanes and alkyl groups are of doubtful origin; the bands from 800 cm^{-1} to 1300 cm^{-1} should be considered as C—C *str* or C—H *def*, and thereafter disregarded.

2.8.3   ALKENES (CHART 1(iii))

Most of the useful assignments are shown on the correlation chart for alkenes.

The similarity to aromatic absorptions is immediately apparent, and it is often difficult to detect an alkene group *within an aromatic molecule* by infrared spectroscopy: conjugation of the double bond with phenyl lowers C=C *str* to around 1630 cm^{-1}.

Out-of-plane C—H *def* bands are easily seen, and are usually strong enough for their overtone bands to be substantial: since $2 v$ lies in the region 1800–2000 cm^{-1}, where few other absorptions appear, the overtone band may be prominent. *Trans* alkenes can often be distinguished from the *cis* isomer by the former's C—H *def* band around 970 cm^{-1}; the corresponding band for *cis* isomers is often around 700 cm^{-1}, but it is not so certainly identified as that of the *trans* isomer.

Tetrasubstituted alkene groups such as $R_2C=CR_2$ show either no C=C *str* band or, at best, a weak band: they do show a Raman band, however (see section 2S.3).

The infrared spectrum of limonene (figure 2.14) shows alkene C—H *str* and C=C *str* : the C—H *def* absorption around 900 cm^{-1} shows an overtone band around 1800 cm^{-1}.

Cumulative double bonds, as in allenes, give rise to strong absorptions around 2000 cm^{-1}: these are thought to be the antisymmetric C=C *str*, etc., and a few of the more common examples are: C=C=C (allenes),

73

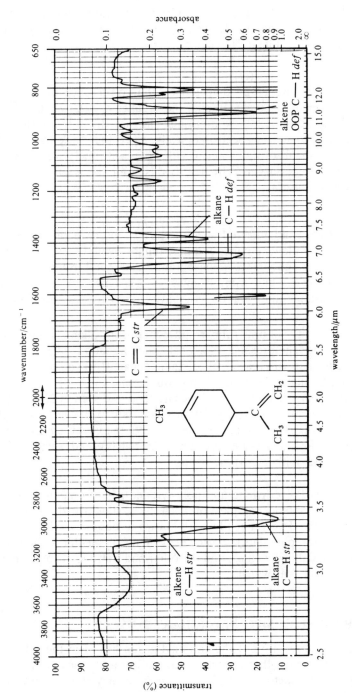

Figure 2.14 *Infrared spectrum of limonene. Liquid film.*

1950 cm^{-1}; O=C=O, 2350 cm^{-1}; C=C=O (ketenes), 2150 cm^{-1}; —N=C=O (isocyanates), 2250 cm^{-1}; —N=C=S (isothiocyanates), 2100 cm^{-1}; —N$_3$ (azides), 2140 cm^{-1}.

2.8.4 ALKYNES (CHART 1(iv))

Terminal alkynes are easily detected by the co-presence of C—H *str* (strong band near 3300 cm^{-1}) and the weaker C≡C *str* band near 2200 cm^{-1}: see figure 2.2. Nonterminal alkynes are one of the most difficult classes to detect spectroscopically, since they lack the valuable alkyne C—H *str* absorption, and C≡C *str* is extremely weak: Raman spectroscopy is necessary to help identify this group (see section 2S.3) and ^{13}C NMR is also useful.

Example 2.2

Question. How could infrared spectroscopy be used to distinguish the members of the following pairs, and how reliable would these distinctions be? (a) Natural rubber (*cis*-polyisoprene) and butyl rubber (polyiso-butene); (b) the *cis* and *trans* isomers of 3-hexene; (c) 1-hexyne and 3-hexyne? Assume that you do not have access to any authenticated spectra.

Model answer. (a) Natural rubber still contains an alkene group; butyl rubber does not. The presence of a weak/medium band for C=C *str* will identify the former with reasonable certainty: chart 1(iii) and section 2.8.3. (b) The out-of-plane C—H *def* bands for *cis* and *trans* alkenes appear, respectively, near 970 and 700 cm^{-1}: chart 1(iii) and section 2.8.3. This is usually fairly reliable, especially if both spectra are available for comparison. (c) 1-Hexyne is a terminal alkyne and both the C≡C *str* and alkyne C—H *str* bands are easily assigned: chart 1(iv) and section 2.8.4. 3-Hexyne shows neither of these bands.

Exercise 2.2

(a) Polyethylene and polystyrene are both manufactured in film form: how could samples of each be distinguished by infrared spectroscopy? (b) Two other films are known to be polyacrylonitrile and poly(ethylene terephthalate), respectively. One of the films showed an absorption band at 2200 cm^{-1}, and the other had a strong absorption near 1710 cm^{-1}: which was which?

2.9 CARBONYL COMPOUNDS (CHART 2)

The stretching absorption of the carbonyl group has been the subject of more study than any other absorption in the infrared; consequently, our knowledge of the factors that give rise to frequency shifts is extensive. These factors have been discussed in section 2.3.

While the presence of C=O *str* absorption is almost always easily detected on the spectrum, identification of the carbonyl-containing function is not always feasible simply by noting the C=O *str* frequency. As ever, the functions that are easiest to detect are those that give rise to the greatest number of characteristic absorptions. Chemical and/or NMR evidence is particularly helpful in the case of aldehydes, quinones, esters, carboxylic acids and amides.

The C=O *str* band is always strong, and, consequently, its overtone at $2v$ is often visible around 3400 cm^{-1}: this overtone band can be clearly seen in figures 2.4 and 2.15.

2.9.1 ALDEHYDES AND KETONES (INCLUDING QUINONES) (CHART 2(i))

Aldehydes are usually distinguishable from ketones, etc., by their C—H *str* absorptions: in both classes the absorptions below 1500 cm^{-1} are of little diagnostic value, one possible exception being the case of methyl ketones (see figure 2.4), which give rise to a very strong characteristic absorption just below 1400 cm^{-1} (noted on chart 1(ii)). Enols are easily identified by the broad H-bonded O—H *str* absorption, and by the very low C=O *str* frequency—as low as 1580 cm^{-1} in acetylacetone (which is about 85 per cent enolic). A sharp strong C—O *str* band also appears near 1200 cm^{-1}.

It is important to note the narrow limits of C=O *str* frequencies for these compounds, and an industrious attempt must be made to relate this frequency to the detailed environment of the carbonyl group. A good example would be the cyclic ketones, where ring size can be gauged accurately from the position of the C=O *str* band. Such assignments must be in accord with deductions concerning the carbon skeleton, and with ^{13}C NMR evidence.

2.9.2 ESTERS AND LACTONES (CHART 2(ii))

In addition to accurately known C=O *str* frequencies, the C—O *str* band(s) for esters and lactones can be highly informative. Thus, most alkanoate esters show one C—O *str* band, while aryl and αβ-unsaturated carboxylate esters show two. (See bands at 1120 cm^{-1} and 1280 cm^{-1} in figure 2.15.)

As indicated in chart 2(ii), the position of the C=O *str* band for saturated esters is near 1740 cm^{-1}.

If the carbonyl group is in conjugation with a double bond or an aromatic ring (such as in the acrylates, CH$_2$=CH—CO—O—, or in the benzoates, Ph—CO—O—), the frequency is lowered to near 1720 cm^{-1}: chelation of the C=O group to a nearby OH group (such as in the salicylates) lengthens and weakens the C=O bond even further, and the stretching frequency moves to near 1680 cm^{-1}.

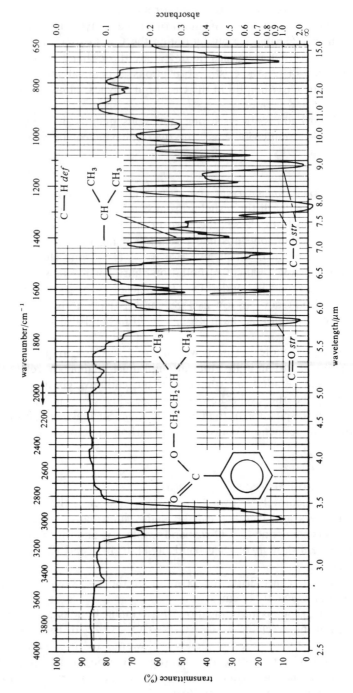

Figure 2.15 *Infrared spectrum of isoamyl benzoate. Liquid film.*

In   contrast,   for   esters   containing   the   groupings
—CO—O—CH=CH— or —CO—O—Ar (such as vinyl esters or
phenyl esters), the C=O *str* band appears at much higher frequency, near
1770 cm^{-1}. For the special case of βγ-unsaturated γ-lactones, where ring
strain also has an effect, the frequency is near 1800 cm^{-1}.

^{13}C NMR spectroscopy offers additional help in classifying esters, from
the chemical shifts of the carbonyl carbon atoms.

2.9.3 CARBOXYLIC ACIDS AND THEIR SALTS (CHART 2(iii))

Carboxyl groups are one of the easiest functions to detect by infrared
spectroscopy because of the co-presence of C=O *str* with the exceedingly
broad O—H *str* absorption centered around 3000 cm^{-1}. The infrared
spectrum of benzoic acid (figure 2.7) is typical. The broad O—H *str* band
corresponds to the dimer structure discussed in section 2.3: O—H *def* in
the dimer gives rise to a characteristic band near 950 cm^{-1}, seen in the
benzoic acid spectrum in figure 2.7.

Carboxylic acid salts are most accurately represented with the
resonance-stabilized carboxylate anion, which contains two identical
C$\doteq$O bonds, and therefore gives antisymmetric and symmetric stretching
absorptions. Very reliable proof of identity involves converting the salt to
the free acid, when the true C=O *str* absorption appears at higher
frequency than C$\doteq$O *str*.

Many carboxylate salts contain water of crystallization, which produces
broad absorption around 3400 cm^{-1}. It is usually possible nevertheless to
separate this from N—H *str* bands in the specific case of ammonium salts.

Example 2.3

Question. How might the following pairs of isomers be distinguished
from their infrared absorption spectra, assuming you have both spectra
for each pair? (a) Propiophenone, PhCOCH$_2$CH$_3$, and phenylacetone,
PhCH$_2$COCH$_3$;   (b)   propiophenone   and   3-phenylpropanal,
PhCH$_2$CH$_2$CHO; (c) 2,5-hexanedione and 2,4-hexanedione; (d) methyl
benzoate and phenyl acetate.

Model answer. (a) Propiophenone is a monoaryl ketone; in phenylacetone
the carbonyl group has alkyl groups on both sides, and in infrared terms it
is considered a dialkyl ketone. The C=O *str* band for the former will be at
lower frequency than for the latter: chart 2(i) and section 2.9.1. (b) Apart
from the position of the C=O *str* bands (lower frequency in the former,
since it is a monoaryl ketone), the aldehyde C—H *str* bands near
2800 cm^{-1} will characterize the latter. (c) 2,4-Hexanedione is a β-diketone
and will be highly enolic. Its spectrum will show broad O—H *str*
absorption and a very-low-frequency C=O *str* band: chart 2(i) and section
2.9.1. 2,5-Hexanedione will show the infrared spectral qualities of a
simple dialkyl ketone. (d) The C=O *str* bands of these two are near

1720 cm^{-1} for the benzoate (conjugation on the C=O group of the ester) and near 1770 cm^{-1} for the aryl acetate (conjugation on the —O— group of the ester): section 2.9.2.

Exercise 2.3   The hydrolysis of an ester, RCOOR, in dilute aqueous KOH produces the alcohol and the potassium salt of the carboxylic acid: the progress of the hydrolysis can be followed by infrared spectroscopy. Suggest how this procedure might be carried out, using a prominent infrared feature as a marker.

Exercise 2.4   An air conditioning filter was badly contaminated by an oily substance, which might have been a mineral oil (engine lubricating oil) or a vegetable oil (cooking oil). The IR spectrum of the contaminant showed a strong absorption near 1720 cm^{-1}: which was the contaminant?

Exercise 2.5   Food is often packaged in clear foil, which may be Cellophane (cellulose acetate) or polypropylene: how could the infrared spectrum distinguish these?

2.9.4 AMINO ACIDS (CHART 2(iv))

Amino acids with the amino group directly attached to an aryl ring exist with the *free* amino and carboxyl functions: all other amino acids (including all naturally occurring members) exist as zwitterions, and therefore show absorptions due to —N—H *str* and —CO$_2^-$ *str*. (Infrared studies historically provided some of the best evidence for the very existence of zwitterions.) On acidification the protonated amino acid then gives rise to true C=O *str* at higher frequency than C=O *str*.

zwitterion

2.9.5   CARBOXYLIC ACID ANHYDRIDES (CHART 2(v))

The characteristic high-frequency *double* band for C=O *str* (near 1800 cm^{-1}) for anhydrides makes them very easy to detect: the splitting of the band is due to Fermi resonance (see section 2.3). A common contaminant is the free acid, with O—H *str* absorption around 3000 cm^{-1}. Interestingly, imides also give rise to two coupled C=O vibrations, their frequencies being 1680–1710 cm^{-1} and 1730–1790 cm^{-1}, respectively.

2.9.6 AMIDES (PRIMARY AND N-SUBSTITUTED) (CHART 2(vi))

Few liquid amides exist. (Exceptions: HCONMe$_2$, HCONEt$_2$.)

The infrared spectrum of benzamide (figure 2.16) shows the very characteristic pair of bands for N—H *str* near 3200 cm^{-1} and 3400 cm^{-1}: the separation between these bands in solid-state spectra (120–180 cm^{-1}) is usually sufficient to identify primary amides.

The two bands are not associated simply with antisymmetric and symmetric N—H *str*, but arise from different degrees of association in amides: indeed, a third band (near 3300 cm^{-1}) is present, but not observed in KBr-disk sampling.

Amides readily self-condense, as do carboxylic acids, but in amides the situation is additionally complicated by (a) the considerable double-bond character of the CO—N bond due to mesomerism which may lead to *cis* and *trans* forms, and (b) the degree of vibrational coupling that takes place (N—H *str* may couple with both N—H *str* and even N—H *def*: N—H *def* couples strongly with C=O *str* and C—N *str*; etc.). These three factors (self-association, *cis–trans* isomerism and vibrational coupling) completely rule out a simple reductive description of amide absorptions: for clarity, and at the risk of oversimplification, the charts show only those absorptions that are important in solid-state spectra. In dilute solution the degree of association changes: N—H *str* moves to higher frequency, while the coupled C=O *str* and N—H *def* vibrations move to lower frequency.

N—H *str* vibrations arise either from monomer, dimer, chains of dimers or polymers as shown: the number of units in the dimer chain or polymer determines the N—H *str* frequencies, which therefore shift with concentration changes. Primary amides in very dilute solution give N—H *str* near 3400 cm^{-1} and 3500 cm^{-1}: secondary, near 3450 cm^{-1}.

C=O *str* and N—H *def*, being coupled vibrations, give rise to bands that are best named simply as amide I and II bands, respectively: these can be seen in figure 2.16 near 1650 cm^{-1}, and this double band is also highly characteristic of amides. The higher-frequency band is predominantly C=O *str* and the lower is predominantly N—H *def*. Other amide bands of mixed assignment are named amide III, IV, V and VI.

Example 2.4

Question. Discuss what features of their infrared spectra could help in distinguishing the following pairs? (a) 4-(Aminomethyl)benzoic acid, H$_2$NCH$_2$C$_6$H$_4$COOH, and 4-aminophenylacetic acid, H$_2$NC$_6$H$_4$CH$_2$COOH; (b) butanamide, CH$_3$CH$_2$CH$_2$CONH$_2$, and *N*-methylpropanamide, CH$_3$CH$_2$CONHCH$_3$.

Model answer. (a) Aryl amino groups are more weakly basic than alkyl amino groups, and, in general, do not protonate in the molecules of amino acids to give zwitterions. Thus, while 4-aminophenylacetic acid shows the separate characteristics of both NH$_2$ and COOH groups (N—H *str* at

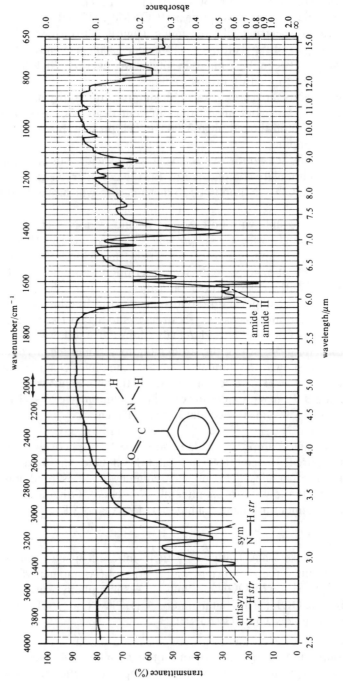

Figure 2.16 *Infrared spectrum of benzamide. KBr disc.*

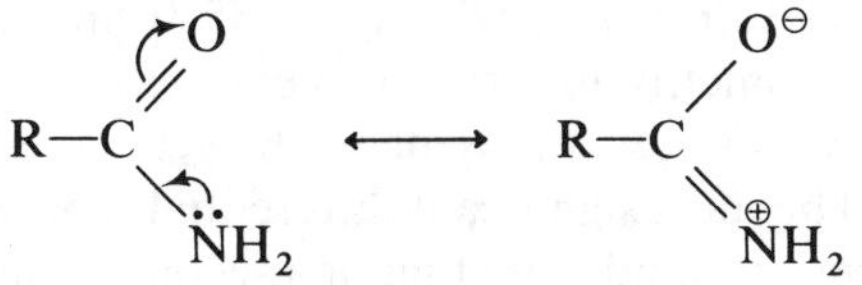

mesomerism in amides

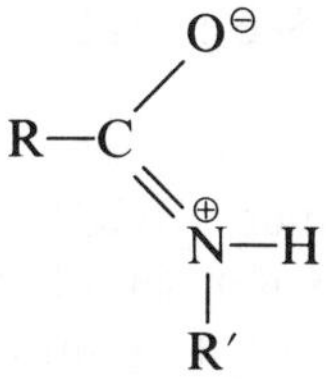

cis monomer

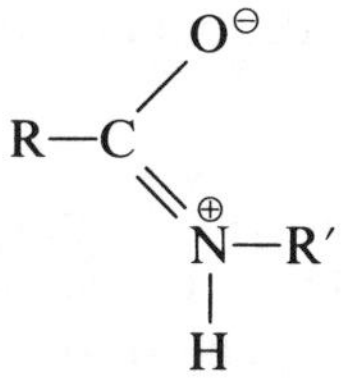

trans monomer

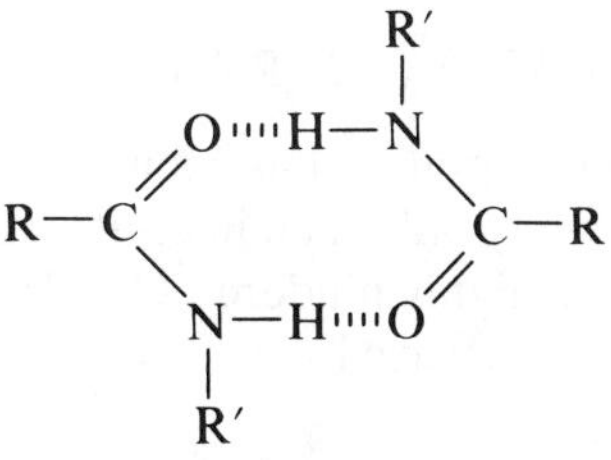

cis dimer

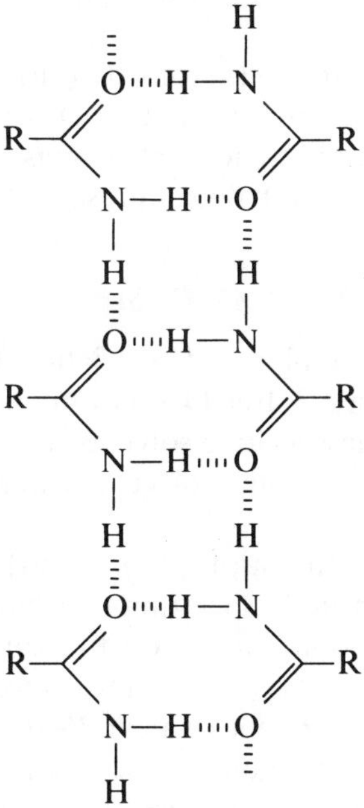

chains of dimers

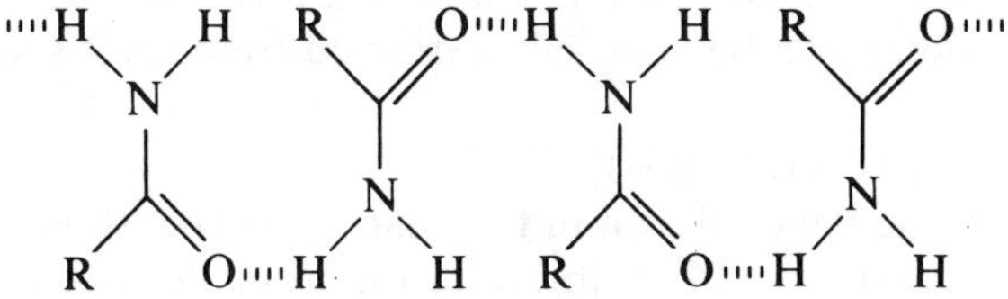

polymer

3300 cm^{-1}, broad OH *str* from 2500 cm^{-1} to 3500 cm^{-1} and C=O *str* near 1700 cm^{-1}), 4-(aminomethyl)benzoic acid exists as a zwitterion and shows the characteristic C=O *str* bands near 1600 cm^{-1}. See chart 2(iv) and section 2.9.4. (b) The most significant differential is between the N—H *str* absorption of a primary amide and that of a secondary amide. They appear near 3300 cm^{-1}; two bands, about 170 cm^{-1} apart, are shown for the primary amide, but only one for the secondary. See chart 2(vi) and section 2.9.6.

Exercise 2.6 The hydrolysis of a primary amide, RCONH$_2$, in dilute aqueous KOH produces ammonia and the potassium salt of the carboxylic acid: the progress of the hydrolysis can be followed by infrared spectroscopy. Which absorption bands will be easiest to use for this procedure? Note the problem of aqueous media.

2.9.7   ACYL HALIDES (CHART 2(vii))
The high-frequency C=O *str* band for acyl halides is reasonably characteristic, but chemical or mass spectral evidence is needed to detect and identify the halogen present.

2.10 HYDROXY COMPOUNDS AND ETHERS (CHART 3)

Hydrogen-bonded association in hydroxy compounds has been discussed in section 2.3; free O—H *str* bands are normally only seen in dilute solutions in nonassociating solvents. (Exceptions are highly hindered O—H groups or certain intramolecularly bonded *o*-hydroxybenzyl alcohols.)

2.10.1   ALCOHOLS (CHART 3(i))
The infrared spectrum of 1-butanol is typical, showing all of the expected absorptions shown on the correlation chart (see figure 2.6): the dotted trace on this spectrum was recorded on a dilute solution in CCl$_4$, and free O—H *str* appears near 3600 cm^{-1}.

Classification of an alcohol as primary, secondary or tertiary can frequently be successful using the bands for the coupled vibrations C—O *str* and O—H *def*.

2.10.2   CARBOHYDRATES (CHART 3(ii))
There is no difficulty normally in classifying a carbohydrate as such, but little further detail can be generally extracted from their spectra.

2.10.3   PHENOLS (CHART 3(iii))
In addition to the bands shown in the charts, phenols show the characteristic C=C *str* and C—H *def* vibrations for aromatic residues (chart 1(i)). The strong O—H *str* band may swamp the weaker C—H *str* band just

above 3000 cm^{-1}. It is difficult to distinguish a phenol from an aryl alcohol from infrared evidence.

Exercise 2.7 The NMR spectra of menthol are discussed in chapter 3. See figure 3.33 for the formula of menthol and predict the major infrared absorption bands from 1600 cm^{-1} to 4000 cm^{-1}.

2.10.4 ETHERS (CHART 3(iv))

Aliphatic ethers show only one characteristic band for C—O *str* (in addition to C—H and C—C vibration bands arising from alkyl or aryl groups in the molecule): while this band is easily identified in *known* ethers, certain identification of an ether in an unknown molecule is difficult, because many other strong bands can appear in the region 1050–1300 cm^{-1}.

In highly unsymmetrical ethers, especially alkyl aryl ethers, the two C—O bonds couple to give antisymmetric and symmetric C—O *str* absorptions: for dialkyl or diaryl members, the symmetric C—O *str* is infrared-inactive and only the antisymmetric C—O *str* is seen.

Note that keto ethers (such as *p*-methoxyacetophenone) may be confused with esters, since identification of both types rests on identifying both C=O *str* and C—O *str* . Epoxides are a special case of cyclic ether, and in addition to their C—O *str* band they have a reasonably characteristic C—H *str* absorption near 3050 cm^{-1}, and two bands around 800 cm^{-1} and 900 cm^{-1}, respectively.

Exercise 2.8 A low-boiling solvent labeled 'Petroleum Ether' showed none of the characteristic C—O stretching absorptions shown by ethers around 1100 cm^{-1}, but the spectrum was very similar to that in figure 2.1: what was the solvent and what is a preferred name for it?

2.11 NITROGEN COMPOUNDS (CHART 4)

The correlation chart contains most of the information necessary to assign principal bands associated with nitrogen functions not already met in section 2.9 (and chart 2).

2.11.1 AMINES (CHART 4(i))

Hydrogen bonding in amines leads to modification of both the symmetric and antisymmetric N—H *str* bands, but shifts on dilution are less than for O—H *str* in alcohols and phenols. The values given on the charts refer to condensed phases (liquid films, KBr disks, etc.) and in dilute solution free N—H *str* can be seen near 3500 cm^{-1}. The infrared spectrum of *o*-toluidine (figure 2.17) shows all the characteristics of a primary aromatic amine spectrum: N—H *str* (two bands only may sometimes be seen),

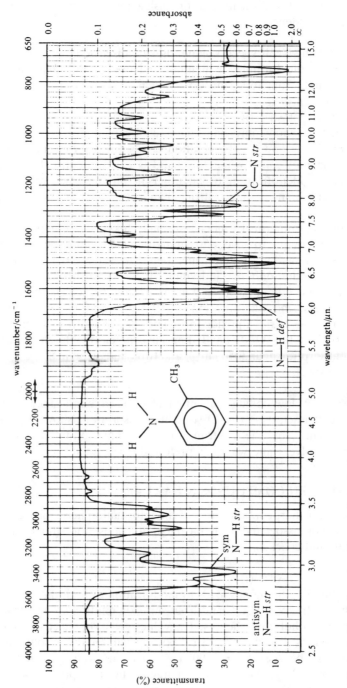

Figure 2.17 *Infrared spectrum of o-toluidine. Liquid film.*

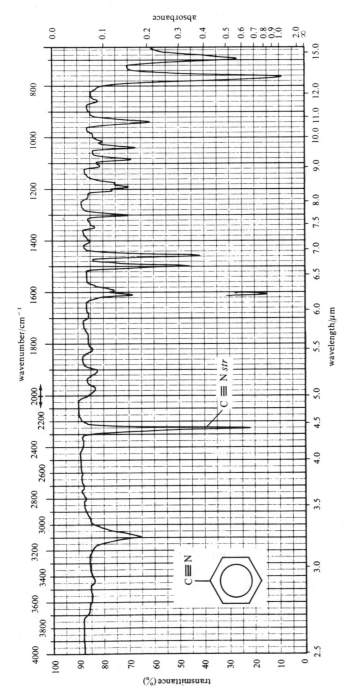

Figure 2.18 *Infrared spectrum of benzonitrile. Liquid film.*

N—H *def* (at 1620 cm^{-1}) and C—N *str* at 1280 cm^{-1} (not identifiable in alkyl amines).

2.11.2   IMINES AND ALDEHYDE-AMMONIAS (CHART 4(ii))
These are both uncommon classes and the chart contains the only useful diagnostic bands for them.

2.11.3   NITRO COMPOUNDS (CHART 4(iii))
The symmetric and antisymmetric N═O *str* bands constitute a reliable method for detecting nitro groups, particularly in aryl nitro compounds (where few other detection methods are simple or reliable).

2.11.4   NITRILES AND ISONITRILES (CHART 4(iv))
Without doubt these two functions are two of the easiest to detect: they each give one strong band, and the spectrum of benzonitrile (figure 2.18) shows clearly C≡N *str* for nitriles at 2250 cm^{-1}, in a spectral region that is usually sparsely populated.

2.12   HALOGEN COMPOUNDS (CHART 5)

The detection of halogen by infrared spectroscopy is totally unreliable. The most useful assignment is C—X *str*, which for most environments appears at frequencies beyond the range of low-cost instruments (that is, below 650 cm^{-1}). The chart nevertheless shows these band positions for the sake of interest and completeness. Mass spectrometry is the method of choice for identification of halogens in organic molecules (chapter 5).

2.13   SULFUR AND PHOSPHORUS COMPOUNDS (CHART 6)

A wide range of sulfur-containing functional grounds is included in the chart, mainly characterized according to the presence of absorptions due to S—H, S═O, SO$_2$, SO$_3$, C—S, C═S or S—S bonds. A much more limited range of phosphorus functions is included.

Very little need be added over the information given in the chart, and only one representative spectrum is given (toluene-*p*-sulfonamide; figure 2.19). The S═O *str* bands are remarkably constant in position, and for sulfones and sulfonic acid derivatives, containing O═S═O groups, the strong antisymmetric and symmetric S═O *str* bands are highly reliable absorptions. Compounds containing the S—H group are usually recognized as such by their smells, before their infrared spectra are even recorded.

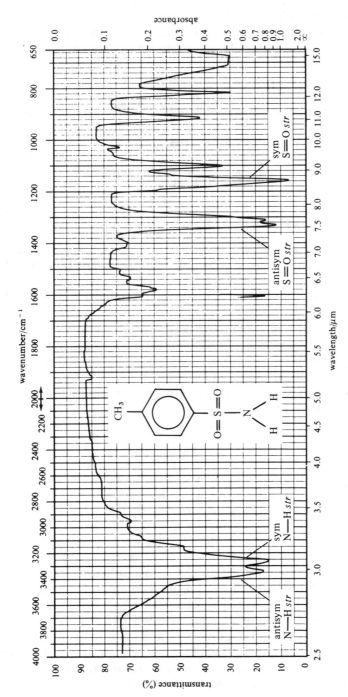

Figure 2.19 *Infrared spectrum of toluene-p-sulfonamide.* KBr *disc.*

All amino sulfonic acids exist as zwitterions, since —SO_3H is sufficiently powerful to protonate aryl amines and alkyl amines alike (contrast aryl amino carboxylic acids). It is very difficult by infrared spectroscopy to distinguish amino sulfonic acids from amine sulfates.

The importance of phosphate and phosphite esters as plasticizers, fuel additives and insecticides warrants their inclusion in the correlation charts. The Wittig reaction with triphenylphosphine produces triphenylphosphine oxide as byproduct: it is useful to know that the unmistakeable P=O *str* band for this compound appears at 1190 cm^{-1}.

SUPPLEMENT 2

2S.1 QUANTITATIVE INFRARED ANALYSIS

Infrared spectra can provide a means of quantitative estimations of components in a mixture, but a number of factors must be taken into account before the accuracy is acceptable. Instrument design is also a factor; ratio recording spectrometers (see section 2.4.4) and Fourier Transform spectrometers (see section 2.4.5) are much superior to optical null machines; pyroelectric detectors perform better than simple thermopiles (see section 2.4.3), and computer software makes various corrections and calculations simpler. The following comments are general, and apply to most instruments (although Fourier Transform instruments have fewer problems associated with sections 2S.1 and 2S.2).

2S.1.1 Absorbance

Quantitative analyses are best carried out using standard solutions, and for solutions the Beer–Lambert law should hold. The *absorbance*, A, of a solution is given by $A = \epsilon c l$, where c is the molar concentration of the solute and l is the path length (in cm): ϵ is the *molar absorptivity*, and is constant for the solute causing the absorbance.

As we have seen in section 2.4, most infrared spectrometers measure band intensities linearly in percentage transmittance, %T, and therefore logarithmically in absorbance, A. Not only does this make it inconvenient to relate band intensity to concentration (which is related linearly with A but inversely and logarithmically with %T), but also slight inaccuracies in the measurement of %T may correspond to very large inaccuracies in the measurement of A. This is particularly true when %T approaches 100, and acceptable accuracy

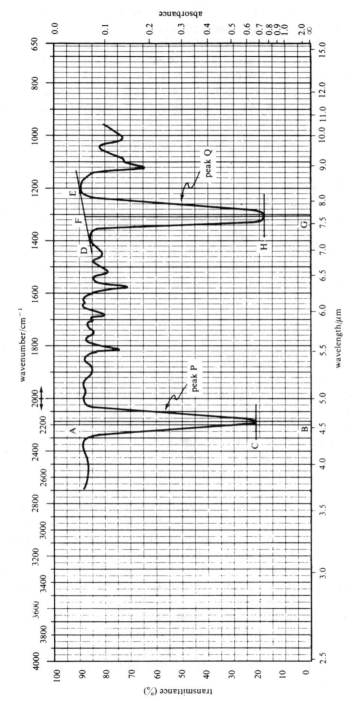

Figure 2.20 *Measurement of absorbance in quantitative infrared analysis.*

in the measurement of A can only be achieved when the band intensity lies between 30% T and 60% T.

Absorbance is defined as $\log(I_0/I)$: in figure 2.20 we can theoretically measure I_0 for peak P as the distance AB, and thus I corresponds to BC. Since, however, there is always background absorption, and the spectrum never reaches 100% T, more accurate results are obtained using the *tangent baseline*, DE. For peak Q, absorbance is given by

$$A = \log(I_0/I) = \log(FG/GH)$$

2S.1.2 Slit widths

The measured absorbance of a peak depends on the slit width at that wavelength—the wider the slit width, the broader and flatter will be the band appearance, because of the intrusion through the slit of neighboring wavelengths at which no absorption is taking place. In practice, reproducibility in the measurement of absorbance can only be achieved if the slit width is less than one-fifth of the band width at half height.

2S.1.3 Path lengths

Solution-cell path lengths cannot accurately be inferred from the spacer thickness, and are best measured by the measurement of *interference fringes*. In figure 2.21 beam 1 is transmitted directly, while beam 2 suffers double internal reflection at the internal surfaces of the cell; the consequent delay in beam 2 leads to reinforcement and cancellation of the resultant beam at appropriate wavelength values, and a scan over the wavelength range of the instrument (through the empty cell) leads to an interference pattern as shown.

The delay in beam 2 is equal to $2t$ (t = cell thickness), and when $2t$ equals an integral number of wavelengths, reinforcement leads to a fringe maximum; at $2t - \frac{1}{2}$ integral number of wavelengths, cancellation leads to a fringe minimum. By choosing two reinforcement maxima ($\bar{\nu}_1$ and $\bar{\nu}_2$) and counting the number of intervening interference fringes between them (n), the cell thickness in this example is given by $t = 0.5n/(\bar{\nu}_1 - \bar{\nu}_2)$.

Interference fringes can also be seen in the spectra of polystyrene film (figure 2.11), between 2000 cm^{-1} and 2800 cm^{-1}: film thickness can be calculated as above, using the modified equation $t = 0.5n/(\bar{\nu}_1 - \bar{\nu}_2)\mu$, where μ is the refractive index of the film material. (For polystyrene $\mu = 1.6$.)

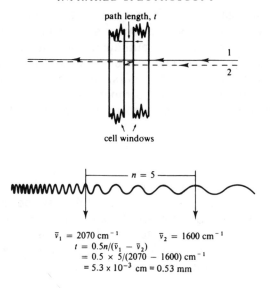

path length, t

cell windows

$n = 5$

$\bar{\nu}_1 = 2070$ cm^{-1} $\bar{\nu}_2 = 1600$ cm^{-1}

$t = 0.5n/(\bar{\nu}_1 - \bar{\nu}_2)$
$= 0.5 \times 5/(2070 - 1600)$ cm^{-1}
$= 5.3 \times 10^{-3}$ cm $= 0.53$ mm

Figure 2.21 *Interference fringes in an empty solution cell. Measurement of path length.*

Exercise 2.9 (a) If the above cell were modified to a path length of 0.1 mm (0.01 cm), how many interference fringes would be generated in the range 1600–2100 cm^{-1}? (b) If the cell were to be modified to a path length of 0.05 mm exactly, over what wavenumber range, starting at 1600 cm^{-1}, would five interference fringes be observed?

Exercise 2.10 Estimate the film thickness used to record the polystyrene spectrum in figure 2.11.

2S.1.4 *Molar absorptivity*
Molar absorptivity, ϵ, can only be measured accurately when all the above parameters are taken into account; coupled with the instrumental difficulty of producing infrared sources and detectors which are linear over the wavelength range, it is clear that accurate ϵ values are rarely achieved.

The most successful quantitative infrared work is done on a routine repetitive analytical basis, using the same machine, the same machine settings of slit widths and scan speeds, etc., and the same sample and reference cells. Even more successful is the preparation of standard solutions of pure authentic specimens, from which a calibration curve of %T against solute concentration can be drawn up for a given set of machine parameters. The concentration of solute in an unknown sample can then be most accurately measured. Computer subtraction of the solvent spectrum is the ultimate choice—see section 1S.1.

Exercise 2.11 Contamination of water sources may arise from spillage of fuels such as diesel fuel: this can be quantified by extracting the fuel from the suspect water with carbon tetrachloride and measuring the C—H stretching absorptions against standard solutions. Devise a procedure for this, specifying the wavenumber to be used and any serious limitations in the method.

2S.2   ATTENUATED TOTAL REFLECTANCE (ATR) AND MULTIPLE INTERNAL REFLECTANCE (MIR)

It is frequently necessary to measure the infrared spectrum of a material for which the normal sampling techniques are inapplicable; examples of such material are polymer films or foams, fabrics, thick pastes, coatings such as paint films or paper glazes, printing inks on metal, etc. The spectrum can be recorded by a modified reflectance technique which depends on the total internal reflectance of light.

Provided the refractive index of the prism, 1 (figure 2.22(a)), is greater than that of the sample medium, 2, and if the angle θ is greater than the critical angle, the infrared light beam 2 will suffer *total internal reflectance* at the interface between 1 and 2. The light beam must, in fact, travel a very short distance (a few μm) along medium 2 before reemerging into 1, and if medium 2 absorbs some of the light, then the reemerging beam will be attenuated (weakened), rather than *totally* reflected. The principle can be easily extended to produce the infrared absorption spectrum of 2, the technique being called *attenuated total reflectance*, ATR.

That part of the radiation which passes through medium 2 (the sample) is referred to as the *evanescent wave*.

Because of the very short path length within medium 2 (the sample) on a single reflection, a very weak spectrum results unless multiple reflectance can be arranged as in figure 2.22(b). This whole assemblage can be inserted into the sample beam of the infrared

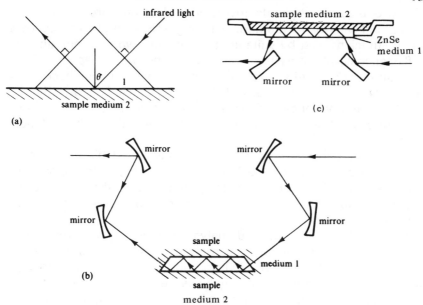

Figure 2.22 *Attachments for* (a) *attenuated total reflectance (ATR) and* (b)
*multiple internal reflectance (MIR) measurements. In a 'circle cell',
medium 1 is rod-shaped, and is placed in a bath containing the
sample (liquid).* (c) *Simplified horizontal ATR cell.*

spectrometer, achieving up to 25 reflections and producing spectra
comparable to normal transmission spectra. This *multiple internal
reflectance* (MIR) is preferred to simple ATR. The alternative name
frustrated multiple internal reflectance (FMIR) is also applied.

The success of the method depends on using a medium 1 which is
transparent to infrared light, and has a high refractive index (μ,
2.5–3.5). Crystals normally used are germanium, zinc selenide or
KRS-5 (mixed ThBr and ThI ($\mu = 2.4$) transmitting to 300 cm^{-1}). The
intensity of the spectrum depends on the efficiency of contact
between the crystal and the sample, on the contact area and on the
angle of reflectance. In double-beam spectrometers, because the ATR
unit increases the path length in the sample beam, a corresponding
increase in the reference beam has to be provided by a second
(blank) ATR unit.

This problem is overcome differently in FTIR instruments, which
work in the single-beam mode. A background spectrum is first
recorded, with the empty ATR cell in place, and then the spectrum
with the sample is recorded; subtraction of background from sample
is then carried out as usual.

Low-energy experiments involving reflectance are much more successful with the FT method, including ATR. A single reflectance is often sufficient, and by suitable modification two different kinds of reflectance can be measured separately. *Specular reflectance* (from the Latin word for mirror) returns from an object directly back along the incident beam, whereas *diffuse reflectance* is that which is scattered at other angles. These phenomena are important in studying surfaces—both their chemical nature and their physical geometry.

The use of reflectance spectroscopy on developed thin-layer chromatography plates allows the infrared spectrum of each spot to be recorded directly from the plate (provided that the TLC support medium is not infrared-active, so that aluminium plates must be used and not glass).

A major difficulty in infrared analysis is in dealing with material in solution, particularly in water, largely because of the strong water absorption bands around $3300 \, cm^{-1}$ and $1600 \, cm^{-1}$. Absorbance subtraction by computer can only be carried out if the water absorption bands do not reach 0% transmittance, which is only the case in extremely short-path-length solution cells (figure 2.12). Using ATR and a modified multiple-reflectance cell, combined with spectral summation (see section 1S.1), the infrared spectra of solutes in very dilute aqueous solution can be measured.

The cell (see figure 2.22(b)) may be in the form of a cylindrical or square rod of zinc selenide inside a stainless steel bath furnished with infrared-transparent windows; the bath is filled to cover the rod, and the infrared beam is then directed into the bath at one end, through the MIR rod cell and out (meanwhile attenuated) at the other end. The term 'CIR' (cylindrical internal reflection) can be used.

In a simpler version the zinc selenide crystal (medium 1, approximate size 5 cm × 1 cm) is horizontal and set as the base of a shallow stainless steel tray, as illustrated in figure 2.22(c). The mirrors which reflect the infrared beam are located below and adjusted to divert the beam up into the crystal, and the sample is simply poured into the tray to cover the crystal. All manner of viscous material—paints, creams or greases—is easily added to and cleaned from the device, and rapid throughput of samples is ensured. Powders, and flexible sheet material such as fabrics or polymer films, can be laid on the crystal and clamped thereto to ensure intimate contact. The application of the method in biology, medicine and the pharmaceutical and chemical industries includes measurement of drugs during manufacture by fermentation methods, following the kinetics of metabolism of dietary constituents, investigation of detergents and cosmetics, or identification of contaminants in waste water, etc.

2S.3 LASER-RAMAN SPECTROSCOPY

2S.3.1 The Raman effect

When a compound is irradiated with an intense exciting light source (mercury arc or laser beam), some of the light is scattered as shown in figure 2.23. If the scattered light is passed into a spectrometer, we can obviously detect in it the frequency of the exciting radiation as a strong line (the so-called *Rayleigh line*), but the Indian physicist C. V. Raman recorded in 1928 that additional frequencies are also present in the scattered beam; these frequencies are symmetrically arrayed above and below the frequency of the Rayleigh line, and the *differences* between the Rayleigh line frequency and the frequencies of the weaker *Raman lines* correspond to the vibrational frequencies present in the molecules of the sample. By measuring these *frequency differences* we can obtain information about the vibrational frequencies within the sample molecules. For example, we may observe Raman lines at ± 750 cm^{-1}, on either side of the Rayleigh line; the molecule therefore possesses a vibrational mode of this frequency. The plot of all the Raman frequency shifts against their intensities is a Raman spectrum, which is very similar to the vibrational spectrum produced in infrared spectroscopy.

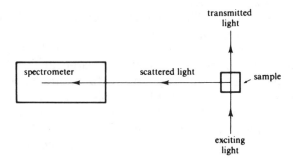

Figure 2.23   *Origin of Raman spectrum in scattered light.*

Raman spectroscopy mainly utilized mercury arcs as the source of the exciting radiation until the development of the laser, and the era of routine laser-Raman work has only now arrived. Not only is laser light more intense and more coherent, but also it enables longer wavelengths to be used for excitation, eliminating many problems of sample decomposition.

State-of-the-art FT laser-Raman spectrometers use a Nd–YAG laser operating at 1064 nm, which minimizes the troublesome fluorescence which arises from many samples if excitation of shorter wavelength is used. The FT technique brings to the recording of Raman spectra all of the advantages discussed for IR spectra—the Fellgett and Jacquinot advantages, wavenumber accuracy, spectra summation and subtraction, short measurement times, etc.—and in a dual instrument, in which switching between the FTIR and FT-Raman modes is carried out by computer, both IR and Raman spectra can be obtained easily and speedily on the same sample.

2S.3.2 Comparison of infrared and Raman spectra

An infrared absorption arises when infrared light (electromagnetic) interacts with a fluctuating *dipole* within the molecule. The Raman effect is not an absorption effect like infrared, but depends on the *polarizability* of a vibrating group, and on its ability to interact and couple with an exciting radiation whose frequency does not match that of the vibration itself. A given vibrational mode within a molecule may lead to fluctuating changes in the dipole (an *infrared-active* vibration), but may not necessarily lead to changes in the polarizability (a *Raman-active* vibration), the converse also being true. Many absorptions that are weak in infrared (for example, the stretching vibrations of symmetrically substituted $C-C$ and $C\equiv C$) give strong absorptions in Raman spectra.

The 'change in polarizability' which is necessary for a Raman signal to be seen is, to a close approximation, the 'ease of imposition of a dipole', but the dipole involved in the Raman signal may not be the same dipole involved in an associated infrared absorption signal. Indeed, in molecules with high symmetry (for example, benzene) the infrared-active vibrations that lead to dipole changes and those that lead to changes in the polarizability are mutually exclusive. The two techniques are therefore complementary, and figure 2.24 shows the infrared spectrum of indene (upper trace) compared with its Raman spectrum (lower trace).

The principal advantages of laser-Raman spectroscopy over infrared are its increased sensitivity (sample sizes as small as several nanograms), the ease of sample preparation (powders or solutions may be used) and the fact that water makes an ideal solvent (since its Raman spectrum consists essentially of only one broad weak band at 3654 cm^{-1}). The application to aqueous biological samples is extensive, and includes drugs analyses in body fluids and studies of aqueous solutions of amino acids. Organometallic complexes can also be easily studied in aqueous solution.

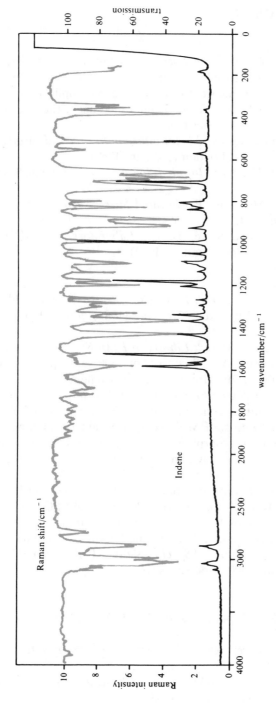

Figure 2.24 *Infrared and laser-Raman spectra of indene compared (infrared spectrum above).*

FURTHER READING

MAIN TEXTS

Bellamy, L. J., *The Infrared Spectra of Complex Molecules*, Methuen, London (Vol. 1, 3rd edn, 1975; Vol. 2, 1980). These two texts are undoubtedly the standard works for organic applications of infrared spectroscopy. Vol. 2 was formerly published under the title *Advances in Infrared Group Frequencies* (1968).

Cross, A. D. and Jones, R. A., *Introduction to Practical Infrared Spectroscopy*, Butterworths, London (3rd edn, 1969).

van der Maas, J. H., *Basic Infrared Spectroscopy*, Heyden, London (2nd edn, 1972).

George, B. and McIntyre, P., *Infrared Spectroscopy*, Wiley, Chichester (1987). ACOL text.

Miller, R. G. J. and Stace, B. C., *Laboratory Methods in Infrared Spectroscopy*, Heyden, London (1972).

SPECTRA CATALOGS

Pouchert, Charles J., *Aldrich Library of Infrared Spectra*, Aldrich Chemical Co. Inc., Milwaukee, 3rd edn (1984). Over 12 000 dispersive spectra.

Pouchert, Charles J., *Aldrich Library of FTIR Spectra*, Aldrich Chemical Co. Inc. 3 volumes (Vol. 3 consists of 6600 vapor phase spectra), also available in digital form from Nicolet Instruments Inc.

Kelly, R. K. (Ed.), *Sigma Library of FTIR Spectra*, Sigma Chemical Co., St. Louis. 2 volumes containing 10 400 FTIR spectra of biochemicals. Also associated with an IBM-PC compatible aid to searching through the library, by input of absorption wavenumbers of unknown compounds.

Merck, E. and Bruker Analytische Messtechnik, *Merck FTIR Atlas*, VCH Verlagsgesellschaft, Weinheim, Germany (1987).

Sadtler Digital FTIR Libraries, Sadtler, Pennsylvania, and Heyden, London (1989). Over 100 000 digitized spectra in IBM-PC compatible form, subdivided into subsets such as polymers, pharmaceuticals, adhesives, solvents, flavors, Georgia State Crime Laboratory Library, etc.

Sadtler Handbook of Infrared Spectra, Sadtler, Pennsylvania, and Heyden, London (1978). 3000 dispersive spectra.

SUPPLEMENTARY TEXTS

Freeman, S. K., *Applications of Laser Raman Spectroscopy*, Wiley, London (1973).

Tobin, M. C., *Laser Raman Spectrosocopy*, Wiley-Interscience, Chichester (1971).

Grasseli, J. G. and Snavely, M. K., *Chemical Applications of Raman Spectroscopy*, Wiley, London (1981).

Griffiths, P. R. and de Haseth, J. A., *Fourier Transform Infrared Spectrometry*, Wiley, Chichester (2nd edn, 1986).

Haris, P. I. and Chapman, D., FTIR in Biochemistry and Medicine, *Chemistry in Britain*, **5**, 1015 (1988).

Long, D. A., *Raman Spectroscopy*, McGraw-Hill, London (1977).

Baranska, H., Labudzinska, A. and Terpinski, J., *Laser Raman Spectrometry*, Ellis Horwood, New York (1987).

Harrick, N. J., *Internal Reflectance Spectrometry*, Wiley, New York (1967).

Several authors, introduction by Long, D. A., The Renaissance of Raman Spectroscopy, *Chemistry in Britain*, **6**, 589 (1989).

Bist, H. D. (Ed.), *Raman Spectroscopy: Sixty Years On*, Elsevier, Amsterdam (1989).

Nuclear Magnetic Resonance Spectroscopy

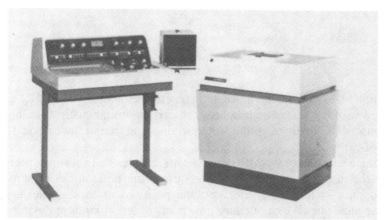

60 MHz NMR spectrometer for proton only.

500 MHz multinuclear NMR spectrometer. The superconducting 11.7 T magnet is on the right.

As is implied in the name, nuclear magnetic resonance (or NMR) is concerned with the magnetic properties of certain atomic nuclei, notably the nucleus of the hydrogen atom—the proton—and that of the carbon-13 isotope of carbon.

Studying a molecule by NMR spectroscopy enables us to record differences in the magnetic properties of the various magnetic nuclei present, and to deduce in large measure what the positions of these nuclei are within the molecule. We can deduce how many different kinds of environments there are in the molecule, and also which atoms are present in neighboring groups. Usually we can also measure how many atoms are present in each of these environments.

The proton NMR spectrum of toluene, $C_6H_5CH_3$, represents an extremely simple example of an NMR spectrum, and it is shown in diagrammatic form in figure 3.1(a).

Toluene has two groups of hydrogen atoms—the methyl hydrogens and the ring hydrogens. *Two signals* appear on the spectrum, corresponding to these *two different chemical and magnetic environments*.

The areas under each signal are in the ratio of the number of protons in each part of the molecule, and measurement will show that the *ratio of these areas is 5 : 3*. The areas are found from the step heights in the integration trace: see section 3.7.

A typical carbon-13 NMR spectrum is shown in figure 3.1(b), and the assignment of each signal to a unique carbon environment in the molecule (*p*-hydroxyacetophenone) is also indicated. Each different carbon atom produces a different signal. Because of molecular symmetry, C-2 and C-6 are equivalent, as are C-3 and C-5; note, therefore, that the signals numbered 2 and 4 in the spectrum are of high intensity, since each corresponds to two carbons. Note also that the other signals are of low intensity, but are unequal, even though each corresponds to a single carbon. The three-line signal from the solvent corresponds to the single carbon atom of deuteriochloroform, $CDCl_3$; it appears as a triplet because of an interaction with the deuterium atom. These features will be discussed later in this chapter.

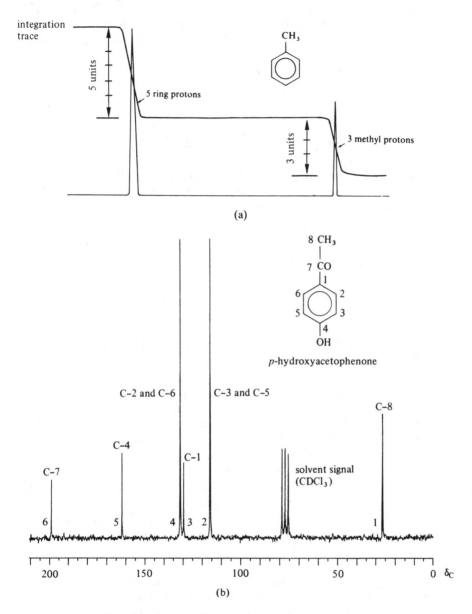

(a)

(b)

Figure 3.1 (a) *Diagrammatic ^{1}H NMR spectrum of toluene, $C_6H_5CH_3$, showing two signals in the intensity ratio 5:3. (b) ^{13}C NMR spectrum of p-hydroxyacetophenone, p-$CH_3COC_6H_4OH$, showing six signals corresponding to the six different carbon environments in the molecule. (20 MHz, in $CDCl_3$.)*

The most appropriate starting point for a study of NMR is the proton: it is the simplest nucleus, and it was the nucleus on which the phenomenon of NMR was first observed. (Felix Bloch and Edward M. Purcell shared the Nobel Prize in physics for this, their early but independent studies being reported nearly simultaneously in 1946.) Carbon-13 NMR is of equal importance, and it will be treated extensively; other magnetic nuclei (such as fluorine-19, phosphorus-31, nitrogen-14, nitrogen-15 and oxygen-17) will be surveyed more selectively.

PROTON NMR SPECTROSCOPY

Much of the theory and practice relevant to proton NMR is appropriate also to all other magnetic nuclei; the development of these topics will center around the proton, but every opportunity will be taken to indicate that the arguments are multinuclear.

3.1   THE NMR PHENOMENON

3.1.1   THE SPINNING NUCLEUS
The nucleus of the hydrogen atom (the proton) behaves as a tiny spinning bar magnet, and it does so because it possesses both electric charge and mechanical spin; any spinning charged body will generate a magnetic field, and the nucleus of hydrogen is no exception.

3.1.2   THE EFFECT OF AN EXTERNAL MAGNETIC FIELD
Like all bar magnets, the proton will respond to the influence of an external magnetic field, and will tend to align itself with that field, in the manner of a compass needle in the earth's magnetic field. Because of quantum restrictions which apply to nuclei but not to compass needles (see section 3.2), the proton can only adopt two orientations with respect to an external magnetic field—either *aligned with the field* (the lower energy state) or *opposed to the field* (the higher energy state). We can also describe these orientations as *parallel* with or *antiparallel* with the applied field.

3.1.3   PRECESSIONAL MOTION
Because the proton is behaving as a *spinning* magnet, not only can it align itself with or oppose an external magnetic field, but also it will move in a characteristic way under the influence of the external magnet.

   Consider the behavior of a spinning top: as well as describing its spinning motion, the top will (unless absolutely vertical) also perform a slower waltz-like motion, in which the spinning axis of the top moves slowly around the vertical. This is *precessional* motion, and the top is said to be *precessing* around the vertical axis of the earth's gravitational field. The precession arises from the interaction of spin—that is, gyroscopic motion—with the earth's gravity acting vertically downward. Only a spinning top will precess; a static top will merely fall over.

As the proton is a spinning magnet, it will, like the top, precess around the axis of an applied external magnetic field, and can do so in two principal orientations, either aligned with the field (low energy) or opposed to the field (high energy). This is represented in figure 3.2, where B_0 is the external magnetic field.

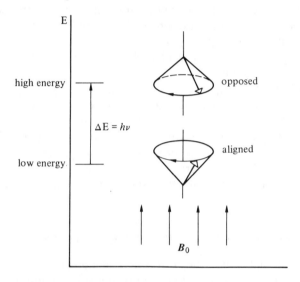

Figure 3.2 *Representation of precessing nuclei, and the ΔE transition between the aligned and opposed conditions.*

3.1.4   PRECESSIONAL FREQUENCY

The spinning frequency of the nucleus does not change, but the speed of precession does. The *precessional frequency*, ν, is directly proportional to the strength of the external field, B_0: that is,

$$\nu \propto B_0$$

This is one of the most important relationships in NMR spectroscopy, and it is restated more quantitatively in section 3.2.

As an example, a proton exposed to an external magnetic force of 1.4 T ($\equiv$ 14 000 gauss) will precess $\approx$ 60 million times per second, so that ν = 60 MHz. For an external field of 2.3 T, ν is $\approx$ 100 MHz, and at 14.1 T ν is $\approx$ 600 MHz. (Strictly, the tesla is a measure of magnetic flux density, not field strength.)

3.1.5 Energy Transitions

We have seen that a proton, in an external magnetic field of 1.4 T, will be precessing at a frequency of ≈ 60 MHz, and be capable of taking up one of two orientations with respect to the axis of the external field—aligned or opposed, parallel or antiparallel.

If a proton is precessing in the *aligned* orientation, it can absorb energy and pass into the *opposed* orientation; subsequently it can lose this extra energy and relax back into the aligned position. If we irradiate the precessing nuclei with a beam of radiofrequency energy of the correct frequency, the low-energy nuclei may absorb this energy and move to a higher energy state. The precessing proton will only absorb energy from the radiofrequency source if the precessing frequency is the same as the frequency of the radiofrequency beam; when this occurs, the nucleus and the radiofrequency beam are said to be *in resonance*; hence the term *nuclear magnetic resonance*.

The simplest NMR experiment consists in exposing the protons in an organic molecule to a powerful external magnetic field; the protons will precess, although they may not all precess at the same frequency. We irradiate these precessing protons with radiofrequency energy of the appropriate frequencies, and promote protons from the low-energy (aligned) state to the high-energy (opposed) state. We record this absorption of energy in the form of an NMR spectrum, such as that for toluene in figure 3.1(a).

3.2 THEORY OF NUCLEAR MAGNETIC RESONANCE

The only nuclei that exhibit the NMR phenomenon are those for which the spin quantum number I is greater than 0: the spin quantum number I is associated with the mass number and atomic number of the nuclei as follows:

Mass number	Atomic number	Spin quantum number
odd	odd or even	$\frac{1}{2}, \frac{3}{2}, \frac{5}{2}, \ldots$
even	even	0
even	odd	$1, 2, 3, \ldots$

The nucleus of ^{1}H, the proton, has $I = \frac{1}{2}$, whereas ^{12}C and ^{16}O have $I = 0$ and are therefore nonmagnetic. If ^{12}C and ^{16}O had been magnetic, the NMR spectra of organic molecules would have been much more complex.

Other important magnetic nuclei that have been studied extensively by NMR are ^{11}B, ^{13}C, ^{14}N and ^{15}N, ^{17}O, ^{19}F and ^{31}P (see sections 3.13–3.15,

3S.5 and 3S.6). Both deuterium (^{2}H) and nitrogen-14 have $I = 1$, and the consequences of this observation will become apparent later (see 'Hetero-nuclear coupling', section 3.9.4).

Under the influence of an external magnetic field, a magnetic nucleus can take up different orientations with respect to that field; the number of possible orientations is given by $(2I + 1)$, so that for nuclei with spin $\frac{1}{2}$ (^{1}H, ^{13}C, ^{19}F, etc.) only two orientations are allowed. Deuterium and ^{14}N have $I = 1$ and so can take up three orientations: these nuclei do not simply possess magnetic *dipoles*, but rather possess electric *quadrupoles*. Nuclei possessing electric quadrupoles can interact with both magnetic and electric field gradients, the relative importance of the two effects being related to their magnetic moments and electric quadrupole moments, respectively.

In an applied magnetic field, magnetic nuclei like the proton precess at a frequency v, which is proportional to the strength of the applied field. The exact frequency is given by

$$ v = \frac{\gamma B_0}{2\pi} $$

where B_0 = strength of the applied external field experienced by the proton,

γ = magnetogyric ratio, being the ratio between the nuclear magnetic moment, μ, and the nuclear angular momentum, I: γ is also called the gyromagnetic ratio.

Typical approximate values for v are shown in table 3.1 for selected values of field strength B_0, for common magnetic nuclei.

Table 3.1 Precessional frequencies (in MHz) as a function of increasing field strength

B_0/tesla	1.4	1.9	2.3	4.7	7.1	11.7	14.1
Nucleus							
^{1}H	60	80	100	200	300	500	600
^{2}H	9.2	12.3	15.3	30.6	46.0	76.8	92
^{11}B	19.2	25.6	32.0	64.2	96.9	159.8	192
^{13}C	15.1	20.1	25.1	50.3	75.5	125.7	151
^{14}N	4.3	5.7	7.2	14.5	21.7	36.1	43
^{15}N	6.1	8.1	10.1	20.3	30.4	50.7	61
^{17}O	8.1	10.8	13.6	27.1	40.7	67.8	81
^{19}F	56.5	75.3	94.1	188.2	288.2	470.5	565
^{31}P	24.3	32.4	40.5	81.0	121.5	202.4	243
(Free electron)	3.9×10^4						

The strength of the signal, and, hence, the sensitivity of the NMR experiment for a particular nucleus, are related to the magnitude of the magnetic moment, μ. The magnetic moments of ^{1}H and ^{19}F are relatively large, and detection of NMR with these nuclei is fairly sensitive. The magnetic moment for ^{13}C is about one-quarter that of ^{1}H, and that of ^{2}H is roughly one-third the moment of ^{1}H; these nuclei are less sensitively detected in NMR. (In contrast, the magnetic moment of the free electron is nearly 700 times that of ^{1}H, and resonance phenomena for free radicals can be studied in extremely dilute solutions; see section 3S.7 for a discussion on electron spin resonance.)

Even with very large external magnetic fields, up to 14.1 T, the energy difference $\Delta E = h\nu$ is very small, even at 600 MHz. Because the difference is so small (of the order of 10^{-4} kJ mol^{-1}), the populations of protons in the two energy states are nearly equal; the calculated Boltzmann distribution for ^{1}H at 1.4 T and at room temperature shows that the lower energy state has a nuclear population only about 0.001 per cent greater than that of the higher energy state. The relative populations of the two spin states will change if energy is supplied of the correct frequency to induce transitions upward or downward.

What happens when protons absorb 60 MHz radiofrequency energy?

Nuclei in the lower energy state undergo transitions to the higher energy state; the populations of the two states may approach equality, and if this arises, no further net absorption of energy can occur and the observed resonance signal will fade out. We describe this situation in practice as *saturation* of the signal. In the recording of a normal NMR spectrum, however, the populations in the two spin states do not become equal, because higher-energy nuclei are constantly returning to the lower-energy spin state.

How can the nuclei lose energy and undergo transitions from the high- to the low-energy state?

The energy difference, ΔE, can be reemitted as 60 MHz energy, and this can be monitored by a radiofrequency detector as evidence of the resonance condition having been reached. Of great importance, however, are two *radiationless* processes, which enable high-energy nuclei to lose energy.

The high-energy nucleus can undergo energy loss (or *relaxation*) by transferring ΔE to some electromagnetic vector present in the surrounding environment. For example, a nearby solvent molecule, undergoing continuous vibrational and rotational changes, will have associated electrical and magnetic changes, which might be properly oriented and of the correct dimension to absorb ΔE. Since the nucleus may be surrounded by a whole array of neighboring atoms, either in the same molecule or in solvent molecules, etc., this relaxation process is termed *spin–lattice relaxation*, where *lattice* implies the entire framework or aggregate of neighbors.

A second relaxation process involves transferring ΔE to a neighboring nucleus, provided that the particular value of ΔE is common to both nuclei: this mutual exchange of spin energy is termed *spin–spin relaxation*. While one nucleus loses energy, the other nucleus gains energy, so that no net change in the populations of the two spin states is involved.

The rates of relaxation by these processes are important, and in particular the rate of spin–lattice relaxation determines the rate at which *net* absorption of 60 MHz energy can occur.

Rather than define the *rate constants* (k) for these first-order relaxation mechanisms, a *spin–lattice relaxation time* (T_1) is defined, being the reciprocal of k. T_1 is related to the half-life of the exponential process by $T_1 = (\text{half-life})/\ln 2$. T_2 is similarly defined as the *spin–spin relaxation time*. A more rigorous treatment of relaxation is given in section 3S.3. If T_1 and T_2 are small, then the lifetime of an excited nucleus is short, and it has been found that this gives rise to very broad absorption lines in the NMR spectrum. If T_1 and T_2 are large, perhaps of the order of 1s, then sharp spectral lines arise.

For nonviscous liquids (and that includes solutions of solids in nonviscous solvents) molecular orientations are random, and transfer of energy by spin–lattice relaxation is inefficient. In consequence, T_1 is large, and this is one reason why sharp signals are obtained in NMR studies on nonviscous systems.

The important relationship between relaxation times and line broadening can be understood qualitatively by using the uncertainty principle in the form $\Delta E \cdot \Delta t \approx h/2\pi$, or, since $E = h\nu$, $\Delta \nu \cdot \Delta t \approx 1/2\pi$. Expressed verbally, the product $\Delta \nu \cdot \Delta t$ is constant, and if Δt is large (that is, the lifetime of a particular energy state is long), then $\Delta \nu$ must be small (that is, the uncertainty in the measured frequency must be small, so that there is very little 'spread' in the frequency, and line-widths are narrow). Conversely, if Δt is small (fast relaxation), then $\Delta \nu$ must be large, and broad lines appear.

The ^{14}N nucleus possesses an electric quadrupole (see page 157), and is therefore able to interact with both electric and magnetic field gradients, which cause the nucleus to tumble rapidly: spin–lattice relaxation is highly effective, and therefore T_1 is small. Since T_1 is small, NMR signals for the ^{14}N nucleus are very broad indeed, and for the same reason the NMR signals for most protons attached to ^{14}N (in N—H groups) are broadened. (See also under 'Heteronuclear coupling', page 155.)

3.3 CHEMICAL SHIFT AND ITS MEASUREMENT

For a field strength of 1.4 T, we have seen that protons have a precessional frequency of ≈ 60 MHz. The precessional frequency of all protons in the same external applied field is not, however, the same, and the precise value for any one proton depends on a number of factors. Historically, this was first observed by Packard in 1951; he was able to detect three different values for the precessional frequencies of the protons in ethanol, and the realization that these corresponded to the three different chemical environments for the protons in ethanol (CH_3, CH_2 and OH) marked the beginning of NMR as a tool of the organic chemist. Because the shift in frequency depended on chemical environment, this gave rise to the term *chemical shift*. An even simpler example is the NMR spectrum of toluene shown in figure 3.1(a). The protons in toluene are in two different chemical environments, and two signals appear on the spectrum, corresponding to two different precessional frequencies (two different chemical shift values).

There are six different *chemical shift values* in the carbon-13 NMR spectrum in figure 3.1(b), etc.

3.3.1   MEASUREMENT OF CHEMICAL SHIFT—INTERNAL STANDARDS

To measure the precessional frequency of a group of nuclei in absolute frequency units is not difficult but is rarely required. More commonly the *differences* in frequency are measured with respect to some reference group of nuclei. For protons and ^{13}C, the universally accepted reference is tetramethylsilane, TMS:

H CH_3 CH_3
| | |
H—Si—H CH_3—Si—CH_3 CH_3—Si—$CH_2CH_2CH_2SO_3^-$ Na^+
| | |
H CH_3 CH_3

silane   tetramethylsilane   sodium salt of
(TMS)   3-(trimethylsilyl)-propanesulfonic acid

TMS is chosen because it gives an intense sharp signal even at low concentrations (having 12 protons in magnetically equivalent positions); the signal arises on the NMR spectrum well clear of most common organic protons (for reasons we shall meet in section 3.4); it is chemically inert and has a low boiling point, so that it is easily removed from a recoverable sample of a valuable organic compound; it is soluble in most organic solvents, and can be added to the sample solution (0.01–1.0 per cent) as an *internal standard*.

TMS is not soluble in water or in D_2O; for solutions in these solvents the sodium salt of 3-(trimethylsilyl)-propanesulfonic acid is used.

The choice of standards against which to measure the precessional frequency of other nuclei is based on the same criteria. TMS is used for ^{13}C; $CFCl_3$ for ^{19}F; an *external* sample of concentrated H_3PO_4 for ^{31}P; and although liquid ammonia (external) is the absolute frequency standard for ^{15}N, nitromethane (CH_3NO_2) is commonly used as a more convenient primary standard. Water is the standard for ^{17}O.

3.3.2 MEASUREMENT OF CHEMICAL SHIFT—THE NMR SPECTROMETER

The basic features of the instrumentation needed to record an NMR spectrum are a magnet, a radiofrequency source and a detection system to indicate that energy is being transferred from the radiofrequency beam to the nucleus. Such an arrangement is shown schematically in figure 3.3.

Magnet strengths and frequencies depend on the various factors discussed above, but typical practical parameters for proton NMR will be used for this discussion.

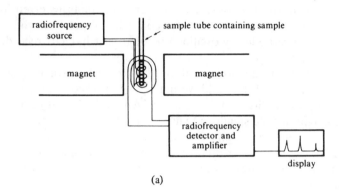

(a)

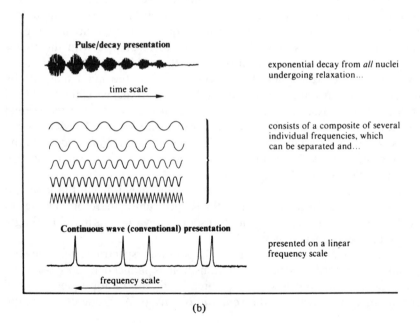

(b)

Figure 3.3   (a) *Basic features of an NMR spectrometer.* (b) *Schematic representation of pulsed NMR; the output in the time domain (the free induction decay, FID) is converted to the frequency domain by Fourier Transform. See also figure 1.6.*

Continuous wave method—CWNMR

The sample is placed in a glass tube (for example, 5 mm diameter) between the pole faces of a magnet (for example, of field strength 1.4 T). A radiofrequency source feeds energy at 60 MHz into a coil wound around the sample tube, and the radiofrequency detector is tuned to 60 MHz. If the nuclei in the sample do not resonate with the 60 MHz source, the detector will only record a weak signal coming directly from the source coil to the detector coil. An increased signal will be detected if nuclei in the sample resonate with the source, since energy will be transferred from the source, via the nuclei, to the detector coil. The output from the detector can be fed to a cathode-ray oscilloscope or to a strip chart recorder after amplification, etc.

Because of chemical shift, not all protons come to resonance at exactly 60 MHz in a field of 1.4 T, and we must have provision in the instrument for detecting these different frequencies. An easy method in theory is to hold the magnetic field steady, and use a tunable radiofrequency source (or 'synthesizer') which can scan over the nearby frequencies until the various nuclei come to resonance in turn as their precessional frequencies are matched by the scanning source. Such an instrument is operating on *frequency sweep*.

It is in practice cheaper and easier to make use of the fact that $v \propto B_0$. By holding the radiofrequency source steady at 60 MHz and varying B_0, we can increase or decrease the precessing frequencies of all the nuclei, until each, in turn, reaches 60 MHz and comes to resonance with the radio-frequency source. In such an instrument the field strength is varied by fixing small electromagnets to the pole faces of the main magnet; by increasing the current flowing through these electromagnets, the total field strength is increased. As the field strength increases, so the precessional frequency of each proton increases until resonance with the 60 MHz source takes place. This design of instrument is operating on *field sweep*, and the majority of low-cost NMR machines use this mode. The variable electro-magnet coils are called *sweep coils*.

As each proton comes to resonance, the signal from the detector produces a peak on the chart; the NMR spectrum is therefore a series of peaks plotted on an abscissa that corresponds to variations in field strength or (since $v \propto B_0$) frequency.

In the above mode of instrument operation, spectra require several minutes to record, because each transition is induced in succession by continuous scan from low field to high field, the radiofrequency signal remaining constant; this is the *continuous wave* (CW) mode of operation.

Pulse techniques and Fourier Transforms—FTNMR

If we could stimulate all transitions *simultaneously*, the same transitions within the nuclear spin states would arise, but the excitation would require a mere fraction of a second to execute. This is precisely what is done in *pulsed NMR*; for proton NMR, the sample is irradiated (at fixed field) with a strong pulse of radiofrequency energy containing all the frequencies over the ^{1}H range (for example, spread around 100 MHz at 2.3 T, etc.). The protons in each environment absorb their appropriate frequencies from the pulse, and these frequencies couple to give beats (similar to the violin analogy discussed in section 1S.2). The presentation at the top of figure 3.3(b) is an interferogram in the *time domain*, and its Fourier Transform renders the same information in the *frequency domain* (see section 1S.2). For this reason the technique of *pulsed NMR* is usually called *pulsed FTNMR* or simply *FTNMR*, thus distinguishing it from *continuous wave NMR, CWNMR*.

The pulse duration may be approximately 10 μs, and when it is switched off, the nuclei undergo relaxation processes, and reemit the absorbed energies and coupling energies. The instrumental problem is that all these reemitted energies are emitted simultaneously as a complex interacting pattern, which decays rapidly. Figure 3.3(b) shows a typical pattern of reemitted decay signals; this output is digitized in a computer, and each *individual* frequency is identified from the interference pattern by the mathematics of Fourier Transforms (see section 1S.2). Each frequency is plotted on a linear frequency scale—the NMR spectrum.

Felix Bloch carried out the NMR experiment in his Nobel prizewinning work, using the CW mode, since contemporary instrumentation did not permit the pulsed mode, and he had no computer to calculate the Fourier Transform. Nevertheless, he predicted the nature of the pulse phenomenon, and distinguished between two ways in which the nuclear magnetization would interact with the radiofrequency field magnetization. *With the RF on*, magnetization would transfer at resonance from the RF to the nuclei by the process of magnetic induction—and he named this *forced induction*. At the *end of a pulse of RF*, however, the nuclei would still retain magnetization from the RF until the end of relaxation, and this Bloch named *free induction*: it is the decay of this free induction that is measured (as the free induction decay, FID) in FTNMR.

Advantages of FTNMR
An entire spectrum can be recorded, computerized and transformed in a few seconds; with a repetition every 2 s, for example, 400 spectra can be accumulated in approximately 13 min, giving twenty times the signal enhancement. Technically, very rapid pulse repetitions are possible, but after excitation by one pulse, the nuclei cannot be reexcited until $5 \times T_1$ (the spin–lattice relaxation time) has elapsed; repetition times up to 60 s may be required for nuclei with long T_1 values.

Three low-sensitivity problems in NMR are now easily overcome: (1) samples at very low concentrations; (2) NMR studies on nuclei with low natural abundance (for example, ^{13}C); and (3) NMR studies on nuclei with low abundance and small magnetic moments (for example, ^{13}C, ^{15}N or ^{17}O).

The first advantage, namely (1), being able to work at very low concentrations in ^1H NMR, is important in biological chemistry, where only microgram quantities of material may be available. The FT method also gives improved spectra for sparingly soluble compounds.

The most dramatic advantage to organic chemists, namely (2) and (3), is the ability to record excellent spectra from 'ordinary' molecules containing only the natural abundance of ^{13}C nuclei, etc.

Carbon-13 NMR spectroscopy is discussed at length in the second part of this chapter. Nitrogen-15 and oxygen-17 NMR are discussed in section 3S.6.

Differences in precessional frequencies are very small indeed, and modern high-resolution instruments can detect differences of the order of 0.001 parts per million (ppm). To obtain these accuracies in resolution, the radiofrequency source and the field strength of the magnet must be accurate to one part in 10^9 or better, and this remarkable accuracy is achieved by a number of means. Much of the accuracy is obtained by having additional contoured electromagnet coils on the pole faces of the main magnet (in addition to the sweep coils); these *Golay coils* can be 'shimmed' to create specifically contoured magnetic fields, which compensate for any inhomogeneity in the main magnet's field. As a further refinement, any remaining magnet inhomogeneity in the horizontal plane can be averaged out if the sample tube is spun about its axis, so that the molecules in different parts of the sample tube experience the same *average* magnetic environment. The sample tube is mounted on a light turbine, and a jet of air is adjusted to provide a steady spinning rate of around 30 Hz.

Adjustment of magnet homogeneity may require weekly or even daily attention to the Golay coils. A standard test of homogeneity is the sharpness and symmetry of the so-called *ringing* pattern associated with the NMR signal and seen under amplification in most sharply resolved proton NMR signals. The ringing pattern is a beat phenomenon caused by the interaction of two close frequencies and has the appearance of a steady exponential decay of pen oscillations. At resonance, the nucleus sets up weak rotating vectors whose frequency alters as the field is swept; interaction of these vectors with the normal relaxation vectors produces the wiggle-beat pattern called ringing. The effect of ringing is seen clearly in figure 3.22: on the high-field side of each peak, the pen has oscillated above and below the base line.

Ringing is only produced in the CW mode, and is not observed in spectra recorded by pulse FTNMR.

The standard test of homogeneity, relating as it does to the resolving power of the instrument, is the narrowness of the NMR signal; this is measured (in Hz) as the width—at half height—of a given signal (for example, one of the narrow lines of the proton NMR spectrum of o-dichlorobenzene). Line-widths as narrow as 0.05 Hz are achievable.

A further specification relates to the sensitivity of the instrument, measured as the ratio between *signal intensity* and the intensity of *background noise* on the spectrum—the signal-to-noise ratio (S/N). Again there are industry norms for calculating the S/N ratio, based for proton NMR on the quartet of the ethylbenzene spectrum (using a 1 per cent solution) and for carbon-13 on the largest aromatic peak in the ethyl-benzene spectrum (using a 10 per cent solution) recorded under very specific instrument parameters. Low-cost spectrometers can reach proton S/N ratios of 55:1. With 600 Mz FTNMR instruments S/N ratios for proton are 600:1, but using a 0.1 per cent ethylbenzene solution (thus correspond-ing to 6000:1 for a 1 per cent solution), while for carbon-13 NMR, the corresponding S/N ratio is 180:1.

Occasionally a strong signal in the spectrum will be flanked by two smaller signals equidistant from the main signal. This may be caused by oscillations set up by irregularities in the spinning of the sample tube; these oscillations will act as modulators, and produce the visible side-bands above and below the main signal frequency (by addition and subtraction of the modulating frequency). These *spinning side-bands* are always symme-trically displayed, but they vary in position with the spinning rate of the sample tube; they are not to be confused with the satellite side-bands caused by ^{13}C nuclei, as discussed on page 156. Spinning side-bands are minimized by the use of high-precision sample tubes, by the avoidance of excessive spinning rates, and by adjustment of the homogeneity controls.

Not all instruments make use of the two sets of radiofrequency coils shown in figure 3.3(a), although these *crossed coils* (so called because they are arranged orthogonally) give high signal-to-noise ratios and are widely used. It is possible to perform the NMR experiment, using only one radiofrequency coil with a radiofrequency bridge arrangement (similar to the Wheatstone bridge) to monitor out-of-balance changes in signal intensities as radiofrequency energy is absorbed by the nuclei.

It is easier to achieve accurate homogeneity over a *small area* of the magnet pole faces, and small sample sizes are advantageous (≈ 0.3 cm^3 of 10 per cent solutions are commonly used in proton work). This considera-tion is offset for nuclei with small magnetic moments, etc., where large sample sizes are demanded for detection sensitivity.

To record the NMR spectra of nuclei other than ^{1}H, using the instru-ment above (field strength 1.4 T), requires a different radiofrequency source appropriate to the nucleus being examined (see table 3.1). For example, with a 1.4 T magnet, ^{19}F spectra can be recorded with a radiofrequency source at 56.5 MHz, and ^{13}C spectra require a 15.1 MHz source.

Instruments with a 1.4 T magnet are usually called 60 MHz instruments, although, strictly, this is only the frequency being used when ^{1}H spectra are being recorded. Similarly, 2.3 T and 14.1 T instruments would be suitably called 100 MHz and 600 MHz instruments, respectively.

Field strengths of 1.4 T and 1.9 T (60 MHz and 80 MHz instruments) can be obtained from either permanent magnets or electromagnets. Field strengths of 2.3 T (100 MHz instruments) are now only obtained commercially from electromagnets. The really high field strengths of 4.7 T and 14.1 T (200 MHz and 600 MHz instruments) use superconducting magnets. When a metal conductor is cooled in liquid helium to a temperature of 4 K, electrical resistance vanishes and the metal becomes a 'superconductor'. An electromagnet immersed in liquid helium and using this principle can reach enormous field strengths, which were hitherto regarded as unattainable.

The photograph at the chapter head shows an NMR spectrometer based on a superconducting magnet system. The can-shaped vessel on the right contains, at its core, the electromagnet, which consists of a solenoid of fine wire, several kilometers in length. The wire is usually either a specially superconducting alloy of niobium and titanium, or, for fields above 10 T, the intermetallic compound Nb_3Sn. The solenoid is surrounded by liquid helium contained in a vacuum-insulated vessel, and this, in turn, is surrounded by a bath of liquid nitrogen (acting as an insulator); the volume of the solenoid is about one-fifth of the volume of the whole container. Field adjustment coils (*shim coils*) are attached to the solenoid (and are therefore at low temperature) and also to the probe housing (and these are therefore at room temperature). To establish the magnetic field, electric current is supplied to the solenoid windings, but when the target value of field is reached, the current supply is shorted out, leaving the electrons to flow continuously through the superconducting wires (with zero resistance). Power consumption in the magnet is zero.

Instruments operating at 600 MHz are now commercially available, and 750 MHz instruments are expected on the market sometime in the 1990s; 'high'-temperature superconducting materials (operating at the temperature of liquid nitrogen rather than that of the more expensive liquid helium) have inevitably a long-term place in these developments.

3.3.3 Measurement of Chemical Shift—Units Used in NMR Spectroscopy

In recording an NMR spectrum (whether by the Fourier Transform method, FTNMR, or on a continuous wave (CW) spectrometer, working either with field sweep or frequency sweep), we measure the differences in chemical shift position between the reference (TMS) and the signals from the compound being examined.

By international convention, an NMR spectrum is always plotted with TMS on the right-hand side, at low frequency, and with high frequency on the left-hand side.

The proton NMR spectrum of benzyl alcohol is shown in figure 3.4: from left to right, the signals correspond to the protons of the aromatic ring, the CH_2 group, the OH group and, lastly, the small TMS internal standard peak.

The protons of TMS are known to come to resonance at exactly 60 MHz when the field strength is 1.409 T (100 MHz at 2.349 T, 600 MHz at 14.092 T, etc.) and we can measure the frequencies of the other signals in relation to TMS: thus, the frequency of the OH signal in the 60 MHz spectrum in figure 3.4 is found to be 144 Hz higher in frequency than TMS and that of the CH_2 group is 276 Hz higher in frequency, etc. If, however, we record the same spectrum at 100 MHz or 600 MHz, then the differences in frequency are *in proportion*—that is, the CH_2 signal will be 2760 Hz higher at 600 MHz, etc. This leads to the desirability of using units which are *field-independent*.

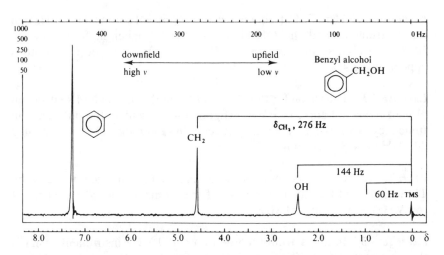

Figure 3.4 ^{1}H *NMR spectrum of benzyl alcohol* (PhCH$_2$OH). (60 MHz *in* CCl$_4$.)

δ *units.* Chemical shift positions are normally expressed in δ (delta) units, which are defined as proportional differences, in parts per million (ppm), from an appropriate reference standard (TMS in the case of proton and carbon-13 NMR).

Since the δ unit is a proportionality, it is a dimensionless number; it is independent of field strength, so that a signal with a δ value of, say, 4.6

derived from a 60 MHz spectrum (1.4 T instrument) such as figure 3.4 will be found at exactly the same value, δ 4.6, on a 100 MHz or on a 600 MHz (14.1 T) instrument. At the bottom of figure 3.4 are shown the δ units—*true for any instrument*—and at the top is shown the frequency scale—*which is only true at 60 MHz*.

The relationship between δ values (in ppm) and frequency (in Hz) is fundamental:

$$\delta_X = \frac{\nu_X - \nu_{TMS}}{\nu_0}$$

where δ_X is the chemical shift (in ppm), ν_X and ν_{TMS} are the frequencies (in Hz) of the signals for X and TMS, respectively, and ν_0 is the operating frequency of the instrument (in MHz).

Example 3.1
Question. The OH signal in figure 3.4 is shown on the spectrum at 144 Hz higher frequency than TMS: how many δ units (ppm) does this correspond to?

Model answer. The TMS frequency is exactly 60 MHz (60 000 000 Hz) on this instrument, and the OH signal is 144 Hz higher in frequency (at 60 000 144 Hz), so the δ value is 144/60 = 2.4. We say 'the OH protons appear at δ 2.4'.

Exercise 3.1   The signal for the CH_2 protons in the proton NMR spectrum of benzyl alcohol appears at δ 4.6 (figure 3.4). Calculate the difference in frequency, expressed in hertz, between this and the TMS signal in a 300 MHz NMR spectrum.

Exercise 3.2   Calculate similarly the difference in frequency between the TMS signal and that for the protons of the aromatic ring (a) in figure 3.4 (60 MHz) and (b) at 300 MHz.

Exercise 3.3   In a spectrum recorded on a 100 MHz instrument, what is the chemical shift position—in δ units—for the CH_2 protons in benzyl alcohol? How many hertz (difference in frequency from TMS) does this correspond to on this spectrum? Calculate similarly for the protons of the aromatic ring.

Exercise 3.4   How many hertz does 1 ppm (one unit on the scale) correspond to on an instrument recording proton spectra at (a) 100 MHz, (b) 250 MHz, (c) 500 MHz?

Two relics of field sweep CWNMR spectroscopy survive in the terminology of the technique. Students coming fresh to NMR spectroscopy should not spend any time learning these (for fear of implanting them), but an awareness is important in reading older literature.

1.   A former convention in the presentation of CW spectra involved field sweep being plotted with low field at the left-hand side of the spectrum, with high field on the right; thus, in figure 3.4 we would say that the CH_2 signal was at lower field than TMS (or *downfield of* TMS) and that the OH signal was at higher field than the CH_2 signal (or *upfield of* CH_2), etc. These parameters have no counterpart in pulsed FTNMR spectra, and since frequency is such a fundamental and unifying property, the terms *upfield* and *downfield* are no longer encouraged, to be replaced by the terms *lower frequency* and *higher frequency*, respectively.

2. τ *units*. To correspond with low field values on the left and high on the right, a scale of units from 0 to 10, also running from left to right, was proposed, thus running counter to the δ scale, so that $\delta = 10 - \tau$. They are only found in older spectra and are not used at all in current practice.

3.4   FACTORS INFLUENCING CHEMICAL SHIFT

3.4.1   ELECTRONEGATIVITY—SHIELDING AND DESHIELDING

Table 3.2 shows the chemical shift positions for CH_3 protons when a methyl group is attached to functions of increasing electronegativity. As the electronegativity of the function is increased, the CH_3 protons come to resonance at higher δ values.

Table 3.2   Chemical shift values for CH_3 protons attached to groups of varying electronegativity

Compound	Chemical shift/δ
$CH_3 - \overset{\displaystyle \mid}{\underset{\displaystyle \mid}{Si}} -$	0.0
CH_3I	2.16
CH_3Br	2.65
CH_3Cl	3.10
CH_3F	4.26

Before we try to explain why, for example, the methyl protons in CH_3F have a higher δ value than those of CH_3Cl, etc., we should restate what this means in instrumental terms.

In pulse FTNMR instruments the field is fixed and a wide pulse of radiofrequency is able to interact with the different precessional frequencies of the groups of protons; in these machines and in CW instruments operating on *frequency sweep* with *fixed field*, the CH_3F signal is at higher frequency than the CH_3Cl signal.

In CW instruments operating on *field sweep*, however, the instrument *frequency is fixed* (for example, at 60 MHz) and the field is increased through the scan until each group of precessing protons comes to resonance with the fixed radiofrequency source at 60 MHz: those protons which come to resonance first (that is, at lowest field values, such as in CH_3F) must possess the highest frequencies at the beginning of the scan, whereas those which require high field values to achieve resonance at 60 MHz (for example, in CH_3Cl) must have inherently lower frequencies of precession.

Why do the protons of the CH_3 group in CH_3F come to resonance at higher frequency than those of CH_3Cl, etc.? The explanation relates to the electron density around the ^1H nuclei.

Hydrogen nuclei are surrounded by electron density, which to some extent *shields* the nucleus from the influence of the applied field B_0, and the extent of this shielding will influence the precessional frequency of the nucleus—the greater the *shielding effect*, the lower the precessional frequency. In a magnetic field the electrons around the proton are induced to circulate, and in so doing they generate a small secondary magnetic field, which acts in opposition (that is, diamagnetically) to the applied field (see figure 3.5(a)). The greater the electron density circulating around the proton, the greater the induced diamagnetic shielding effect and the lower the precessional frequency of the proton. Electronegative groups, such as fluorine in CH_3F, withdraw electron density from the methyl group (inductive effect) and this *deshielding effect* means that the methyl protons experience a greater nett magnetic field, and, hence, precess with higher frequency. Since fluorine is more electronegative than chlorine, its deshielding influence is greater and, hence, the attached protons have higher precessional frequencies (higher δ values).

Silicon is electropositive, and the opposite effect operates in, for example, TMS; silicon pushes electrons into the methyl groups of TMS by a +I inductive effect, and this powerful *shielding effect* means that the TMS protons come to resonance at low frequency (low δ value, defined as zero).

The effect of charged species on chemical shift values is very marked; protons adjacent to N^+ (as in quaternary ammonium ions, R_4N^+) are very strongly deshielded (high δ values), while carbanionic centers act as powerful shielding influences (low δ values).

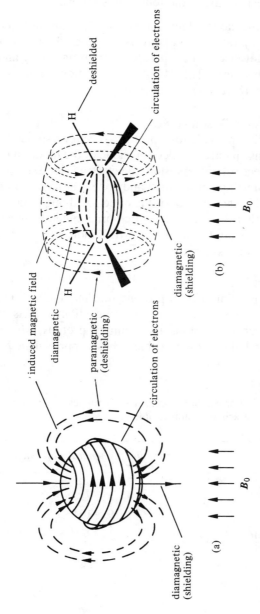

Figure 3.5 *Induced anisotropic magnetic field around* (a) *a hydrogen atom and* (b) *an alkene group.*

121

Many of the data in table 3.4 (page 171) can be explained by electro-negativity considerations, although we shall meet other powerful shielding and deshielding mechanisms in the following sections.

Sign convention. Increased shielding is usually (but not universally) assigned positive (+) values, and conversely. Since, however, increased shielding leads to lower δ values, it is less confusing if (+) and (−) are omitted in diagrams, etc. The IUPAC convention shows (+) where increased frequency (δ values) is observed.

3.4.2   VAN DER WAALS DESHIELDING

In a rigid molecule it is possible for a proton to occupy a sterically hindered position, and in consequence the electron cloud of the hindering group will tend to repel, by electrostatic repulsion, the electron cloud surrounding the proton. The proton will be deshielded and appear at higher δ values than would be predicted in the absence of the effect. Although this influence is small (usually less than 1 ppm), it must be borne in mind when predicting the chemical shift positions in overcrowded molecules such as highly substituted steroids or alkaloids.

3.4.3   ANISOTROPIC EFFECTS

The chemical shift positions (δ) for protons attached to C=C in alkenes is higher than can be accounted for by electronegativity effects alone. The same is true of aldehydic protons and aromatic protons, whereas alkyne protons appear at relatively low δ. Table 3.3 lists approximate δ values for these protons.

Table 3.3   Approximate chemical shift ranges for protons attached to anisotropic groups

Structure	Approximate chemical shift range/δ
$-C\overset{H}{\underset{O}{\Big\langle}}$	9.5–10.0
$\overset{\ }{\underset{\ }{C}}=\overset{H}{\underset{\ }{C}}$	4–8
(benzene ring)—H	6–9
$-C{\equiv}C-H$	1.5–3.5

The explanation is again collated with the manner in which electrons, in this case π electrons, circulate under the influence of the applied field. The effect is complex, and can lead to shifts to higher frequency (downfield shifts, or paramagnetic shifts) or to lower frequency (upfield shifts, or diamagnetic shifts). In addition, the effects are paramagnetic in certain directions around the π clouds, and diamagnetic in others, so that these effects are described as *anisotropic*, as opposed to *isotropic* (operating equally through space).

Alkenes. When an alkene group is so oriented that the plane of the double bond is at 90° to the direction of the applied field (as in figure 3.5(b)), induced circulation of the π electrons generates a secondary magnetic field, which is diamagnetic around the carbon atoms, but paramagnetic (that is, it augments B_0) in the region of the alkene protons.

Where the direction of the induced magnetic field is parallel to the applied field, B_0, the net field is greater than B_0. Protons in these zones come to resonance therefore at higher δ values than expected.

Any group held above or below the plane of the double bond will experience a *shielding effect*, since in these areas the induced field opposes B_0. In α-pinene one of the geminal methyl groups is held in just such a shielded position, and comes to resonance at significantly lower δ (frequency) than its twin. The third methyl group appears at higher δ (frequency), since it lies *in* the plane of the double bond and is thus *deshielded*.

α-pinene

In summary, we can divide the space around a double bond into two categories, as shown in figure 3.6(a). Deshielding occurs in the cone-shaped zones, and in these zones δ values will tend to be higher. Shielding is found outside the cones and protons in these zones are shielded (lower δ values).

Carbonyl compounds. For the carbonyl group a similar situation arises, although the best representation of shielding and deshielding zones is slightly different from the alkene pattern; see figure 3.6(b) Two cone-shaped volumes, centered on the oxygen atom, lie parallel to the axis of the

C=O bond; protons within these cones experience deshielding, so that aldehydic protons, and the formyl protons of formate esters, appear at high δ values. Protons held above or below these cones will come to resonance at lower δ values.

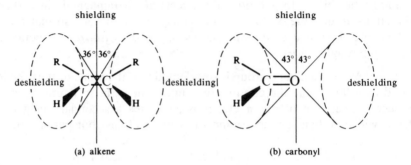

Figure 3.6   *Anisotropic shielding and deshielding around* (a) *alkene groups,* (b) *carbonyl groups. Protons inside the cones are deshielded (have higher δ values).*

Alkynes. Whereas alkene and aldehydic protons appear at high δ values, alkyne protons appear around δ 1.5–3.5. Electron circulation around the triple bond occurs in such a way that the protons experience a *diamagnetic shielding* effect. Figure 3.7 shows how this arises, when the axis of the alkyne group lies parallel to the direction of B_0. The cylindrical sheath of π electrons is induced to circulate around the axis, and the resultant annulus-shaped magnetic field acts in a direction that opposes B_0 in the vicinity of the protons. These protons experience lower values of field; therefore, acetylenic protons appear at low δ values in the spectrum.

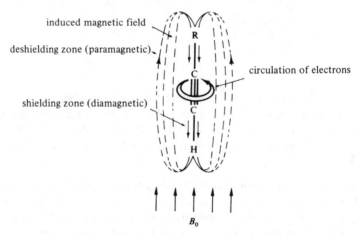

Figure 3.7   *Anisotropic shielding of a proton in an alkyne group.*

Aromatic compounds. In the molecule of benzene (and aromatic compounds in general) π electrons are delocalized cyclically over the aromatic ring. These loops of electrons are induced to circulate in the presence of the applied field, B_0, producing a substantial electric current, called the *ring current*. The magnetic field associated with this electric field has the geometry and direction shown in figure 3.8. (An analogy in the macroworld is a ring of copper wire moved into a magnetic field: electric current flows in the wire, and sets up a magnetic field similar in geometry and direction to that shown for benzene in figure 3.8.)

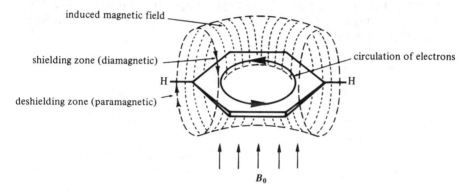

Figure 3.8 *Anisotropic shielding and deshielding associated with the aromatic ring current. The molecules are constantly tumbling, but a nett effect is still present. Protons attached to the ring have high δ values.*

The induced field is diamagnetic (opposing B_0) in the center of the ring, but the returning flux outside the ring is paramagnetic (augmenting B_0).

Protons around the periphery of the ring experience an augmented magnetic field, and consequently come to resonance at higher δ values than would otherwise be so. Protons held above or below the plane of the ring resonate at low δ values. Two examples will illustrate the magnitude of these effects.

In the molecule of toluene the methyl protons resonate at δ 2.34, whereas a methyl group attached to an acyclic conjugated alkene appears at δ 1.95. This is some measure of the greater deshielding influence of the ring current in aromatic compounds (cyclically delocalized π electrons) compared with the deshielding of conjugated alkene groups (having no cyclic delocalization). Indeed, so important is this observation that NMR has become one of the principal criteria used in deciding whether an organic compound has substantial aromatic character (at least in so far as aromatic character relates to cyclic delocalization of $(4n + 2)$ π electrons). The method has been applied successfully to heterocyclic systems and to the annulenes; for example, [18]annulene sustains a ring current, so that the twelve peripheral protons are deshielded and the six internal protons shielded. The outer protons appear at δ 8.9, while the inner protons are at a *lower* frequency than TMS at δ − 1.8. (This is true only around 20 °C; molecular motion makes the spectrum change with variable temperature.)

One of the most dramatic observations in NMR work on aromatic systems involves the dimethyl derivative of pyrene (I), in which the methyl groups appear at δ −4.2, *lower* in frequency than TMS. This shows that the cyclic π electron system around the periphery of the molecule sustains a substantial ring current, and therefore indicates aromatic character in a nonbenzenoid ring system. The methyl groups are deep in the shielding zone of this ring current, and it is for this reason that they appear at such an extraordinary δ value.

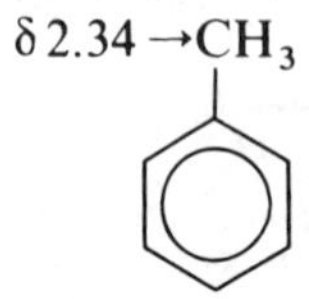

δ 2.34 →CH₃

toluene

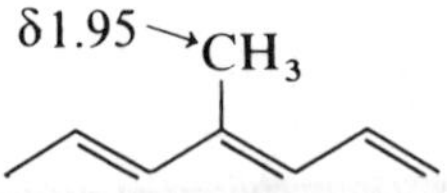

δ 1.95 →CH₃

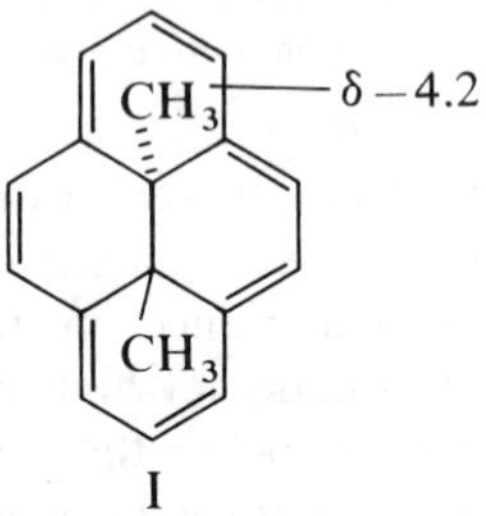

CH₃ — δ −4.2

CH₃

I
dihydrodimethylpyrene

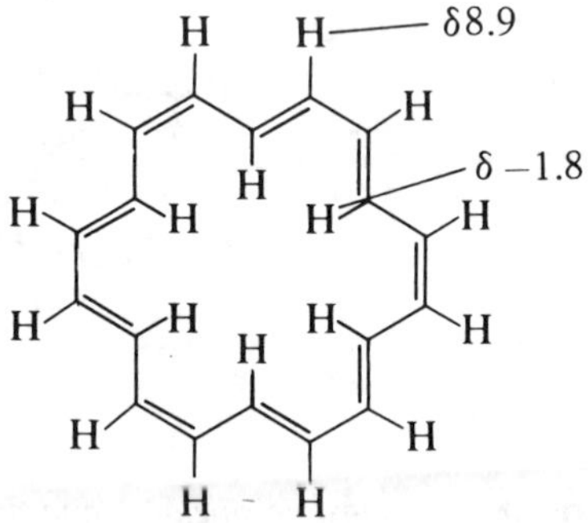

H   H—— δ8.9

δ −1.8

[18]annulene

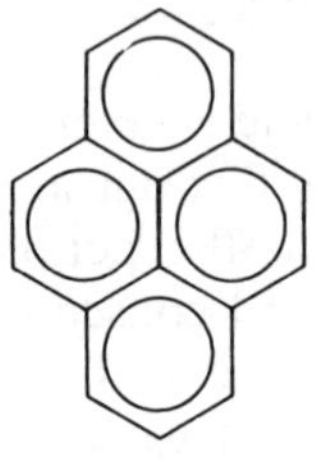

pyrene

Alkanes. The equatorial protons in cyclohexane rings come to resonance about 0.5 ppm higher than axial protons, and this is attributed to anisotropic deshielding by the σ electrons in the βγ bonds, as shown in figure 3.9. The effect is small compared with the anisotropic influence of circulating π electrons, but is readily observed in rigid systems and also in mobile systems at low temperature.

Simple electronegative (inductive) effects operate only along a chain of atoms, the effect weakening with distance, but magnetic anisotropy operates through space irrespective of whether the influenced group is directly joined to the anisotropic group. For this reason the stereochemistry of molecules must be carefully studied to predict whether magnetically anisotropic groups are likely to have an influence on the chemical shift of apparently distant protons.

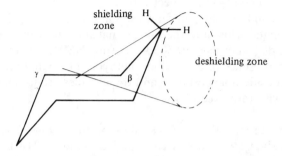

Figure 3.9 *Anisotropic shielding and deshielding in cyclohexanes.*

3.5   CORRELATION DATA FOR PROTON NMR SPECTRA

3.5.1   USE OF CORRELATION TABLES

Although chemical shift data have been rationalized to a large extent, using the factors discussed above, much of the application of NMR to organic chemistry is what one might call 'explained empiricism'.

Predicting the ^{1}H NMR spectrum of an organic compound begins with predicting the chemical shift positions for the different hydrogens in the molecule. Figure 3.10 is a useful chart containing approximate proton classifications, and all chemists working with NMR are thoroughly familiar with these allocations.

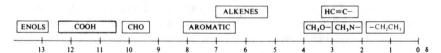

Figure 3.10 *Approximate chemical shift positions for protons in organic molecules.*

Tables 3.4–3.9 later in the chapter contain more detailed figures, and a few examples will serve to illustrate the use of the tables.

We shall predict only the approximate chemical shift positions as a drill exercise; single-line signals are not always obtained, and section 3.9 deals with this important complicating feature.

Example 3.2
Question. Predict the chemical shift positions for the protons in methyl acetate, CH_3COOCH_3.

Model answer. Table 3.4 shows (first column of data) that CH_3 attached to —COOR appears at δ 2.0, and that CH_3 attached to —OCOR appears at δ 3.6. Therefore, the ^1H NMR spectrum of methyl acetate shows two signals, at δ 2.0 and δ 3.6, respectively.

Acetic acid and its esters all show the same signal at ≈ δ 2.0; all methyl esters of aliphatic acids show the same signal at δ 3.6 (with methyl esters of aromatic acids appearing at δ 3.9). Using NMR, we can, therefore, diagnose the presence of acetates or methyl esters in an organic molecule with some considerable certainty and these values are usually accurate to ±0.2 ppm, with rare exceptions.

Exercise 3.5 Predict the chemical shift positions for those protons attached to carbon atoms in (a) *N*-methylacetamide, $CH_3CONHCH_3$; (b) acetone, CH_3COCH_3; (c) dimethyl ether, MeOMe; and (d) dimethylamine, MeNHMe. (See section 3.6.2 for protons attached to nitrogen.)

Example 3.3
Question. Predict the chemical shift positions for the protons in ethyl acrylate, $CH_2{=}CHCOOCH_2CH_3$.

Model answer. Table 3.4 shows (second column of data) that the CH_2 group of an alkyl chain (RCH_2) adjacent to —OCOR appears at δ 4.1.

The first column of data in the same table shows that the terminal CH_3 group on an alkyl chain (CH_3—R) appears at δ 0.9, but the —OCOR group nearby, on the β-carbon, has a minor deshielding effect, as shown in table 3.5. The value given is + 0.4 ppm; therefore, the CH_3 group appears, not at δ 0.9, but at δ 1.3.

Ethyl esters, of whatever acid, all have the same chemical shift positions for the CH_2 and CH_3 groups—namely around δ 4.1 and δ 1.3, respectively.

The δ values for the alkene protons are shown in table 3.7, and three different chemical shift positions are predicted.

alkene δ values   4-nitroanisole

Exercise 3.6 Predict the chemical shift positions for the protons in (a) butanone, $CH_3COCH_2CH_3$; (b) ethyl propanoate (propionate), $CH_3CH_2COOCH_2CH_3$; (c) vinyl propanoate, $CH_3CH_2COOCH{=}CH_2$; and (d) *n*-propyl acetate, $CH_3COOCH_2CH_2CH_3$.

Example 3.4

Question. Predict the chemical shift positions for the protons in 4-nitroanisole.

Model answer. Table 3.4 shows that CH_3 attached to —OAr appears at δ 3.7.

In the aromatic ring of 4-nitroanisole there is considerable symmetry, so that the protons marked *a* are in chemically equivalent environments, and protons *b* likewise. Chemical shifts in the benzene ring are shown in table 3.9: taking the protons of benzene itself as reference (at δ 7.27), the —NO_2 group shifts *ortho* protons to higher frequency by 1.0 ppm, so that protons *a* appear at δ 8.27. The methoxyl group (—OR) moves the protons *ortho* to it to lower frequency by 0.2 ppm, so that the *b* protons appear at δ 7.07.

(Although the accuracy of the tables does not always justify it, one should also compute the influence of the —NO_2 group on the *b* protons *meta* to it, and of the MeO— group on the *a* protons *meta* to it. Thus, the *a* protons are *ortho* to —NO_2 (+ 1.0 ppm) and *meta* to —OR (− 0.2 ppm), and should appear at δ (7.27 + 1.0 − 0.2) = δ 8.07. Similarly, the *b* protons are *ortho* to —OR (− 0.2 ppm) and *meta* to —NO_2 (+ 0.3 ppm), and should appear at δ (7.27 − 0.2 + 0.3) = δ 7.37.)

In practice one would expect to find the *a* protons lying somewhere around δ 8.2, and the *b* protons between δ 7.1 and δ 7.4. Where two *para*-substituents are in conjugation (as in 4-nitroanisole) the errors are greatest.

Exercise 3.7   Predict the chemical shift positions for the protons attached to carbons in (a) acetanilide, $C_6H_5NHCOCH_3$; (b) methyl benzoate C_6H_5COOMe; (c) dimethyl phthalate, the dimethyl ester of benzene-1,2-dicarboxylic acid; and (d) phenacetin (aceto-*p*-phenitidide) *p*-$CH_3CH_2OC_6H_4NHCOCH_3$ (see section 3.6.2 for protons attached to nitrogen).

Example 3.5

Question. Predict the chemical shift positions for the protons in methyl phenoxyacetate, $PhOCH_2COOCH_3$.

Model answer. As we saw in example 3.2, methyl esters of acids with an aliphatic residue on the carboxyl group show the methyl signal at δ 3.6. Table 3.9 shows that the aromatic protons are all moved to lower frequency by the alkoxy group by approximately 0.2 ppm (−0.2 for —OR); the aromatic protons should come to resonance around δ 7.07.

The methylene group, flanked by two electronegative groups, is more difficult to deal with, but Shoolery has computed the relationships listed in table 3.6. For the methylene group in methyl phenoxyacetate, we take the base value of δ 1.2 and add 2.3 ppm for —OPh and add 0.7 ppm for —COOR, giving δ 4.2. The Schoolery rules are usually sufficiently accurate for an organic chemist's purposes, particularly for methylene groups, although the error in predicting the δ values for methine protons frequently exceeds 0.5 ppm.

Exercise 3.8   Predict the chemical shift positions for the protons in (a) diethyl malonate, $CH_2(COOCH_2CH_3)$; (b) methyl cyanoacetate, $NCCH_2COOCH_3$; (c) benzyl methyl ether, $PhCH_2OMe$; and (d) chloroform, trichloromethane.

3.5.2 INFLUENCE OF RESTRICTED ROTATION

The NMR spectrum of *N,N*-dimethylformamide, $HCONMe_2$, recorded around room temperature, shows *two* signals for the methyl groups, although it might have been expected that the two methyl groups would be in magnetically equivalent environments. Dimethylformamide is represented by the two resonance forms shown below, and the result of conjugation between the carbonyl group and the nitrogen nonbonding pair is to increase the double-bond character of the C—N bond sufficiently to restrict the rotation at room temperature: one methyl group is *cis* to

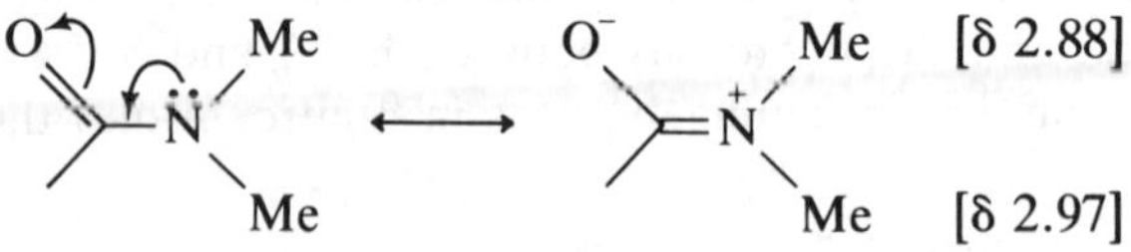

N, N-dimethylformamide

oxygen, the other is *trans*, and anisotropy of the carbonyl group is sufficient to influence the chemical shift position for the *cis* group. Using a heated probe in the NMR instrument (see section 3S.2), the spectrum of *N,N*-dimethylformamide can be recorded at high temperature (≈ 130 °C), and this spectrum shows only one signal for the methyl groups; at elevated temperatures rotation around the C—N bond is so rapid that each methyl group experiences the same *time-averaged* environment.

The four protons in 1,2-dibromoethane ($BrCH_2CH_2Br$) are chemically indistinguishable, and one might also suppose that they are magnetically equivalent; we might predict the same chemical shift position for all four. However, the Newman diagrams (below) show clearly that protons in the different conformations are not in magnetically equivalent environments. For example, the ringed proton is flanked by H and Br in the first conformer, but by H and H in the second. If the NMR spectrum is recorded at low temperature, the rapid molecular rotations around the C—C bond are 'frozen' sufficiently for these differences to be detected. At room temperature rotation is so rapid that each proton experiences the

rotational isomers of 1,2-dibromoethane

cyclohexane ring-flips

same time-averaged environment and only one sharp signal appears in the NMR spectrum. (See also section 3.8.5.)

Similar effects are found in fluxional molecules and in structures where steric effects can intervene; examples are cyclic structures (cyclohexane ring-flips, etc.), bridged structures (as in α-pinene, page 123), spirans, etc.

The energy barriers involved in these rotational changes can be measured by NMR, using a variable-temperature sample probe, and noting the temperature at which the spectrum changes from that of the mixed conformers to that of the time-averaged situation. Some recent applications of this technique are discussed in section 3S.2. The theory in all cases relates back to $\Delta t \cdot \Delta \nu \approx 1/2\pi$: we can only resolve the *individual* resonances if the molecule stays in one position longer than Δt; otherwise we see only the time-averaged environment.

3.6   SOLVENTS USED IN NMR

3.6.1   CHOICE OF SOLVENT FOR PROTON NMR SPECTRA

To satisfy the condition that nonviscous samples give the sharpest NMR spectra (section 3.2), it is usually necessary to record the spectra of organic compounds in solution; if the compound itself is a nonviscous liquid, the neat liquid can be used. Choice of solvent is not normally difficult, provided that a solubility of about 10 per cent is obtainable, but it is clearly an advantage to use aprotic solvents (which do not themselves give an NMR spectrum to superimpose on that of the sample). The following solvents are commonly used, many of which are normal organic solvents in which hydrogen has been replaced by deuterium: Note that most FTNMR instruments maintain frequency accuracy by 'locking' simultaneously to the deuterium NMR frequency, and in these instruments a deuterium-containing solvent is essential.

CCl_4	carbon tetrachloride
CS_2	carbon disulfide
$CDCl_3$	deuteriochloroform (chloroform-d)
C_6D_6	hexadeuteriobenzene (benzene-d_6)
D_2O	deuterium oxide (heavy water)
$(CD_3)_2SO$	hexadeuteriodimethylsulfoxide (DMSO-d_6)
$(CD_3)_2CO$	hexadeuterioacetone (acetone-d_6)
$(CCl_3)_2CO$	hexachloroacetone

A more comprehensive list, including physical properties, is given in table 3.19.

For deuteriated solvents the isotopic purity should ideally be as high as possible, but since greater isotopic purity means greater cost, most users compromise their ideals. Isotopic purity approaching 100 per cent can often be obtained, but, for example, a small $CHCl_3$ peak at δ 7.3 in 99 per cent $CDCl_3$ will cause no difficulty or ambiguity unless accurate integrals for aromatic protons (also around δ 7.3) are demanded.

3.6.2 SOLVENT SHIFTS—CONCENTRATION AND TEMPERATURE
       EFFECTS—HYDROGEN BONDING

The solvents listed above vary considerably in their polarity and magnetic susceptibility. Not surprisingly, the NMR spectrum of a compound dissolved in one solvent may be slightly different from that measured in a more polar solvent, and it is important in all NMR work to quote the solvent used. The NMR signals for protons attached to carbon are, in general, shifted only slightly by changing solvent, except where significant bonding or dipole–dipole interaction might arise: the NMR spectrum for chloroform dissolved in cyclohexane appears at δ 7.3, but in benzene solution the signal is moved by the exceptionally large amount of −1.56 ppm (to δ 5.74). Benzene is behaving as a Lewis base to chloroform, and considerable charge transfer is responsible for altering the electron density around the chloroform proton, with concomitant shift in the signal to a lower δ value. The benzene ring-current also contributes to this shift.

In contrast, NH, SH and, particularly, OH protons all have their NMR signals substantially moved on changing to solvents of differing polarity. This effect is largely associated with hydrogen bonding, and it is noted even when different concentrations are used in the same solvent.

At low concentrations intermolecular hydrogen bonding is diminished in simple OH, NH and SH compounds: since hydrogen bonding involves electron-cloud transfer from the hydrogen atoms to a neighboring electronegative atom (O, N or S), the hydrogen experiences a net deshielding effect when hydrogen bonding is strong, and is less deshielded when hydrogen bonding is diminished. Thus, at high concentrations (strong

deshielding

Intermolecular H bonding raises δ values: δ values are lowered with increased dilution or temperature

hydrogen bonding, strong deshielding) OH, NH and SH protons appear at higher δ than in dilute solutions.

Table 3.8 lists the chemical shift positions for protons subject to hydrogen bonding, and it can be seen that the range within which they come to resonance is wide (δ 0.5–4.5 for simple alcohols).

Increased temperature also reduces intermolecular hydrogen bonding, so the resonance positions for these protons are temperature-dependent (higher temperatures mean lower δ values).

Intramolecular hydrogen bonding is unchanged by dilution and the NMR spectrum from such systems is virtually unaltered by varying concentration. Salicylates and enols of β-dicarbonyl compounds are examples of such systems: chelates, such as the salicylates, show the OH resonance at very high δ (10–12), and enol OH appears even higher (δ 11–16).

Carboxylic acids are a special case of hydrogen bonding because of their stable dimeric association, which persists even in very dilute solution; carboxylic OH appears betwen δ 10 and δ 13, usually nearer δ 11–12.

salicylates   enol of β-diketone   carboxylic acid dimer

Intramolecular H bonding: δ values largely unaffected by concentration changes

Example 3.6
Question. Predict the chemical shift positions for the protons in (a) benzyl alcohol, PhCH₂OH; and (b) acetic acid (ethanoic acid), CH₃COOH.

Model answer. (a) Table 3.9 is used to predict the δ values for aromatic protons, but the substituent —CH$_2$OH is not listed: it is a legitimate approximation to consider the substituent to be an alkyl group, R, and to ignore the more distant effect of the OH group. Thus, the *ortho*, *meta* and *para* protons are shifted, respectively, by −0.15 ppm, −0.1 ppm and −0.1 ppm: the ring protons will therefore all appear near δ 7.17. The CH$_2$ protons are calculated from table 3.6 to appear at δ (1.2 + 1.3 + 1.7) = δ 4.2. The OH proton (table 3.8) may appear anywhere between δ 0.5 and δ 4.5, dependent on hydrogen bonding—which, in turn, depends on concentration and temperature. The observed values are shown in figure 3.4 to be at δ 7.3, δ 4.5 and δ 2.4, respectively, in CCl$_4$ solution. (b) From table 3.4 the δ value for CH$_3$ attached to —COOH is 2.1. From table 3.8 the carboxyl proton appears at δ 10–13.

Exercise 3.9 Predict the chemical shift positions for the protons in (a) succinic acid, HOOCCH$_2$CH$_2$COOH; and (b) paracetamol (*p*-acetamidophenol), HOC$_6$H$_4$NHCOCH$_3$.

3.7 INTEGRALS IN PROTON NMR SPECTRA

We mentioned very briefly in the introduction to this chapter that the area under each NMR signal in the spectrum is proportional to the number of hydrogen atoms in that environment. The two peaks in the toluene spectrum (figure 3.1) have relative areas 5:3 corresponding to C$_6$H$_5$ and CH$_3$, respectively.

Measurement of the peak areas is carried out automatically on the NMR spectrometer by integration of each signal, and the integral value is indicated on the spectrum in the form of a continuous line in which steps appear as each signal is measured: step height is proportional to peak area. These *integral traces* are shown diagrammatically in figure 3.1 and also on the spectra in problems 6.2(i) to 6.2(v) (pages 347–349).

Note that only *relative* peak areas are recorded, and thus only the *ratio* between protons in each environment is given. The step heights should be measured to the nearest 0.5 mm, and by proportionality the best-fit *integer ratio* calculated.

Additionally, a computer printout of peak frequencies and areas is presented.

Accuracy. The accuracy of the integrator on the instrument will be specified by the manufacturer, and is commonly within 2 per cent for several consecutive scans of a standard spectrum (for example, that of ethylbenzene), but a number of factors militate against this accuracy applying to the spectra of specific compounds. The net absorption of radiofrequency energy in the NMR experiment depends on the relaxation time, T_1, on the rate of scan and on the intensity of the radiofrequency source. Not all protons have the same relaxation times (see section 3.2), and slight deviations from exact integer ratios can be found in the signal intensities. If the irradiating frequency is too intense in relation to the scanning rate, saturation of the signals may arise and lead to low integral values. A spectrum with a noisy baseline, perhaps because a weak signal has to be greatly amplified, may not give an accurate integral trace, since noise in the baseline will be integrated along with the resonance signals. Broad peaks tend also to give less accurate integrals than do sharp peaks. Note that if an integration trace is recorded on the proton NMR spectrum of a mixture, then, in general, this permits a quantitative estimate to be made of the relative amounts of each component in the mixture. This is a valuable quality control procedure in industrial laboratories.

3.8   SPIN–SPIN COUPLING—SPIN–SPIN SPLITTING

3.8.1   The Splitting of NMR Signals in Proton NMR Spectra

The ^{1}H NMR spectrum of *trans*-cinnamic acid is reproduced in figure 3.11.

The aromatic protons, five in number, give rise to the peaks at δ 7.4 and δ 7.55, and the carboxyl proton is at δ 12.5; both of these signals we might have predicted from the discussions in sections 3.4 and 3.5.

What we would not have predicted from chemical shift data alone is that proton H_A appears as *two* lines on the spectrum (centered on δ 6.45), and proton H_X appears as *two* lines (centered on δ 7.8). We say that each signal is *split into a doublet*. Note that the separation between the two H_A lines is the same as the separation between the two H_X lines.

Look now at the spectrum of 1,1,2-trichloroethane (figure 3.12). The signal from proton H_A appears as a *triplet*, while that from protons H_X is a *doublet*.

The number of lines (multiplicity) observed in the NMR signal for a group of protons is not *related to the number of protons in that group; the multiplicity of lines is related to the number of protons in neighboring groups.*

For example, protons H_X in figure 3.12 have only *one* neighboring proton, and H_X appears as a two-line signal (doublet); proton H_A has *two* neighbors and the signal is split into three lines (triplet).

(n + 1) rule. *The simple rule is: to find the multiplicity of the signal from a group of protons, count the number of neighbors (n) and add 1. (Exceptions to the rule are discussed in section 3.10).*

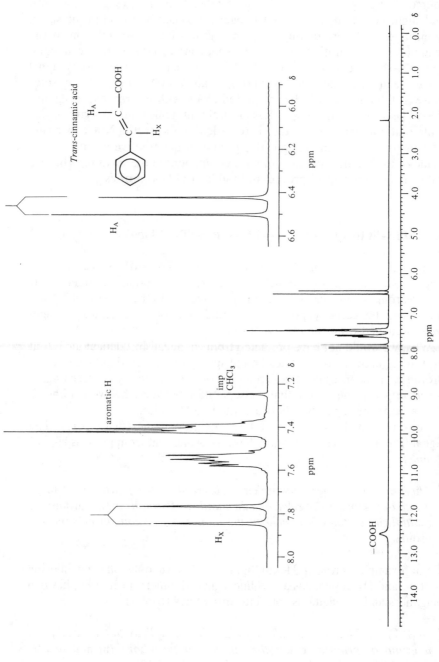

Figure 3.11 ¹H *NMR spectrum of trans-cinnamic acid (200 MHz in CDCl₃). Lower trace, full spectrum; upper traces, expanded × 6.*

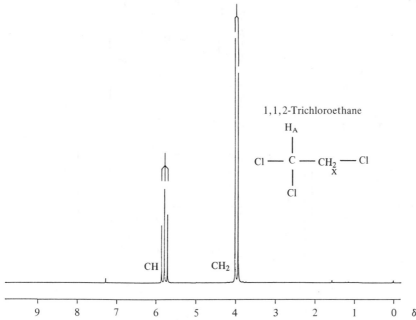

Figure 3.12 ^{1}H *NMR spectrum of 1,1,2-trichloroethane.* (80 MHz *in* CDCl$_3$.)

Splitting of the spectral lines arises because of a *coupling* interaction between neighbor protons, and is related to the number of possible spin orientations that these neighbors can adopt. The phenomenon is called either *spin–spin splitting* or *spin–spin coupling*.

Exercise 3.10   Predict the multiplicities of the signals in the proton NMR spectra of (a) 1,1-dichloro-2,2-dibromoethane; (b) 1,1-dichloro-2-cyanoethane; (c) 1,1-dichloroethane; and (d) 2-chloropropanoic acid, CH$_3$CH(Cl)COOH (ignore the COOH proton).

3.8.2   THEORY OF SPIN–SPIN SPLITTING
The diagram in figure 3.13 represents two vicinal protons similar to the alkene protons in cinnamic acid, H$_A$ and H$_X$. These protons, having different magnetic environments, come to resonance at different positions in the NMR spectrum; they do not give rise to single peaks (singlets) but doublets. The separation between the lines of each doublet is equal: this spacing is called the *coupling constant, J*.

Why is the signal for proton A split into a doublet? A simplistic explanation is that the resonance position for A depends on its total magnetic environment; part of its magnetic environment is the nearby proton X, which is itself magnetic, and proton X can have its nuclear

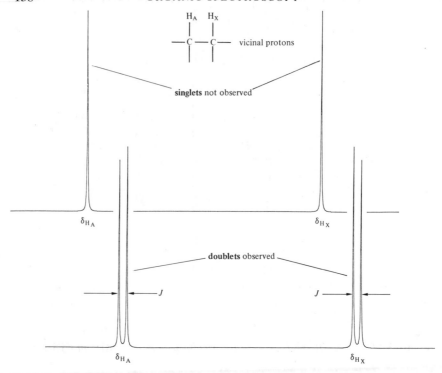

Figure 3.13   *Splitting in the signals of two vicinal protons.*

magnet either *aligned with* proton A or *opposed to* proton A. Thus, proton A can either *increase* the net magnetic field experienced by A (X aligned) or *decrease* it (X opposed); in fact, it does both. The two spin orientations of X create two different magnetic fields around proton A: in roughly half of the molecules the spin orientation of X creates a shielding field around proton A, and in the other half a deshielding field. Therefore, proton A comes to resonance, not once, but twice, and proton A gives rise to a doublet.

Similarly, proton A is a magnet having two spin orientations with respect to X, and A creates two magnetic fields around X. Proton X comes to resonance twice in the NMR spectrum.

This mutual magnetic influence between protons A and X is not transmitted through space, but via the electrons in the intervening bonds. The nuclear spin of A couples with the electron spin of the $C-H_A$ bonding electrons; these, in turn, couple with the $C-C$ bonding electrons and then with the $C-H_X$ bonding electrons. The coupling is eventually transmitted to the spin of the H_X nucleus. This *electron-coupled* spin interaction operates strongly through one bond or two bonds, less strongly through three bonds, and, except in unusual cases, rather weakly through four or more bonds. This point is more rigorously developed in the following section and in supplement 3S.1.

We can represent the possible spin orientations of coupling protons as in figure 3.14. Proton A can 'see' proton X as aligned (parallel ↑) or opposed antiparallel ↓); these two spin orientations correspond effectively to two different magnetic fields. Therefore, proton A comes to resonance twice. The same argument explains why proton X appears as a doublet.

The A and X protons of cinnamic acid give rise to this characteristic pair of doublets, caused by two protons undergoing spin coupling: such a spectrum is called an AX spectrum. Since the probability of the two spin orientations of A and X arising is equal in molecules throughout the sample, the two lines in each doublet are of equal intensity. (However, see section 3.10.)

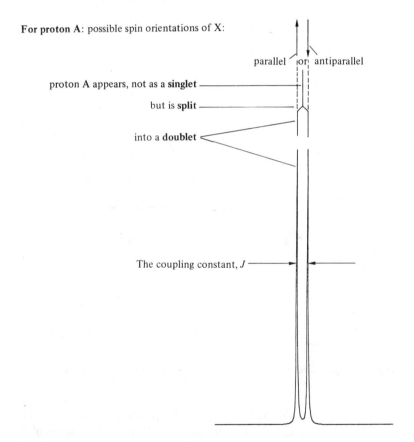

For proton A: possible spin orientations of X:

parallel or antiparallel

proton A appears, not as a **singlet**

but is **split**

into a **doublet**

The coupling constant, J

Figure 3.14 *Simulation of spin coupling between a proton A and one neighboring proton X.*

Figure 3.15 represents the coupling that arises in the triplet signal in the NMR spectrum of 1,1,2-trichloroethane.

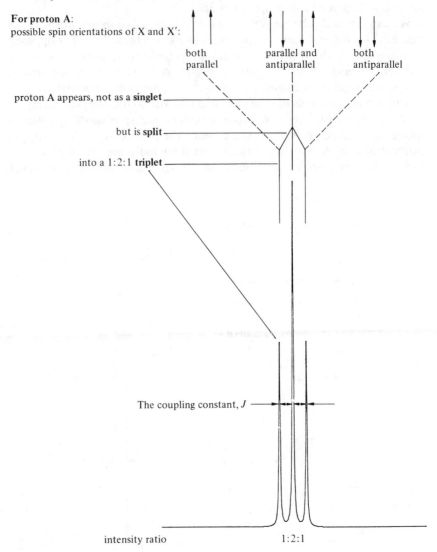

Figure 3.15 *Simulation of spin coupling between a proton* A *and two neighbor protons* X *and* X'.

When proton A 'sees' the two neighboring protons X and X', A can 'see' *three* different possible combinations of spin: (1) the nuclear spins of X and X' can both be parallel to A (↑ ↑); (2) both can be antiparallel to A

($\downarrow$ $\downarrow$); (3) one can be parallel and the other antiparallel, and this can arise in two ways—X parallel with X' antiparallel ($\uparrow$ $\downarrow$) or X antiparallel with X' parallel ($\downarrow$ $\uparrow$). Three distinct energy situations, (1), (2) and (3), are created, and therefore proton A gives rise to a triplet. The probability of the first two energy states arising is equal, but since the third state can arise in two different ways, it is twice as likely to arise; the intensity of the signal associated with this state is twice that of the lines associated with the first two states, and we see in the spectrum of 1,1,2-trichloroethane that the relative line intensities in the triplet are 1:2:1.

The spectrum of 1,1,2-trichloroethane, consisting of the characteristic doublet and triplet of two protons coupling with one proton, can be called an AX_2 spectrum.

The relative line intensities predicted by the above spin coupling mechanism are 1:1 in doublets and 1:2:1 in triplets. Real spectra almost always depart from this first-order prediction in a characteristic manner: the doublet and triplet are slightly distorted, the inner lines being a little more intense than the outer lines. This 'humping' toward the center can be seen in the AX coupling in the cinnamic acid spectrum (figure 3.11) and in the other spectra reproduced later in the chapter. While not all coupling multiplets 'hump' toward the center, it is a reliable enough occurrence to help in deciding which are the coupling multiplets in complex spectra.

3.8.3 MAGNITUDE OF THE COUPLING—COUPLING CONSTANTS, J

The coupling constant, J, is a measure of the interaction between nuclei, and we have stated earlier that the interaction is transmitted through the intervening electrons: for two nuclei, there are four energy levels involved in the NMR transitions, and their relative positions are governed by the internuclear spin coupling as shown in figure 3.16. When there is zero coupling between A and X, both X transitions are equal, as are both A transitions: each nucleus gives rise to one line absorption.

When coupling takes place, we can suppose that the energy levels are altered by $\pm E$, the energy of interaction—then the transitions no longer remain equal: the X transition splits into an X' line (transition energy $X + 2E$) and an X'' line ($X - 2E$). The A transition is likewise split into A' ($A + 2E$) and A'' ($A - 2E$). The spacings between X' and X'' and A' and A'' are equal, and have magnitude $4E$: the coupling constant, J, is therefore equal to $4E$, and is a measure of nuclear interaction, *which is wholly independent of any external magnetic field*. The units of J are energy units, usually Hz; and it is easy to see that J can have sign as well as magnitude. The sign of J is hardly ever of importance to organic chemical applications of NMR; in practical terms, it is simpler to interpret signal multiplicity in more extensive coupling systems by utilizing the $(n + 1)$ rule and the concept of x nuclei 'seeing' y neighboring nuclei, etc.

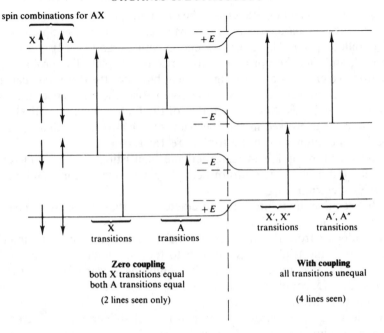

spin combinations for AX

X', X''
transitions

A', A''
transitions

X
transitions

A
transitions

Zero coupling
both X transitions equal
both A transitions equal

(2 lines seen only)

With coupling
all transitions unequal

(4 lines seen)

Figure 3.16 *Spin coupling as the energy of nuclear interaction. Only those transitions are allowed which involve one (and only one) change in nuclear spin; thus* ↓ ↑ → ↑ ↑ *is allowed, but* ↓ ↑ → ↑ ↓ *is forbidden. Example shown is for J positive (antiparallel spins of lower E): for negative J the ±E changes in energy levels are reversed (parallel spins of lower E).*

3.8.4   MORE COMPLEX SPIN–SPIN SPLITTING SYSTEMS

The two examples of spin coupling above, the AX and AX_2 cases, show virtually undistorted multiplets, with the multiplicity following the $(n + 1)$ rule, and signal intensities almost exactly 1:1 and 1:2:1. Such spectra are described as *first-order* spectra. We can extend this first-order treatment successfully to more complex systems, before considering non-first-order spectra (in section 3.10).

The spectrum of 2-chloropropanoic acid (figure 3.17) contains a doublet and a quartet (J, 8 Hz), corresponding to the coupling of one proton with three neighbors; this is an AX_3 spectrum.

The methyl protons have one neighbor, and therefore appear as a doublet.

The methine proton has three neighbors, on the methyl group, and therefore $(n + 1)$ is 4. For this methine proton, we must consider the various ways in which the spin orientations of the methyl protons can be grouped together, and we find that four arrangements are possible: (1) the methine proton can 'see' all three spins of the methyl protons parallel

(↑ ↑ ↑); (2) alternatively, two spins may be parallel with one antiparallel (and there are three ways in which this can arise—↑ ↑ ↓ or ↑ ↓ ↑ or ↓ ↑ ↑); (3) then two spins can be antiparallel with one parallel (arising in three ways—↓ ↓ ↑ or ↓ ↑ ↓ or ↑ ↓ ↓); (4) lastly, all three spins can be antiparallel (↓ ↓ ↓).

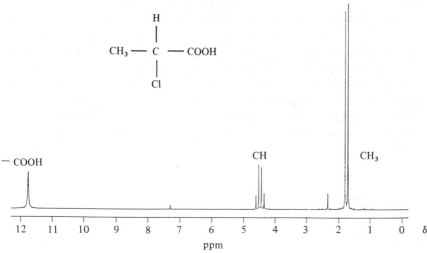

Figure 3.17 ^{1}H *NMR spectrum of 2-chloropropanoic acid. (80 MHz in CDCl$_3$. Note CHCl$_3$ impurity peak in the CDCl$_3$ at δ 7.3.)*

Four different energy combinations are produced; therefore, the me-thine proton comes to resonance four times, appearing as a quartet. The relative probabilities of these states arising are in the ratio 1:3:3:1, and the line intensities in the quartet have the same ratio.

One can go further and predict the theoretical line intensities for quintets, sextets, etc., and find that the ratios are the same as the coefficients in the binomial expansion. Pascal's famous triangle serves to remind:

						singlet
1						singlet
1 1						doublet
1 2 1						triplet
1 3 3 1						quartet
1 4 6 4 1						quintet
1 5 10 10 5 1						sextet

The outer lines in substantial multiplets are of such low intensity that they may be all but unobservable, unless that part of the spectrum is rerun at increased intensity.

The triplet and quartet observed in the spectrum of ethyl bromide (figure 3.18) is an A_2X_3 case, which is one of the easiest of systems to identify in a proton spectrum. All isolated ethyl groups produce a similar spectrum, the chemical shift positions of the CH_2 protons being dependent on substituent (see example 3.3, page 128). The presence of such ethyl groups (as in ethyl esters, ethyl ketones, ethyl ethers, ethanol, etc.) is quite unequivocally identifiable from the ^1H NMR spectrum.

In the ^1H NMR spectrum of 1-nitropropane (figure 3.19) there are three groups of protons, each group coupling with its near neighbors, so that the central methylene group couples both with the methyl protons and with the terminal methylene protons. The methyl group appears as a triplet, and the terminal methylene group also appears as a triplet, since both couple with the central methylene group: $(n + 1) = (2 + 1) = 3$. (The methyl protons show no coupling with the terminal methylene protons, since coupling over four σ bonds is rarely observed; but see *long-range coupling*, section 3.9.1.)

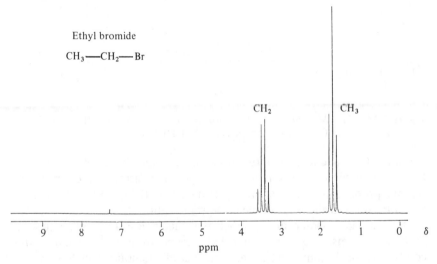

Ethyl bromide

$CH_3 \longrightarrow CH_2 \longrightarrow Br$

CH_2

CH_3

Figure 3.18 ^1H *NMR spectrum of ethyl bromide.* (80 MHz *in* $CDCl_3$.)

The central methylene group can be dealt with by considering the successive coupling, first with the methyl group, and then with the terminal methylene protons. The methyl protons split the central methylene signal into a quartet; the two terminal methylene protons should now split *each line of this quartet* into a triplet, giving twelve lines in all. In the spectrum only six lines are observed, showing that considerable overlapping of the predicted twelve lines has taken place. In fact, the two coupling constants involved (J_{CH_3,CH_2} and J_{CH_2,CH_2}) are equal, and an easier way to consider the central methylene group is to add the *total* number of neighbors with which it couples (CH_3 and CH_2) and then apply the $(n + 1)$ rule. There are five coupling neighbors: therefore, the multiplicity is 6.

This simplified approach only succeeds when the two coupling constants are equal. If the two coupling constants had been different, it might have been possible to observe all twelve predicted lines; the spectrum of ethanol gives us an opportunity to see this degree of coupling; see figure 3.22 and section 3.8.5.

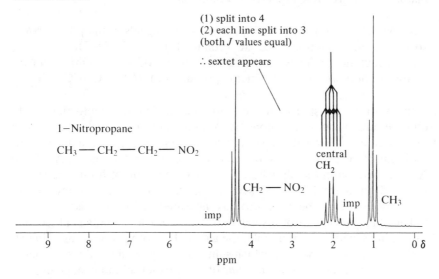

Figure 3.19 ^1H *NMR spectrum of 1-nitropropane.* (80 MHz *in* $CDCl_3$. *Note signals from impurity, 2-nitropropane.*)

Example 3.7

Question. Predict the multiplicities of the signals in the proton NMR spectra of (a) 1,3-dichloropropane, $ClCH_2CH_2CH_2Cl$; (b) 1,1,3,3-tetrachloropropane, $Cl_2CHCH_2CHCl_2$; and (c) di-isopropyl ether, $(CH_3)_2CH—O—CH(CH_3)_2$.

Model answer. It is first necessary to determine how many different chemical shift positions appear in each molecule. (a) Only two chemical shifts are seen (both terminal CH_2 groups have the same value). The terminal CH_2 groups both appear as triplets (since they each have two coupling neighbors—the middle CH_2). The middle CH_2 appears as a quintet, since it has four neighbors—and because the coupling constant to all four is equal. (b) Again only two values of chemical shift appear. The terminal CH groups appear as a triplet as above, but so does the middle CH_2 group (since it has two coupling neighbors, with J being the same in each case). (c) Two chemical shift positions appear here also, since both methine protons are equivalent, as are all twelve methyl protons. The methine protons both appear as septets (coupling equally with all six neighbors) and the CH_3 groups all appear in the same doublet signal (one coupling neighbor). No coupling is observed across the C—O—C bonds.

Exercise 3.11   Predict the multiplicities of the signals in the proton NMR spectra of (a) ethyl acetate (ethanoate), $CH_3COOCH_2CH_3$; (b) diethyl malonate, $CH_2(COOCH_2CH_3)$; (c) butanone.

Exercise 3.12   Predict the multiplicities of the signals in the proton NMR spectra of (a) isopropyl acetate (ethanoate), $CH_3COOCH(CH_3)_2$; (b) 3-methylbutanone; (c) isopropyl methyl ether, $(CH_3)_2CHOMe$; and (d) methyl isobutyrate (methyl 2-methylpropanoate), $(CH_3)_2CHCOOCH_3$.

In unsaturated systems (and aromatic systems) it is frequently possible to observe three groups of protons, A, M and X, each of which couples with the other two. For such a system to be first-order, the chemical shift positions of the protons must be relatively well separated, just as A, M and X are separated in the alphabet. The NMR spectrum of furan-2-aldehyde (furfural) in figure 3.20 shows such an AMX system; the coupling pattern involves interaction between protons separated by four bonds.

The three nuclear protons each give rise to a four-line signal, so that twelve lines in all are observable; the aldehydic proton appears as a singlet.

Proton A couples with X, which splits the A signal into a doublet; but A also couples with M, so that *each line* of A is further split in two, giving four lines in all; see figure 3.20. The signal for proton A shows two splittings, J_{AM} and J_{AX}, and is therefore a double doublet (rather than a quartet).

Similarly, the M signal is split into two by coupling with X, and each line is further split into two by coupling with A; two coupling constants are again seen, J_{AM} and J_{MX}, and M appears as a double doublet.

Lastly, the X signal is split into a double doublet by two successive couplings, J_{MX} and J_{AX}.

A full analysis of an AMX spectrum involves identifying all three J values, J_{AM}, J_{AX} and J_{MX}. Note that each J value appears in two different multiplets, and that each multiplet contains two different J values.

The NMR spectrum of many vinyl compounds (for example, vinyl acetate) show AMX coupling systems.

Figure 3.21 shows the AMX coupling system.

Example 3.8
Question. Estimate the coupling constants in the proton NMR spectrum of 1-nitropropane, shown in figure 3.19.

Model answer. Since this is an 80 MHz spectrum, 1 ppm corresponds to 80 Hz (1 part per million of 80 million). Careful measurement of any of the line separations, and scaling this with respect to the dimensions of the ppm

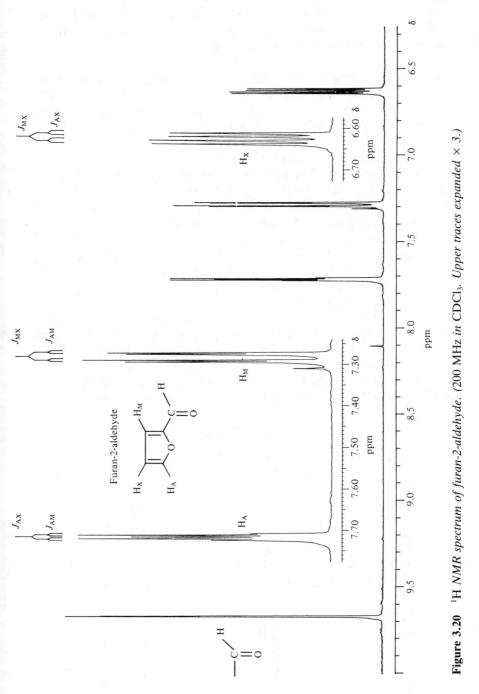

Figure 3.20 ^1H *NMR spectrum of furan-2-aldehyde.* (200 MHz in CDCl$_3$. *Upper traces expanded* × 3.)

axis, shows J to be approximately 7 Hz. Note that the middle CH_2 group *only* appears as a simple sextet because coupling to both of its terminal neighbors has substantially the same J value.

Exercise 3.13 Measure the three AMX coupling constants in the 200 MHz proton NMR spectrum of furan-2-aldehyde (figure 3.20). What would these couplings measure if the spectrum had been recorded at (a) 500 MHz or (b) 600 MHz? Recall that at 200 MHz 1 ppm = 200 Hz.

3.8.5 CHEMICAL AND MAGNETIC EQUIVALENCE IN NMR

We saw in section 3.5.2 that protons which are chemically indistinguishable, in terms of their synthetic reactivity, may nevertheless give rise to more complex NMR spectra than this might have implied. Having studied the separate phenomena of chemical shift and spin–spin coupling, we can now introduce a general set of criteria which govern these complications and therefore must be borne in mind in the analysis of many NMR spectra. The necessary definitions will be mainly presented with respect to proton NMR spectra, but the factors involved apply equally to the NMR spectra of any magnetic nuclei.

Consider that formula I below represents the most stable conformation of the molecule of $XCH_2 — CH_2Y$: because of symmetry, H_A and $H_{A'}$ will have the same chemical shift values, as will H_B and $H_{B'}$. Now H_A will undergo spin–spin coupling to H_B, but (assuming no rotation around the $C — C$ bond) this will be different from the coupling of $H_{A'}$ to H_B. We say that H_A and $H_{A'}$ are *chemically equivalent* but are *magnetically non-equivalent*.

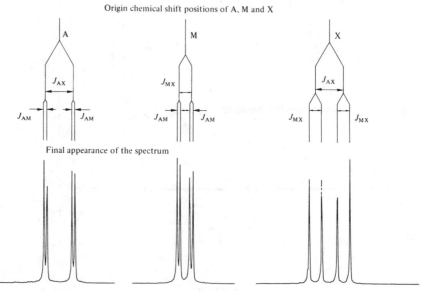

Figure 3.21 *Coupling constants in an AMX system.*

Two protons are defined as being chemically equivalent *if, by virtue of symmetry within the molecule, their electronic environments are indistinguishable and, therefore, they possess the same value of chemical shift.*

I   II   III   IV

Two protons are defined as being magnetically equivalent *if each couples equally to a third neighboring proton; otherwise they are magnetically nonequivalent.*

The labels A, B, C, and so on, are allocated to each separate group of chemically equivalent nuclei; if two protons H_A are magnetically non-equivalent, this is indicated by the use of primes (thus, H_A and $H_{A'}$). Similarly, protons, H_B and $H_{B'}$ must be chemically equivalent but magnetically nonequivalent.

Molecules II and III contain equivalent and nonequivalent groups of nuclei labelled to show the differences between chemical and magnetic equivalence. All *para*-substituted benzene derivatives of this type (II, where group X is different from group Y) show NMR spectra with recognizable features: see section 3.10. The two fluorine atoms in III are subject to the same considerations, and are chemically equivalent but magnetically nonequivalent.

The different implications of chemical and magnetic equivalence are not made especially clear in the conventional choice of terminology which has been acquired in the development of NMR practice, since the terms *chemical* and *magnetic* have so many other connotations outside NMR spectroscopy. It is certainly important to recognize that, fundamentally,

chemical equivalence means simply *chemical shift equivalence,* and *magnetic equivalence* means *coupling equivalence.*

Note the additional corollaries that two protons which are chemically nonequivalent must also be magnetically nonequivalent, but that two protons can be magnetically nonequivalent while still being chemically equivalent.

Deceptive simplicity. As a result of chance, it occasionally arises that the chemical shifts of two protons are equal or very nearly so, even though they do not have the same chemical environment. This is called *accidental equivalence*; if they couple to a third proton, the coupling pattern will be affected and one example is in the molecule of 2,5-dichloronitrobenzene, IV, where H_B and H_C have near-identical chemical shifts. The proton NMR spectrum of this molecule is superficially first-order, with a doublet near δ 7.5 for H_B and H_C (because they couple with H_A) and a triplet near δ 7.9 for H_A (coupling with H_B and H_C). The expected coupling constants are J_{AB} (*meta*) 2–3 Hz, and near zero for J_{AC} (*para*): the observed spacings in both the triplet and the doublet do not correspond to either of these values, but are the *average* of them, at 1.6 Hz.

Although in appearance the spectrum is AX_2 in type, nevertheless it is strictly a special case of an ABX spectrum. Such a spectrum is said to exhibit *deceptive simplicity*.

3.8.6 PROTON-EXCHANGE REACTIONS

If the ^{1}H NMR spectrum of ethanol is recorded on a slightly impure sample and then compared with the spectrum of a high-purity sample, we become faced with yet another problem; figure 3.22 shows these two spectra. The spectrum of the commercial-grade sample is easily explained by use of the multiplicity predictions of an A_2X_3 case, as for ethyl bromide (figure 3.18), but clearly the OH proton is not involved in coupling with the CH_2 group. The spectrum for the pure sample does show this coupling, and we can explain the multiplicity, using the arguments developed for the 1-nitropropane spectrum (figure 3.19).

The CH_3 group is a triplet because of coupling to CH_2. The OH signal is a triplet because of coupling to CH_2. The CH_2 is split into a quartet by the CH_3 group, and each line is further split into two by the OH proton. There are two different coupling constants involved (8 Hz and 6 Hz), so all eight predicted lines are reasonably clear.

Why is the OH coupling not observed in the spectrum recorded on the contaminated sample of ethanol?

Exchange of the OH protons among ethanol molecules is normally so rapid that one particular proton does not reside for a sufficiently long time on a particular oxygen atom for the nuclear coupling to be observed:

$$R—O—H + R—O—H^* \rightleftharpoons R—O—H^* + R—O—H$$

We saw earlier (page 109) that $\Delta t \cdot \Delta v \approx 1/2\pi$. Here Δt is the time needed to resolve accurately the multiplicity in the CH_2 and OH groups brought about by their coupling and Δv is the coupling constant. Provided that the *residence time* of a particular proton on oxygen is sufficiently long (longer than Δt), we can record the coupling. If there is rapid proton exchange, the residence time will be shorter than Δt and the coupling will not be resolved. The rate of exchange is related to the coupling constant (6 Hz in this case), and if we are unable to resolve the coupling, the rate of exchange must be greater than 6 s^{-1}. The exchange is acid-catalyzed and base-catalyzed, and

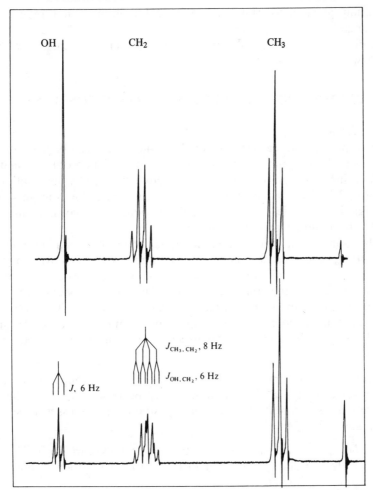

Figure 3.22 ^1H *NMR spectrum of ethanol. Top: sample with acidic impurities. Bottom: pure sample, showing* OH—CH_2 *coupling (recorded at 100 MHz). Chemical shifts:* CH_3, δ 1.2; CH_2, δ 3.7; OH, *near* δ 5, *but variable.*

only in samples that are acid- and base-free is the residence time sufficiently long for the coupling to be observed between OH and CH_2.

Rapid proton exchange occurs in carboxylic acids, phenols, amines and thiols, etc. (see table 3.8), so that, in general, no coupling is observable between the protons on these functions and their neighbours.

In the case of alcohols it is relatively easy to see the OH coupling if the sample is pure, or if the spectrum is recorded with a low-temperature sample (when the exchange is retarded), or if the sample is dissolved in a highly polar solvent such as dimethylsulfoxide, Me_2SO, when strong solvation presumably stabilizes individual molecules and reduces the exchange.

An important corollary of this process is deuterium exchange, which is dealt with in section 3.9.5.

3.9 FACTORS INFLUENCING THE COUPLING CONSTANT, J

3.9.1 GENERAL FEATURES

As was pointed out earlier, the coupling constant, J, can have positive or negative values; initial interpretation of the spectra can profitably omit this factor, since the sign of J cannot be directly extracted by observation.

Table 3.10 lists the commonest proton–proton coupling constants found in organic molecules, and some general points are worth highlighting before beginning a general discussion.

The number of bonds intervening between the coupling nuclei is important, since the coupling is transmitted via the electrons of these bonds. It is a convenient notation to indicate this number as a superscript to the symbol for the coupling constant. Thus, direct coupling (as in the coupling of a proton with an attached carbon-13 nucleus, $^{13}C-^1H$) would be a one-bond coupling, ^1J; the coupling between protons on a CH_2 group would be symbolized by ^2J; that between protons on adjacent carbons as ^3J; and so on.

Geminal coupling, involving protons on $-CH_2-$ groups, is strong, ^2J being typically 10–18 Hz, but it will only be observed where the *gem* protons have different chemical shift positions, as discussed in sections 3.9.2 and 3.10.

Vicinal coupling (three bonds separating the protons) varies from $^3J = 0$ to $^3J = 12$ Hz in rigid systems, but in freely rotating carbon chains (alkyl groups) it is usually around 8 Hz.

Long-range coupling in alkane systems (extending over more than three bonds—that is, ^4J and longer) is usually vanishingly small, but is observed within rigid systems where the W-shaped zig-zag of bonds is near to being coplanar, as indicated by the heavy bonds in the formulae at the bottom of table 3.10.

Trans coupling in alkene groups (^3J, 11–19 Hz) is stronger than *cis* coupling (J, 5–14 Hz). Typical values, 16 Hz and 8 Hz, respectively.

Aromatic coupling depends on whether the coupling protons are *ortho*, *meta* or *para* to each other, and in simple cases the coupling constant is definitive in deciding the orientation; thus, $^3J_{ortho}$, 7–10 Hz; $^4J_{meta}$, 2–3 Hz; $^5J_{para}$, 0–1 Hz.

Allylic coupling, as in allyl chloride, is the most likely four-bond coupling to be met in nonaromatic molecules and is very small (^4J, 0–2 Hz). The analogous coupling in aromatic systems (for example, between the methyl protons and the *ortho* protons in the ring) is not normally large enough to be measured, although it has been observed in certain polynuclear aromatic hydrocarbons, such as 2-methylpyrene, in which considerable double-bond character exists in the intervening aromatic $C\!\!=\!\!\!=\!\!C$ bond.

$$CH_2\!\!=\!\!CHCH_2Cl$$
allyl chloride

J_{AX}, allylic coupling, 0–2 Hz

2-methylpyrene

Other magnetic nuclei present in the molecule (^{14}N, ^{19}F, etc.) may increase the complexity of proton spectra (see section 3.9.4), while substitution of deuterium for hydrogen may lead to simplification (see sections 3.9.4 and 3.9.5).

Exercise 3.14   Estimate the coupling constant for the alkene protons in cinnamic acid, whose proton NMR spectrum is in figure 3.11; does this confirm that this is the *trans* isomer?

Exercise 3.15   Measure the AMX coupling constants for the vinyl protons of vinyl acetate: see figure 3.29. Do the measured coupling constants accord with the principle that *cis* coupling constants are smaller than *trans*? (First verify the three chemical shift positions from table 3.7.)

3.9.2   FACTORS INFLUENCING GEMINAL COUPLING

The electronegativity of an attached substituent alters the values of *gem* coupling, but not always predictably. In groups such as $-CH_2-X$ the *gem* coupling will range from 12 Hz to 9 Hz as the electronegativity of X is increased. These couplings cannot usually be measured directly, because the two protons will have identical δ values unless they are diastereotopic, but in the derived $-CHD-X$ the *gem* coupling between H and D ($H-C-D$) can be measured; $J_{H,H}$ can then be calculated from the equation $J_{H,H} = 6.53 J_{H,D}$ (see section 3.9.4).

The magnitude of J_{gem} also varies with the $C—\hat{C}—C$ bond angle, being of greatest magnitude (10–14 Hz) in the strain-free cyclohexanes and cyclopentanes. With increasing angular strain the value of J_{gem} drops, being 8–14 Hz in cyclobutanes and 4–9 Hz in cyclopropanes.

3.9.3 Factors Influencing Vicinal Coupling

The electronegativity of attached substituents alters the value of vicinal coupling, as it does that of geminal coupling. In qualitative terms, the more electronegative the substituent the smaller the value of J_{vic}, so that in unhindered ethanes the value is ≈ 8 Hz and in halogenoethanes it is lowered to 6–7 Hz. Where there is restricted rotation, the angle subtended by the electronegative substituent at the $C—C$ bond also has an effect on J_{vic}, and other constraints which alter the angles $H—\hat{C}—C$ and $C—\hat{C}—H$, particularly the presence of small rings, will influence J_{vic}.

The factor that is ostensibly most easy to predict in its influence on J_{vic} is the dihedral angle, ϕ, between the two vicinal $C—H$ bonds; the equations due to Karplus give frequent agreement with the observed values.

Karplus's equations

$$\phi \text{ between } 0° \text{ and } 90°: J_{vic} = 8.5 \cos^2 \phi - 0.28$$
$$\phi \text{ between } 90° \text{ and } 180°: J_{vic} = 9.5 \cos^2 \phi - 0.28$$

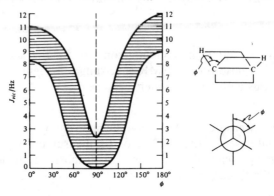

Figure 3.23   *Variation of vicinal coupling constant, J_{vic}, with dihedral angle ϕ (graphical presentation of Karplus's equations).*

It is more convenient to express this graphically (see figure 3.23), and indeed the reliability of the method is not sufficiently high to justify accurate calculations.

To summarize the Karplus rules, the largest vicinal couplings arise with protons in the *trans* coplanar positions ($\phi = 180°$). Vicinal couplings for *cis* coplanar protons are almost as large ($\phi = 0°$). Very small couplings arise between protons at 90° to each other.

As an example of the successful application of these rules, we can consider the protons in chair cyclohexanes: diaxial protons have coupling constants around 10–13 Hz (somewhat larger than predicted) and this collates with their 180° orientation; diequatorial protons, or those with axial/equatorial relationship, have coupling constants around 2–5 Hz, corresponding to about 60° orientation.

$$H_a$$
$$H_e$$
$$H_e$$
$$H_a$$

3.9.4 HETERONUCLEAR COUPLING

The presence in an organic molecule of magnetic nuclei other than hydrogen can introduce additional complications into the spectrum, since these nuclei will take up spin orientations with respect to the applied field and may cause spin–spin splitting of the proton signals in the same way as neighbor protons do.

As a simple example, the spectrum of 2,2,2-trifluoroethanol is shown in figure 3.24. The CH_2 group does not appear as a singlet, but couples with the three vicinal ^{19}F atoms; since ^{19}F has $I = \frac{1}{2}$, each fluorine nucleus can take up two spin orientations in exactly the same manner as hydrogen.

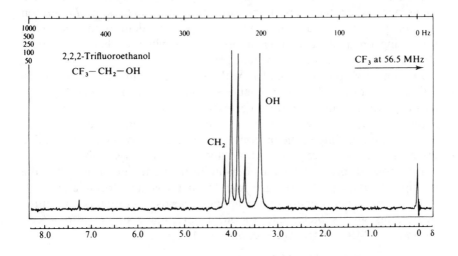

Figure 3.24 ^1H *NMR spectrum of 2,2,2-trifluoroethanol* (60 MHz in CDCl$_3$). *The* CH_2 *group is split into a quartet by the three neighbor* ^{19}F *atoms.*

The splitting of the CH_2 group by CF_3 in CF_3CH_2OH is analogous to the splitting of the CH_2 group by CH_3 in CH_3CH_2OH; the coupling constants are about the same (8 Hz and 11 Hz, respectively, for H—H and H—F coupling).

The signal for ^{19}F itself is about 60 000 ppm lower in frequency than TMS; if we wish to observe this at the same field strength (1.4 T), we need a radiofrequency source at 56.5 MHz, the precessional frequency for ^{19}F. Couplings with ^{19}F and ^{31}P are dealt with in more detail in section 3S.5.

Carbon-13 coupling. Only 1.1 per cent of carbon atoms are magnetic (the ^{13}C isotope) but these nuclei split the protons in ^{13}C—H groups into doublets; these weak signals can be seen sitting symmetrically astride any strong ^{12}C—H signals as *satellites*, separated by the coupling constant $J_{C—H}$ (about 120 Hz).

Deuterium coupling. We saw earlier in section 3.2 that the number of orientations which any magnetic nucleus can adopt in a magnetic field is $(2I + 1)$, where I is the spin quantum number. Thus, for protons and carbon-13 nuclei two orientations arise, but for deuterium (with $I = 1$) three orientations are allowed, corresponding, respectively, to $-I$, 0 and $+I$: figure 3.25 shows these orientations in relation to the applied field B_0. Any group of protons coupled to one deuterium nucleus will 'see' three spin orientations for the deuteron, and therefore experience three different nett fields. Orientation (a) will augment B_0 (thus raising the precessional frequency of the attached protons); orientation (c) will diminish B_0 (proton frequency reduced); and in orientation (b) the deuterium nucleus is precessing on a plane cutting across B_0 and will not change the field strength (proton frequency unchanged).

In summary, protons coupled with one deuterium nucleus come to resonance at three different frequencies—that is, the proton signal appears as a triplet; the line separations correspond to $J_{H,D}$.

Since the populations of the three deuterium spin states are very nearly equal, the probability of the protons 'seeing' these is also equal, so the line intensities are in the ratio 1:1:1 (and not 1:2:1, as in AX_2 triplets).

If a group of protons is coupled to more than one deuterium, then the multiplicity of the proton signal is found from the general formula $2nI + 1$.

Thus, two (equal) deuterium couplings give rise to quintets, and three deuteriums give septets, and so on. Deuterium coupling also affects carbon-13 NMR spectra: see section 3.13.4.

If we record the ^1H NMR spectrum of a ⟍CHD group, the proton signal will appear as a 1:1:1 triplet. (We can only see the deuterium NMR signal itself if we substitute an appropriate radiofrequency probe at 9.2 MHz, the precession frequency of ^2H at 1.4 T: the deuterium signal will be a doublet, because of coupling to ^1H.) See also section 3.13.4.

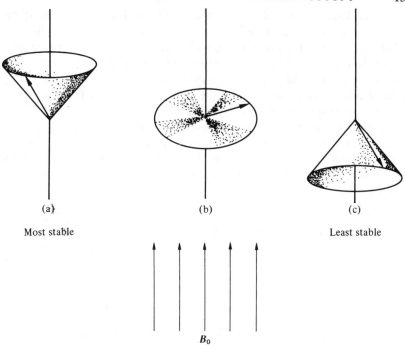

(a)   (b)   (c)

Most stable   Least stable

B_0

Figure 3.25 *Precession modes for a nucleus with spin quantum number = 1. It can precess around a conical path either aligned with the field (a), opposed to the field (c) or across the field on a plane as in (b).*

Nitrogen coupling. The ^{1}H spectrum for N—H groups might be expected to show splitting similar to the deuterium case, since ^{14}N also has $I = 1$ and thus three spin orientations, equally populated. As we saw in section 3.2, page 109, we must also recognize the efficient spin–lattice relaxation for protons attached to ^{14}N. Because of its electric quadrupole, nitrogen can flip rapidly among its spin states, thus remaining in each for very short times. Since T_1 is short, the ^{1}H signal is usually broadened, and the ^{14}N spin splitting of the proton resonance is not resolved. (As in the cases of fluorine and deuterium, an appropriate radiofrequency probe—4.3 MHz at 1.4 T—is needed to observe the ^{14}N NMR. See section 3S.6.) N—H proton exchange also eliminates the coupling: see page 150.

3.9.5 DEUTERIUM EXCHANGE
If deuterium oxide, D_2O, is used as solvent for NMR work, the D_2O exchanges with labile protons such as OH, NH and SH. The mechanism is the same as that for proton exchange discussed in section 3.8.6, page 150. In effect, because of the rapidity of the exchange, ROH becomes ROD, RCOOH becomes RCOOD, $RCONH_2$ becomes $RCOND_2$, etc.:

$$ROH + D—O—D(D_2O) \rightleftharpoons ROD + H—O—D$$

Peaks previously observed for the labile protons disappear (or are diminished) and a peak corresponding to H—O—D appears around δ 5.

This technique of *deuteriation* is widely used to detect the presence of OH groups, etc., and is easily carried out. The NMR spectrum can be recorded conventionally in a solvent other than D_2O and then a few drops of D_2O are shaken with the sample and the spectrum is rerun. Peaks diminish or disappear for OH, NH signals, etc.

The method can be extended to detect reactive methylene groups, such as those flanked by carbonyl. The spectrum is recorded normally, and then a small amount of D_2O and sodium deuteroxide is added to the sample tube; base-catalyzed deuterium exchange occurs only on reactive sites—for example,

When the spectrum is rerun, signals from protons previously on these sites disappear.

3.10 NON-FIRST-ORDER SPECTRA

In general, first-order spectra will only arise if the separation between multiplets (the chemical shift difference between signals, expressed in Hz) is much larger than the coupling constant, J; if $\Delta v \geqslant 6\,J$, then fairly unperturbed spectra will arise.

For example, in the spectrum of 1,1,2-trichloroethane (figure 3.12) the coupling constant, J, is 8 Hz, while the difference between the chemical shift positions of CH and CH_2 is $\delta(5.7-3.9) = 1.8$ ppm, which corresponds to 144 Hz on an 80 MHz spectrum.

If the signals from coupling protons are closer together on the spectrum, and the chemical shift difference is small, distortion of the signals arises. Figure 3.26 shows a simple example of this in the AX case (for example, cinnamic acid, figure 3.11). As the signals move together, the inner peaks become even larger at the expense of the outer peaks, and the positions of the lines also change: the origin chemical shift positions are no longer found at the mid-points of the doublets, but lie approximately at the 'center of gravity' of the doublets. Such spectra are usually called AB spectra, indicating that the chemical shift values are closer than in AX cases.

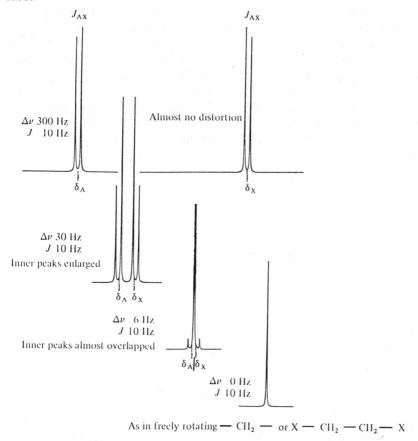

Figure 3.26 *Effect of ratio $\Delta v{:}J$ on doublet appearance (computer simulated). For AB cases, the origin chemical shift positions can be calculated from $v_A - v_B = [(v_1 - v_4)(v_2 - v_3)]^{1/2}$.*

In the ultimate case of two protons having the same chemical shift, so that $\Delta v = 0$ (for example, the *gem* protons of many CH_2 groups), the interaction between spins, although present at the nuclear level, is not observable in the NMR spectrum and a singlet peak is produced. See section 3S.2.2.

As the chemical shift values of coupling protons approach equality, the energy levels involved in the transitions among spin states become closer: new transitions frequently become possible, and additional lines may appear on the spectrum. Such interactions also give rise to changes in the expected Boltzmann distributions among spin states, and the predicted first-order line intensities are distorted, as in the AB diagram (figure 3.26). Population changes in the spin states can also be brought about deliberately, as discussed in section 3S.1 (double irradiation techniques).

In the AX_2 case of 1,1,2-trichloroethane (figure 3.12) only five lines appear, as predicted by first-order $(n + 1)$ rules. If the chemical shift values of A and X are closer together, the spectrum is non-first-order (and should be called AB_2); altogether nine lines may appear, the ninth line often being vestigial. The precise appearance of the spectrum depends on the ratio $\Delta v:J$, as shown in figure 3.27. Origin chemical shift positions are no longer easily extracted with exactitude, but they again lie near enough to the centers of gravity of the multiplets for most structural organic analyses.

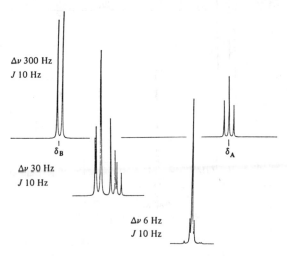

Figure 3.27   *Some possible appearances of AB_2 spectra. These computer simulated spectra can be varied until a match is achieved with an observed spectrum.*

For three coupling environments, the first-order appearance is AMX (see above). If two of the origin chemical shift positions are closer than in a true AMX case, then the ABX system results, which in many cases, like AMX, exhibits twelve lines. Chemical shift positions and coupling constants cannot easily be extracted from the spectrum directly, but sufficient accuracy for structural work can be achieved by treating a clear twelve-line ABX spectrum as an AMX case.

If all three coupling environments are close together in chemical shift terms, ABC systems arise, in which case up to fifteen lines can be observed. No attempt should be made to extract coupling constant data from such spectra, since none of the line spacings represents a true J value: chemical shift data should be evaluated on an approximate basis only.

Many non-first-order spectra can be simplified by recording at higher fields (see section 3.11), and in other cases a successful analysis can be achieved by computer programs for simulation of the spectrum. Trial chemical shifts and coupling constants are supplied to the program and subsequently adjusted until a good replication of the observed spectrum is obtained.

Aromatic systems. The majority of aromatic coupling systems are non-first-order, and many examples can be seen in perusing spectra catalogs (for example, the Aldrich catalogs). It nevertheless serves a useful purpose to describe the most frequently encountered cases, beginning with single-line spectra. The characteristic appearance of a selection of simple systems is shown in figure 3.28. Scale expansion will show further small splittings.

Unsplit signals (or nearly so) are produced by several monosubstituted benzenoid derivatives, provided that the substituent has no strong shielding or deshielding effect (for example, toluene, I): this 'accidental equivalence' is discussed in section 3.8.5. Compounds with identical *para* substituents, whatever their electronegativity, give single-line spectra because of the molecular symmetry; all four protons in the ring are magnetically equivalent (for example, *p*-dinitrobenzene, II).

| I | II | III | IV | V |

Unsymmetrical *para* substitution, if the two substituents have different shielding influences, can give rise to almost AB simplicity for the aromatic protons (for example, *p*-chloroaniline, III). Such a system is easily recognized in the spectrum of *p*-chlorophenol (see figure 3.28) and can be analyzed as for the AB case, to which it approximates; but because each chemically equivalent proton couples with the protons *ortho* and *para* to it, they are magnetically nonequivalent, and additional transitions become possible. Additional lines appear, characteristically within each doublet (see figure 3.28) and the system is more correctly described as the non-first-order AA'BB' system, as discussed in section 3.8.5.

A single substituent that is either strongly shielding or deshielding (for example, $-COCH_3$, in acetophenone, IV, or $-COOH$, in benzoic acid) usually causes the *ortho* protons to move to lower or higher δ values with respect to the *meta* or *para* protons. We usually observe a two-proton complex multiplet separated from a three-proton complex multiplet (see figure 3.28).

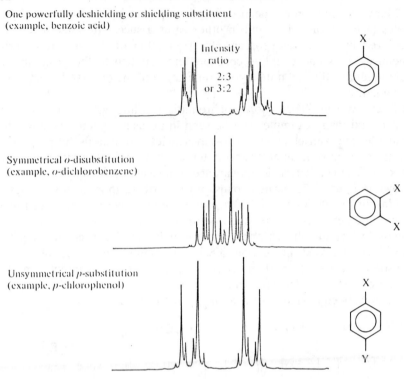

One powerfully deshielding or shielding substituent
(example, benzoic acid)

Intensity
ratio

2:3
or 3:2

Symmetrical *o*-disubstitution
(example, *o*-dichlorobenzene)

Unsymmetrical *p*-substitution
(example, *p*-chlorophenol)

Figure 3.28 *The appearance of some common aromatic coupling systems in* ^{1}H *NMR spectra.*

Identical groups *ortho* to each other produce considerable symmetry in the molecule, and the NMR spectrum is usually complex but symmetrical about the mid-point of the multiplet (see figure 3.28). Such spectra are also AA'BB' systems, but with more complexity than the simplest of the corresponding *para* systems (for example, *o*-dichlorobenzene, or diethyl phthalate, V).

Highly substituted rings may produce very simple spectra, and in a great number of cases it is possible to use the coupling constants to confirm orientation, since $J_{ortho} > J_{meta} > J_{para}$.

Heterocyclic aromatic systems are, for the most part, governed by similar rules to those governing benzenoid compounds. Clear first-order spectra should be analyzed completely, and complex multiplets should be treated with caution.

Polynuclear aromatic hydrocarbon systems invariably give very complex spectra, the chemical shift positions ranging from δ 7 to δ 9.

Vinyl and allyl systems. Many vinyl systems give rise to AMX coupling (see figure 3.29), especially where the substituent is an oxygen function, as in vinyl acetate. To understand the coupling, we should recall that H_A and H_M are *trans* and *cis*, respectively, to the substituent, and this often means that they have widely different chemical shift positions. The *trans* coupling, J_{MX}, is usually larger than the *cis* coupling, J_{AX}; J_{AM}, being a *gem* coupling, is also significant. In the absence of a strongly electronegative substituent, vinyl spectra will have ABC complexity and only approximate chemical shift data should be extracted.

Altogether four proton environments couple together in allyl systems, and this produces extremely complex multiplets. A very large, three-proton multiplet may extend almost from δ 4.5 to δ 6.5, generated by the alkene protons. The methylene protons may appear superficially as a doublet because of coupling to H_A, but the small allylic couplings (0–2 Hz) to protons H_B or H_C may also be observed under closer examination. The chemical shift positions for the CH_2 group can be predicted fairly reliably from table 3.6. The typical appearance of an allyl coupling system is shown in figure 3.29.

vinyl protons   allyl protons

Virtual coupling. The methyl group of a long alkyl chain ($...CH_2CH_2CH_2CH_3$) should appear as a simple triplet in the NMR spectrum, but more commonly it appears as a very blurred triplet indeed. The broadening of the CH_3 signal is attributed to interaction with the numerous nonneighbor CH_2 groups, even though there is no true coupling with these. Strong coupling only occurs with the adjacent CH_2 group, but this group serves to link spin interactions with the other methylenes to the terminal methyl. Such a higher-order effect is termed *virtual coupling*, and is responsible for a number of distortions in what one might have expected to be first-order spectra.

For virtual coupling to arise in protons P, they must be strongly coupled to their neighbor protons Q, which must, in turn, be strongly coupled to their neighbors R (that is, both J_{PQ} and J_{QR} must be $\gg$ 0); Q and R must have nearly the same chemical shift, and the coupling between P and R must also be zero.

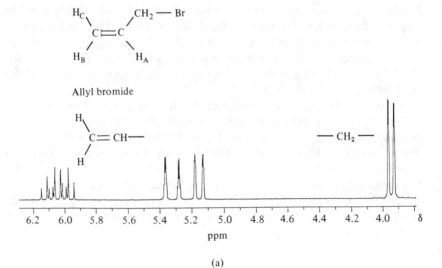

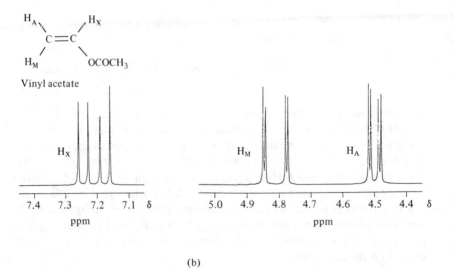

Figure 3.29 *The appearance of representative vinyl and allyl coupling systems in* *¹H NMR spectra: vinyl acetate, allyl bromide, at 200 MHz in CDCl₃.*

3.11   SIMPLIFICATION OF COMPLEX PROTON NMR SPECTRA

If a non-first-order spectrum is obtained at 60 MHz, a number of techniques can be applied to simplify the complexity and enable more accurate analyses of chemical shift and coupling constant data to be made.

3.11.1   INCREASED FIELD STRENGTH

We have seen that first-order multiplicity is usually produced when $\Delta v \geqslant 6 J$. Chemical shift positions (when measured in Hz) are field-dependent: the methyl resonance in acetates appears at δ 2.0, or 2 ppm higher frequency than TMS. In a 60 MHz instrument (1.4 T) 2 ppm corresponds to 120 Hz, while in a 100 MHz instrument (2.3 T) 2 ppm corresponds to 200 Hz. At 14.1 T (600 MHz) 2 ppm corresponds to 1200 Hz.

However, coupling constants are independent of field strength, so that the ratio $\Delta v : J$ is effectively increased as the field strength is increased from 1.4 T to 14.1 T. If two coupling multiplets overlap at 1.4 T, we can pull the multiplets apart by increasing the magnetic field. The further we pull the multiplets apart, the more likely is the spectrum to approach first-order, since we are in effect increasing Δv with respect to J.

If a compound gives a complex NMR spectrum at 60 MHz, it will further be improved by recording at 200 MHz or 600 MHz, although the degree of improvement depends on the particular chemical shift differences and coupling constants involved.

Figure 3.30(a) shows simulations of the proton NMR spectrum of 4-chlorobutanoic acid: at 60 MHz the central methylene group is unresolved, but the resolution is progressively improved when the field strength is increased. At field strength corresponding to 200 MHz (4.7 T) the central methylene group is separated, but still shows some non-first-order effects; at 500 MHz it is first-order.

Figure 3.30(b) shows the proton NMR spectrum of menthol recorded at 80 MHz, 200 MHz, 360 MHz and 600 MHz. At 80 MHz the spectrum is non-first-order, and hopelessly overlapped, while at 600 MHz much of the spectrum is capable of being interpreted as for first-order spectra. These spectra are presented in uniform frequency scale for emphasis: note that in this format the 600 MHz spectrum is 7.5 times as wide as the 80 MHz spectrum, and the full resolving power of such a high-field instrument is not apparent until the spectrum is plotted on a very wide piece of paper—certainly wider than this book page.

3.11.2   SPIN DECOUPLING OR DOUBLE RESONANCE (DOUBLE IRRADIATION)

Multiplicity of signals arises because neighboring protons have more than one spin orientation (low-energy or parallel, and high-energy or antiparallel): proton A in figure 3.31 appears as a doublet because of the two spin

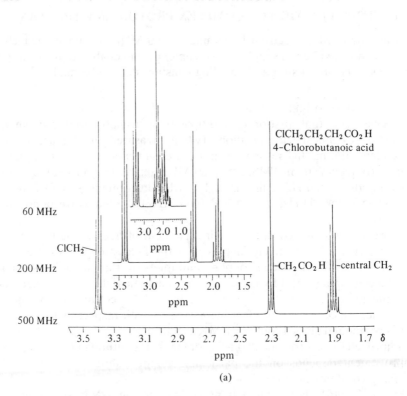

Figure 3.30 shows "ClCH₂CH₂CH₂CO₂H / 4-Chlorobutanoic acid" label with 60 MHz, 200 MHz, 500 MHz spectra.

Figure 3.30 *The advantages of increased field strength in spectrum simplification. (a) Simulation proton NMR spectra for 4-chlorobutanoic acid. (b) Proton NMR spectra of menthol at 80, 200, 360 and 600 MHz (in CDCl₃).*

orientations of X. If we irradiate X with the correct radiofrequency energy, we can stimulate rapid transitions (both upward and downward) between the two spin states of X, so that the lifetime of a nucleus in any one spin state is too short to resolve the coupling with A. If proton A 'sees' only one time-averaged view of X, then A will come to resonance only once, and not twice. By the same argument, if we irradiate proton A with the correct radiofrequency energy to cause it to undergo rapid transitions between its two spin states, proton X will only 'see' one time-averaged view of A, and appear only as a singlet.

We have seen earlier that $\Delta t \cdot \Delta v \approx 1/2\pi$: for a pair of coupled protons, the time needed to resolve the two lines of a doublet (Δt) is related to the separation between the lines—that is, the coupling constant (Δv). For the above example, we shall be able to resolve the H_A doublet, provided that each spin state of H_X has a lifetime greater than Δt. Double irradiation

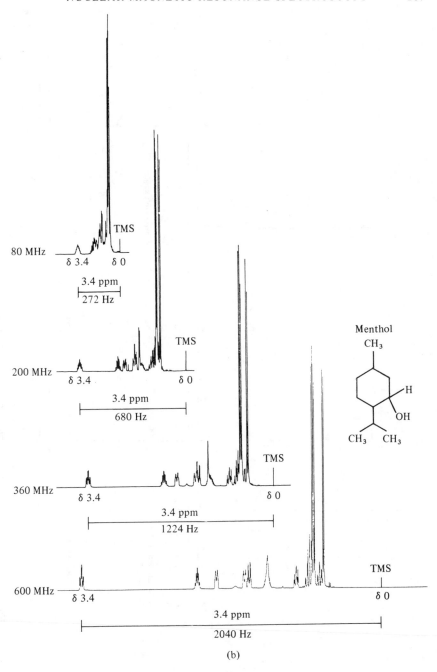

(b)

Fig. 3.30 (*continued*)

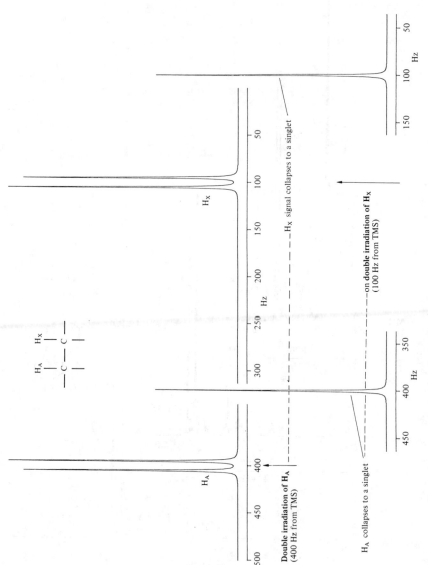

Figure 3.31 *Double irradiation (spin decoupling) of an AX spectrum.*

shortens this lifetime to less than Δt, and consequently we are unable to resolve the H_A doublet—which appears, therefore, as a singlet. See also section 3.8.4.

To perform this operation, we require, in addition to the basic NMR instrument, a *second* tunable radiofrequency source to irradiate proton X at the necessary frequency (that is, its precession frequency), while recording the remainder of the spectrum as before. Since we are making simultaneous use of two radiofrequency sources, the technique is called *double resonance* or *double irradiation*: since the nuclear spins during the process are 'less coupled' than before, we also call it *spin decoupling*.

For the method to be successful, the chemical shift positions for the coupling multiplets should be no closer than ≈ 1 ppm. Decoupling of non-first-order spectra can frequently lead to first-order spectra, provided that this condition is met.

Other applications of the double resonance method are discussed in section 3S.1.

3.11.3 LANTHANIDE SHIFT REAGENTS—CHEMICAL SHIFT REAGENTS

The proton NMR spectrum of 6-methylquinoline is reproduced in figure 3.32. The lower spectrum is the normal record; the upper spectrum was recorded after the addition of a soluble europium(III) complex to the solution, and the spectrum is pulled out over a much wider range of frequencies, so that it is simplified almost to first-order. The paramagnetic europium(III) ion complexes with the quinoline and induces enormous shifts to higher frequency in the quinoline resonances. The use of europium and other lanthanide derivatives as *chemical shift reagents* or *lanthanide shift reagents* has extended the applicability of detailed NMR studies to very complex molecules.

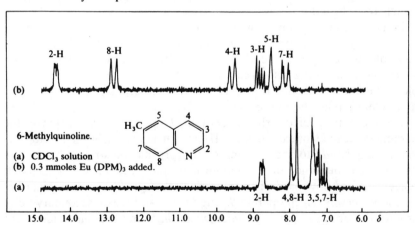

Figure 3.32   *Effect of a lanthanide shift reagent (Eu(DPM)$_3$) on the ^{1}H NMR spectrum of 6-methylquinoline. (Ring protons only; 60 MHz in CDCl$_3$.)*

For the method to succeed, the organic molecule must be able to donate nonbonding pairs to the europium ion, so that we are concerned principally with the following functional classes: amines; alcohols; ketones and aldehydes; ethers and thioethers; esters; nitriles; and epoxides.

The lanthanide complex should be soluble in common NMR solvents for wide applicability, and those most frequently used are complexes with two enolic β diketones, *di*pivaloyl*methane* (DPM) and hepta*fluorodimethyl-octanedione* (FOD). Deca*fluoroheptanedione* (FHD) has no protons and is also soluble in CCl_4.

$$R_1 = R_2 = -CMe_3 \equiv Eu(DPM)_3$$
$$R_1 = -CMe_3, R_2 = -C_3F_7 \equiv Eu(FOD)_3$$
$$R_1 = R_2 = -C_2F_5 \equiv Eu(FHD)_3$$

The FOD complexes are more soluble than the DPM variants, although their hygroscopic nature can be troublesome. The fully deuterated derivative of FOD is also commercially available, and this eliminates the signals from ligand protons.

In general, europium complexes produce shifts to higher δ, while praseodymium complexes produce shifts to lower δ. Ytterbium, erbium and holmium compounds tend to give greater shifts, but in these last two some line broadening also occurs and complicates the analysis of multiplets: the line broadening is associated with the paramagnetic ion's ability to accelerate relaxation processes.

The mechanism of the shift is twofold. Unpaired electron spin in the paramagnetic ion (for example, Eu(III)) is partially transferred *through the intervening bonds* to the protons of the organic substrate; this is true *contact shift*, but since this is seldom important, the name 'contact shift reagent' is not now used. The spinning paramagnetic ion also generates magnetic vectors which operate *through space* and create secondary fields around the protons; this is *pseudocontact shift*, and predominates in the case of the lanthanide ions.

3.12 TABLES OF DATA FOR PROTON NMR

Protons of NH, OH *and* SH *groups* (whose δ values are listed in table 3.8) show special characteristics. All are removed by deuteriation (see page 158) and all are affected by solvent, temperature and concentration (see page 132). Signals for ROH protons will appear as a singlet or as a multiplet, depending on whether coupling to neighbor protons is observed (see page 150). Primary amines with concentrated sulfuric acid are completely protonated (to RNH_3^+): proton exchange is suppressed, and the signal changes because of coupling to ^{14}N, with $J_{NH} \approx 50$ Hz (see page 157).

In some cases (such as $CH_3NH_3^+$) the signal shows 1:1:1 triplets for J_{NH}, together with slightly broadened multiplicity from other nearby protons (such as broad quartets from the CH_3 splitting in $CH_3NH_3^+$, with $J \approx 5$ Hz).

In other cases (such as many aryl amines) the broadening is so marked that the NH signal almost disappears into the baseline.

Secondary and tertiary amines with concentrated sulfuric acid give a sharpened line at low field, because ^{14}N relaxation is so rapid that no N—H coupling is observable.

Table 3.4 δ values for the protons of CH_3, CH_2 and CH groups attached to groups X, where R = alkyl and Ar = aryl

X	CH_3X	$R'CH_2X$	$R'R''CHX$
—R	0.9	1.3	1.5
—CH^b ⟨ CH$_2^a$ / O	1.3	a 3.5	b 3.0
＼ = ＼	1.7	1.9	2.6
＼ =—=—=, etc. (i.e., end-of-chain)	1.8		
＼ =—=—=, etc. (i.e., in-chain)	2.0	2.2	2.3
＼ =N—	2.0	—	—
—≡	2.0	2.2	—
—COOR, —COOAr	2.0	2.1	2.2
—CN	2.0	2.5	2.7
—CONH$_2$, —CONHR, —CONR$_2$	2.0	2.0	2.1
—COOH	2.1	2.3	2.6
—COR	2.1	2.4	2.5
—COAr	2.5	2.8	2.9

Table 3.4 (*continued*)

—SH, —SR	2.1	2.4	2.5
—NH$_2$, —NHR, —NR$_2$	2.1	2.5	2.9
—I	2.2	3.1	4.2
—CHO	2.2	2.2	2.4
—Ph, Ar	2.3	2.6	2.9
—NHAr	2.5	3.1	—
—Br	2.6	3.3	4.1
—NHCOR, —NRCOR	2.9	3.3	3.5
—Cl	3.0	3.4	4.0
—OR	3.3	3.3	3.8
$\overset{+}{\text{—N}}$R$_3$	3.3	3.4	3.5
—OH	3.4	3.6	3.8
—OCOR	3.6	4.1	5.0
—OAr	3.7	3.9	4.0
—OCOAr	3.9	4.2	5.1
—NO$_2$	4.3	4.4	4.6
—F	4.3	—	—

Table 3.5 Influence of functional group X on the chemical shift position (δ) of CH$_3$, CH$_2$ and CH protons β to X

	For β-shifts, add the following to the δ values given in table 3.4		
X	CH$_3$—C—X	CH$_2$—C—X	CH—C—X
—C=C	0.1	0.1	0.1
—COOH, —COOR	0.2	0.2	0.2
—CN	0.5	0.4	0.4
—CONH$_2$	0.25	0.2	0.2
—CO—, —CHO	0.3	0.2	0.2
—SH, —SR	0.45	0.3	0.2
—NH$_2$, —NHR, —NR$_2$	0.1	0.1	0.1
—I	1.0	0.5	0.4
—Ph	0.35	0.3	0.3
—Br	0.8	0.6	0.25
—NHCOR	0.1	0.1	0.1
—Cl	0.6	0.4	0
—OH, —OR	0.3	0.2	0.2
—OCOR, —OCOAr	0.4	0.3	0.3
—OPh	0.4	0.35	0.3
—F	0.2	0.4	0.1
—NO$_2$	0.6	0.8	0.8

Table 3.6 δ values for the protons of CH_2 (and CH) groups bearing more than one functional substituent (modified Shoolery rules)

Note:

for $H_2C\begin{smallmatrix}X^1\\ \\X^2\end{smallmatrix}$ $\delta CH_2 = 1.2 + \Sigma a$

Less accurate for $HC\begin{smallmatrix}X^1\\-X^2\\X^3\end{smallmatrix}$ but use $\delta CH = 1.5 + \Sigma a$

X	a	X	a
—≡	0.75	—Ph	1.3
—≡	0.9	—Br	1.9
—COOH, —COOR	0.7	—Cl	2.0
—CN, —COR	1.2	—OR, —OH	1.7
—SR	1.0	—OCOR	2.7
—NH$_2$, —NR$_2$	1.0	—OPh	2.3
—I	1.4		

Table 3.7 δ values for H attached to unsaturated and aromatic groups

H—≡—R	1.8*	$H_A \quad H_X$	A, 4.5
H—≡⊤OH	2.4*		M, 4.8
		$H_M \qquad OCO—$	X, 7.2
H—≡—≡—, etc.			
H—≡—Ph	2.9*	H	
H—≡—CO—	3.2*		
$H_2C{=}C\begin{smallmatrix}R\\ \\R'\end{smallmatrix}$	4.6	${=}O$	5.8
$H_2C{=}{-}{=}$, etc.	4.9	H	6.0
$\begin{smallmatrix}H_a \qquad OR'\\ \\R \qquad H_b\end{smallmatrix}$	a, 4.5–5.0	${=}O$	
	b, 6.5–7.5		6.2
		H	*(continued)*

* Alkyne proton signals removed on deuteriation, and δ values increased by a trace of pyridine.

Table 3.7 (*continued*)

Structure	Shift	Structure	Shift
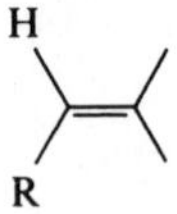 (H)(R)C=C(CH₃)₂	5.3	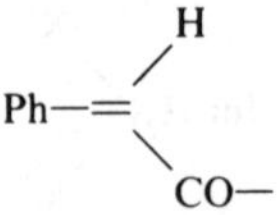 Ph—CH=C(H)(CO—) (*cis or trans*)	6.6
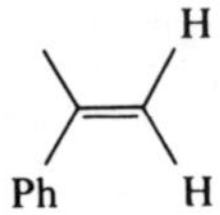 (CH₃)₂C=C(H)(H)	5.0	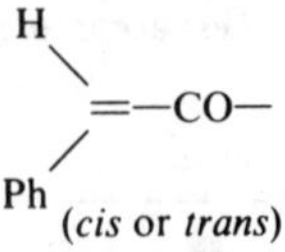 (H)(Ph)C=C=CO— (*cis or trans*)	7.8
(Ph)(H)C=C(H)(H)	5.3	N—C(H)=O	7.8
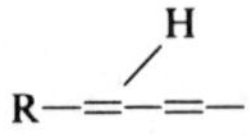 R—=—=— (in-chain)	6.2	ROC(H)=O	8.0
(ring)=(H)	5.6	RC(H)=O	9.6
(CH₃)₂C=C(H)(OR)	6.8	PhC(H)=O	9.9
(CH₃)₂C=C(H)(Ph)	7.0		
Ph—H	7.27 (see table 3.9)		

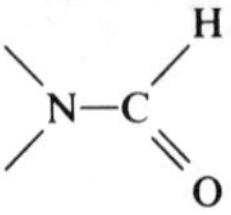

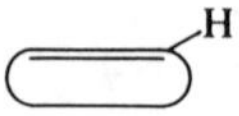

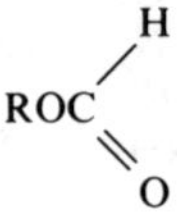

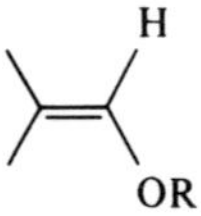

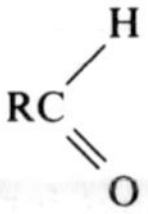

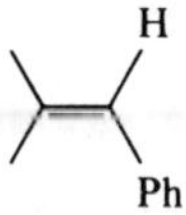

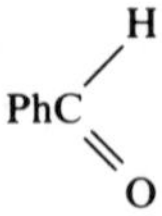

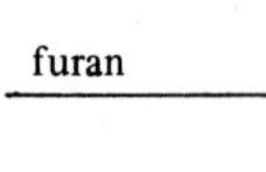

α 7.7
β 7.5

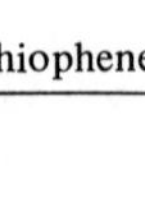

β 6.3
α 7.4
furan

γ 7.4
β 7.0
α 8.5

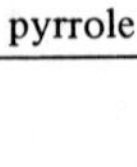

β 7.1
α 7.2
thiophene

β 4.5
α 6.2

β 6.1
α 6.5
pyrrole

Table 3.8 δ values for the protons of OH, NH and SH groups*

ROH	0.5–4.5	higher for enols (11.0–16.0)
		lines often broadened
ArOH	4.5	raised by hydrogen bonding to ≈ 9.0
		chelated OH, ≈ 11.0
RCOOH	10.0–13.0	
RNH_2, RNHR′	5.0–8.0	lines usually broadened
$ArNH_2$, ArNHR′	3.5–6.0	occasionally raised; lines usually broadened
$RCONH_2$, RCONHR′	5.0–8.5	lines frequently very broad, and even unobservable
RCONHCOR′	9.0–12.0	lines broadened
RSH	1.0–2.0	
ArSH	3.0–4.0	
=NOH	10.0–12.0	often broadened

*See notes on page 171.

Table 3.9 Shifts in the position of benzene protons (δ 7.27) caused by substituents

Substituent	ortho	meta	para
—CH_3, —R	−0.15	−0.1	−0.1
—=	+0.2	+0.2	+0.2
—COOH, —COOR	+0.8	+0.15	+0.2
—CN	+0.3	+0.3	+0.3
—$CONH_2$	+0.5	+0.2	+0.2
—COR	+0.6	+0.3	+0.3
—SR	+0.1	−0.1	−0.2
—NH_2, —NHR	−0.8	−0.15	−0.4
—$N(CH_3)_2$	−0.5	−0.2	−0.5
—I	+0.3	−0.2	−0.1
—CHO	+0.7	+0.2	+0.4
—Br	0	0	0
—NHCOR	+0.4	−0.2	−0.3
—Cl	0	0	0
—NH_3^+	+0.4	+0.2	+0.2
—OR	−0.2	−0.2	−0.2
—OH	−0.4	−0.4	−0.4
—OCOR	+0.2	−0.1	−0.2
—NO_2	+1.0	+0.3	+0.4
—SO_3H, —SO_2Cl, —SO_2NH_2, etc.	+0.4	+0.1	+0.1

Table 3.10 Proton–proton spin–spin coupling constants

Function	J_{ab}/Hz
H_a, C, H_b (*gem*)	10–18, depending on the electronegativities of the attached groups (commonly around 10 in cyclohexanes, etc.)
CH_a—CH_b (*vic*)	depends on dihedral angle: see section 3.9.3 for cyclohexanes, etc.
C=C with H_a, H_b	1–4
C=C (*cis*) with H_a, H_b	5–14 (commonly around 6–8)
H_a, C=C (*trans*), H_b	11–19 (commonly around 14–16)
C=C, $\overset{\mid}{C}$—H_a, H_b	4–10
$\overset{\mid}{C}$—H_b (*cis* or *trans*), H_a—C=C	0–2 (for aromatic systems, 0–1)
C=CH_a—CH_b=C	10–13
H_a (ring) H_b	*ortho*, 7–10 *meta*, 2–3 *para*, 0–1

Long-range (4J) couplings

H_e ... H_e	$^4J_{ee}$	1–2 Hz
H_a—H_b ... H_c	$^4J_{ac}$	8 Hz
	$^4J_{bc}$	0.2 Hz
H_a ... H_b	$^4J_{ab}$	18 Hz

CARBON-13 NMR SPECTROSCOPY

In a 1.9 T field the precession frequency of ^{13}C is 20 MHz, that for ^1H being 80 MHz and ^{12}C being nonmagnetic. In principle, therefore, it is not difficult to observe ^{13}C NMR. The magnetic moment of ^{13}C is about one-quarter that of ^1H, so that signals are inherently weaker, but the overwhelming problem is that the natural abundance of ^{13}C is only 1.1 per cent. The problem in simple molecules can be overcome by synthesizing ^{13}C-enriched samples, but this is of little value in complex molecules.

3.13   NATURAL ABUNDANCE ^{13}C NMR SPECTRA

In practice, routine 'natural abundance' ^{13}C NMR spectra are recorded by the pulsed FT method discussed in section 3.3.2, with the sensitivity enhanced by summation of several spectra (commonly a few hundreds to several thousands, depending on the solubility of the compound, the amount available and the number of carbon atoms in the molecule).

An example is shown in figure 3.33 of the ^{13}C spectrum of menthol, and for comparison the ^1H spectrum of menthol is also shown.

3.13.1   RESOLUTION
Each of the ten lines in the carbon-13 NMR spectrum in figure 3.33 represents one carbon atom of menthol, and two immediate differences from the ^1H spectrum are apparent; the ^{13}C spectrum is much simpler, and much more highly resolved.

The chemical shift range in the ^1H spectrum is only $\approx$ 4 ppm (320 Hz in this 80 MHz spectrum), while the range in the ^{13}C spectrum is $\approx$ 80 ppm (1600 Hz in this 20 MHz spectrum). Expressed otherwise, the chemical shift differences in the ^{13}C spectrum are about 20 times those shown in the ^1H spectrum, and this is typical in all other molecules.

3.13.2   MULTIPLICITY
Both ^{13}C and ^1H have $I = \frac{1}{2}$, so that we should expect to see coupling in the spectrum between (a) ^{13}C—^{13}C and (b) ^{13}C—^1H. The probability of two ^{13}C atoms being together in the same molecule is so low that ^{13}C—^{13}C couplings are not usually observed. Couplings from ^{13}C — ^1H interaction have already been discussed (page 156) and these couplings should be observed in the ^{13}C spectra. However, these couplings make the ^{13}C spectra extremely complex, and they have been eliminated by decoupling. The proton-coupled (or non-decoupled) spectrum is shown in figure 3.33.

3.13.3 ^1H DECOUPLING—NOISE DECOUPLING—BROAD BAND DECOUPLING
To eliminate the complicating effects of the proton couplings in the ^{13}C spectra, we must decouple the ^1H nuclei by double irradiation at *their* resonant frequencies (80 MHz at 1.9 T, etc.). This is an example of

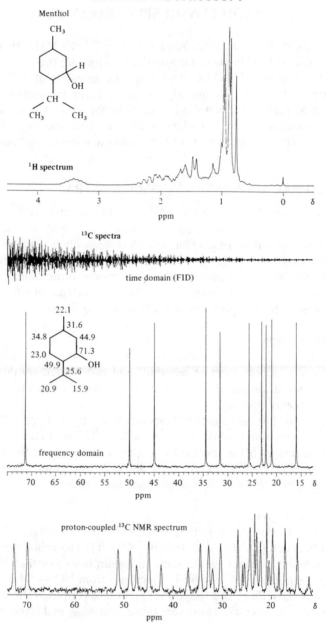

Figure 3.33 *Comparison of the proton and ^{13}C NMR spectra of menthol, recorded at the same field strength, 1.9 T (80 MHz and 20 MHz, respectively, in CDCl$_3$). The proton-coupled (i.e. non-decoupled) ^{13}C NMR spectrum is shown at the bottom.*

heteronuclear decoupling (see page 155), but we do not wish merely to decouple specific protons; rather, we wish to double irradiate *all* protons simultaneously while recording the ^{13}C spectrum. A decoupling signal is used that has all the ^{1}H frequencies spread around 80 MHz, and is therefore a form of radiofrequency noise; spectra derived thus are ^{1}H-*decoupled*, or *noise-decoupled*. Most ^{13}C spectra are recorded in this way; see the menthol spectrum in figure 3.33.

The alternative name *broad band decoupled* spectra simply takes cognizance of the fact that a wide spread of decoupling radiofrequency can be produced by several electronic techniques, other than by simple noise modulation.

The convenient notation ^{13}C-$\{^1$H] can be used to identify proton-decoupled carbon-13 NMR spectra; in the same way ^{31}P-$\{^1$H$\}$ spectra are phosphorus-31 NMR spectra with all proton coupling to phosphorus removed by broad band (or noise) decoupling, and ^{15}N-$\{^1$H$\}$ corresponds for nitrogen-15, etc.

3.13.4   DEUTERIUM COUPLING

Deuteriated solvents (such as deuteriochloroform ($CDCl_3$), deuterio-benzene (C_6D_6), deuterioacetone (CD_3COCD_3) or hexadeuterio-dimethylsulfoxide (CD_3SOCD_3)) give rise to carbon-13 signals which are split by coupling to deuterium. The multiplicity is calculable from the general formula $2nI + 1$ and deuterium has $I = 1$, so that in molecules with one deuteron attached to each carbon (as in $CDCl_3$ and C_6D_6) the carbon-13 signal from the solvent is a 1 : 1 : 1 triplet; this is seen in figures 3.1(b) and 3.34. For CD_3 groups (as in CD_3COCD_3 and CD_3SOCD_3), the solvent gives rise to a septet with line intensities 1 : 3 : 6 : 7 : 6 : 3 : 1; see the insert in figure 3.34. See also section 3.9.4.

3.13.5   NOE SIGNAL ENHANCEMENT

Since decoupling can interfere with (and thereby shorten) relaxation times, the nuclear Overhauser effect (see section 3S.1.3) may operate and lead to signal enhancement of certain ^{13}C peaks. The line intensities in the ^{13}C spectrum of menthol are not all equal, because of these relaxation effects.

It turns out that the major relaxation route for a ^{13}C nucleus involves dipolar transfer of its excitation energy to the proton(s) directly attached to it; there is a corollary that maximum nuclear Overhauser effect operates on CH_3, CH_2 and CH carbons, whereas no enhancement arises for quaternary carbons (and this includes those carbons on aromatic rings with substituents attached). It happens also that these non-proton-bearing carbons have long relaxation times and also tend to give low-intensity signals for this reason (unless special steps are taken to ensure otherwise).

These dual influences ensure easy identification of such carbons: in figure 3.1(b), for example, there are three signals of lower intensity, assigned, respectively, to the two substituted ring carbons and to the carbonyl carbon.

3.13.6   QUANTITATIVE MEASUREMENT OF LINE INTENSITIES

The number of nuclei in any environment (measured by integration of peak areas) in proton NMR is routine and quite accurately quantitative, but this is not so in routine carbon-13 NMR spectra. As we have seen above, there are two main reasons for this.

The nuclear Overhauser effect tends to *increase* the line intensities for those carbons bearing protons, and to leave the quaternary carbons unaltered. To eliminate the nuclear Overhasuer effect requires a special pulse sequence, which is described in section 3S.3, but it is not usually routinely applied.

In the pulsed FT mode used for normal ^{13}C work, the pulses are applied with only short delays between each successive pair; carbon nuclei with long relaxation times will not have fully relaxed after one pulse before the next pulse is applied. The signals are therefore slightly saturated (see section 3.2) and of lower intensity. It is the quaternary carbons which tend to have long relaxation times, so that they show lowered intensities; in contrast to this, proton-bearing carbons not only have shorter relaxation times, *but also experience the enhanced line intensities caused by the nuclear Overhauser effect.*

To avoid this saturation effect would involve longer delays between pulses; because T_1 is a measure of an exponential process, it would be necessary to wait for approximately $5T_1$ before relaxation is complete. Since this would counteract the main asset of the FT method—speed—it is not done unless quantitative information is essential.

Interestingly, small symmetrical molecules (such as the solvents used in NMR, $CDCl_3$, C_6D_6, etc.) also tend to have carbons with long relaxation times; this is one of the reasons for the observation that the solvent peaks in ^{13}C NMR spectra are of low intensity.

Paramagnetic ions may be added to the sample to supply the fluctuating electromagnetic vectors which catalyze the relaxation processes for the excited carbon nuclei; this leads to improvement in the quantitative line heights. Typical paramagnetic species are chromium acetylacetonate $(Cr(ACAC)_3)$, or the shift reagents discussed in section 3.11.3 (which, of course, cause shifts in the δ values except for the Gd complexes).

3.13.7   OFF-RESONANCE PROTON DECOUPLING

Fully proton-decoupled carbon-13 NMR spectra offer two main advantages over fully coupled spectra (sometimes called *non-decoupled* spectra): removal of coupling multiplicity makes the spectrum simpler in appearance and ensures almost no confusing overlap in adjacent signals, but there is a sensitivity bonus in addition. As an example, the methyl carbon in *p*-hydroxyacetophenone (figure 3.1(b)) would appear in a non-decoupled spectrum as a quartet (intensity ratio 1 : 3 : 3 : 1) because of the three attached and coupling protons and, when this is decoupled, the whole of the signal intensity appears as a single line (of intensity 8 relative to the outside lines of the quartet). The fact that the signal is a quartet proves that it arises from a methyl group, and unfortunately this valuable piece of information is lost in the fully decoupled ^{13}C-$\{^1$H$\}$ NMR spectrum. There are several techniques which allow this information to be retained; the simplest (but not the best) consists of carrying out the proton decoupling by irradiation of the sample with radiofrequency which is not quite exactly that of the protons but is a few hundred hertz displaced. The consequence of this *off-resonance decoupling* is an incomplete collapse of the multiplicity, and vestigial quartets remain from methyl carbons, with triplets from CH_2, doublets from CH and singlets from fully substituted carbons. More elaborate procedures (which allow the separate plotting of subspectra, respectively, from CH_3, CH_2 and CH carbons) are discussed in section 3S.3 and these are used in preference to off-resonance decoupling.

It is convenient to annotate signals in ^{13}C-$\{^1$H$\}$ spectra to indicate multiplicity, with the abbreviations q, t, d and s for quartet, triplet, doublet and singlet, respectively, as in figure 3.34.

3.14 STRUCTURAL APPLICATIONS OF ^{13}C NMR

Differentiation among alternative organic structures has a long history in ^{1}H NMR and it is substantially extended by ^{13}C NMR. Increased shift resolution (compared with ^{1}H spectra) is often sufficient in itself to lead to correct structural assignment, but the use of correlation data for chemical shift positions and the calculation of multiplicity in non-decoupled spectra both have their contributions to make. Figure 3.35 shows the approximate chemical shift positions for common organic functional groups; the shifts are measured in ppm from TMS as standard.

Example 3.9
Question. There are three isomeric ethers with the molecular formula $C_4H_{10}O$: name them, and state how many signals will arise in the carbon-13 NMR spectrum of each.

Model answer. The three ethers are diethyl ether (I), methyl propyl ether (II) and methyl isopropyl ether (III). Only in methyl propyl ether are all four carbons in different environments, so this ether shows four signals in its spectrum. In diethyl ether each ethyl group is equivalent, so that only two different environments (and, hence, signals) are present. The two methyl groups of the isopropyl group are equivalent, so methyl isopropyl ether gives rise to three signals in the spectrum.

$$CH_3CH_2OCH_2CH_3 \qquad CH_3OCH_2CH_2CH_3 \qquad CH_3OCH\begin{cases} CH_3 \\ CH_3 \end{cases}$$

I   II   III

IV   V   VI

Figure 3.1(b) shows the ^{13}C-$\{^1H\}$ NMR spectrum of p-hydroxy-acetophenone, p-$CH_3COC_6H_4OH$; the fact that it is the *para* isomer is easily confirmed from the spectrum, since only six nonequivalent carbons are present in the molecule (C-2 and C-6 are equivalent, as are C-3 and C-5). Both the *ortho* and *meta* isomers would have given spectra with eight signals, from the eight nonequivalent sites in each molecule.

(In section 3.15.1 we shall predict the chemical shifts for the carbons in these isomers, thus definitively identifying each isomer.)

Exercise 3.16 State the number of nonequivalent carbon environments (a) in *o*-dinitrobenzene and in its *m*- and *p*-isomers; (b) in *o*-dimethoxybenzene and its *m*- and *p*-isomers; and (c) in the three possible structures, IV, V and VI, for the dimer of cyclooctatetraene. (The actual dimer showed four signals in its ^{13}C NMR spectrum, so which is the correct structure?)

Exercise 3.17 State the number of nonequivalent carbon environments in (a) the three isomeric methyl esters of chlorobenzoic acid (*o*-, *m*- and *p*-) and (b) the three isomers of hydroxybiphenyl, PhC_6H_4OH, (2-hydroxy-, 3-hydroxy- and 4-hydroxybiphenyl).

3.15   CORRELATION DATA FOR ^{13}C NMR SPECTRA

While it is possible to offer reasonable rationales for proton NMR chemical shifts (section 3.4), the explanation of carbon-13 NMR chemical shifts is much less self-consistent, despite extensive studies; happily, predictions based on the tables of empirical data which follow are very reliable.

It is usually very difficult to deduce *a priori* the structure of an organic molecule from its ^{13}C NMR spectrum; indeed, this would be at variance with experimental experience, where much other information is often simultaneously available—both chemical and spectroscopic (IR, UV, MS and proton NMR spectra). Proof of structure usually involves hypothesizing what the likely structures for the compound are, and then using the tables to predict *for each of these possibles* the appearance of the ^{13}C NMR spectrum, and that structure which gives the best fit with observed values is likely to be correct.

Some general features should be given consideration.

sp³ hybridized carbons

Figure 3.35 shows that sp^3 carbons come to resonance in the range δ 0–80; within this overall range, it is worth noting that the carbons of C—O bonds, C—N bonds and C—S bonds appear in the narrower ranges indicated. Exceptions abound, usually as a consequence of influential electronic or steric effects. An extreme and interesting exception is the signal for the carbon atom in tetraiodomethane, CI_4, at δ −300 (that is, at 300 ppm *lower* frequency than TMS).

sp² hybridized carbons

Alkene and aromatic carbon atoms give signals in overlapped areas of the spectrum (δ 80–150 and δ 110–140, respectively)—a fact which can make their distinction less clear than in the proton NMR spectrum. The great diversity of C=O groups is mirrored in their significantly differing shift positions (see table 3.17). A less common sp^2 class (not shown in figure 3.35) is in the C=N group of aromatic imines, often called *Schiff's bases*; the range is δ 130–150. (Aliphatic imines are unstable and tend to decompose or polymerize.)

sp hybridized carbons

For the sp carbons of alkynes, nitriles and isonitriles, the shift ranges are usefully narrow (see figure 3.35).

Each main class of carbon environment (sp^3, sp^2 and sp) will be discussed, showing how the effects of further substitutions can be predicted.

The first steps in deducing the structure of an organic compound, using the ^{13}C NMR spectrum, are:

1. Count the number of signals in the spectrum; this is the number of nonequivalent carbon environments in the molecule. (Identify and discount the signal(s) from solvent; see table 3.19.)

2. Use figure 3.35 to assign signals approximately to the regions δ 0–80, δ 80–150 and δ 160–220 (carbonyl carbons).

3. Note the intensities of the peaks: non-proton-bearing carbons give lower intensity signals, and groups of two or more equivalent carbons give higher intensity signals.

4. Take account of any multiplicity information (q, t, d or s).

5. Use the Correlation Tables (section 3.16.1) to predict the chemical shifts of all carbons in each putative structure.

3.15.1 USE OF THE CORRELATION TABLES

There are two principal predictable influences which we can quantify in determining the chemical shift positions of any carbon atom:

1. The number of other carbon atoms attached to it (and whether these are CH_3, CH_2, CH or C groups).
2. The nature of all other substituents attached (or nearby along a chain of other carbon atoms).

It is imperative to compute 1 before 2.

Example 3.10

Question. Predict the chemical shift positions for the carbons in 3-heptanone (butyl ethyl ketone), I.

Model answer. To do this we must first know the δ values for butane, II, and ethane, III: these are listed in table 3.11 and are shown in the formulae. *Only thereafter* can we predict the influence of the carbonyl substituent on each of these moieties; the influence of C=O on alkane carbons is given in table 3.15.

3-heptanone, I   butane, II   ethane, III

For C-1 we take the base value for ethane (δ 5.7) and note from table 3.15 that a carbonyl group, COR, β to it increases the value of the chemical shift by 2 ppm. The predicted value is therefore δ 7.7 (*ca* δ 8).

For C-2 we again take the base value for ethane (δ 5.7) but the carbonyl group is α to this carbon in 3-heptanone, so the increment is 30 ppm. The predicted value is therefore δ 35.7 (*ca* δ 36).

For C-4 we take the base value for the terminal carbon in butane and add to this (still table 3.15) 30 ppm, giving a predicted shift position of δ 43.

For C-5 the base value of δ 25 is increased by a β-carbonyl group to δ 27.

For C-6 the carbonyl group is in the γ position; for reasons that are not totally clear, but may correlate with molecular geometry, γ shifts are commonly negative—as here, where the base value of δ 25 is decreased (by −3) to δ 22.

For C-7 the effect of the carbonyl group is vanishingly small.

Note that the point of attachment of C=O on butane and ethane (C-2 and C-4) is CH_2CO; hence, the increment in each case is 30. This is notwithstanding the fact that the terminal carbon in each parent hydrocarbon is CH_3. (See also example 3.11.)

Thus, the predicted δ values (to the nearest integer) for the sp³ carbons in 3-heptanone are, from C-1 on, as follows: 8, 36, 43, 27, 22 and 13; the observed values are 8, 36, 42, 26, 23 and 14.

Table 3.17 lists the chemical shift for the C=O carbon of a dialkyl ketone at δ 205–218; it is observed at δ 211.

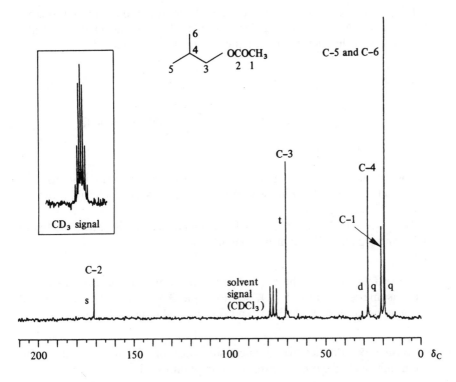

Figure 3.34   ¹³C *NMR spectrum of isobutyl acetate. (20 MHz in* CDCl₃, *broad band proton decoupled.) Multiplicities (s,d,t,q) come from off-resonance data. Insert: Appearance of the septet signal from solvents containing the* CD₃ *group, such as acetone-d₆ or DMSO-d₆* (CD₃SOCD₃).

Example 3.11

Question. Predict the chemical shift positions for the carbons in (a) *sec*-butyl acetate, IV, and (b) isobutyl acetate, V.

Model answer. (a) The starting point again is butane, whose δ values are given in table 3.11. On this occasion the functional group is attached to the C-2 of butane; although this is a CH₂ group in butane itself, it is CH in *sec*-butyl acetate, so we therefore use the increment 50 (not 52) in table 3.15.

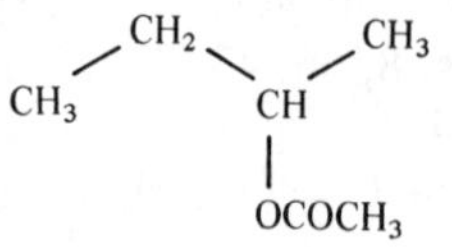

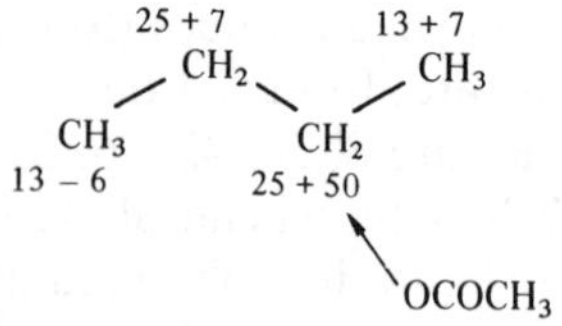

$$25 + 7 \qquad 13 + 7$$

CH$_3$ CH CH$_2$ OCOCH$_3$ → structures

sec-butyl acetate, IV

(b) The starting point for isobutyl acetate is isobutane (methylpropane, VI), the δ values of which are shown in table 3.11 and in the formula, together with the increments for the substitution of OCOCH$_3$.

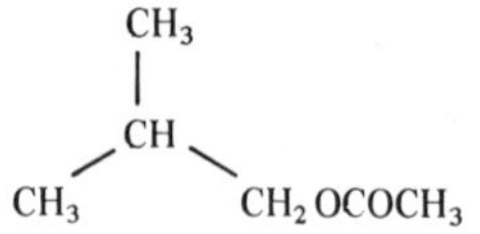

isobutyl acetate, V 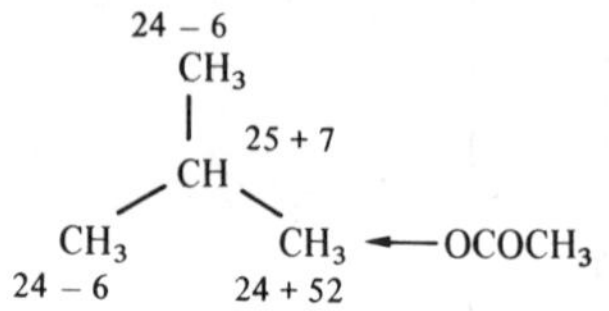   isobutane, VI

The predicted values for the carbonyl carbons in both isomers is δ 169–176 (table 3.17) and likewise for the isolated acetate methyl groups it is δ 20–22 (table 3.18).

The ^{13}C-$\{^1$H$\}$ NMR spectrum for isobutyl acetate is shown in figure 3.34, and signal assignment is straightforward.

Exercise 3.18   Predict the chemical shift positions for the carbons in the three isomeric ethers C$_4$H$_{10}$O; see example 3.5 above. Use ethane and propane as models (table 3.15) and find methyl ethers in table 3.18.

Exercise 3.19   Predict the chemical shift positions for the carbons in the three unbranched isomeric carbonyl compounds C$_5$H$_{10}$O.

Example 3.12
Question. Predict the chemical shift positions for the carbons in butyl acrylate, VII.

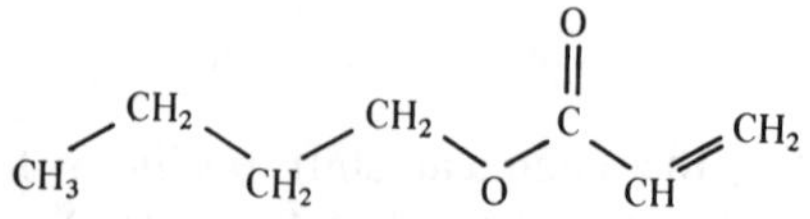

butyl acrylate, VII

Model answer. The δ values for the butyl group are predicted in a similar way to example 3.10, and the ester C=O carbon predicted from table 3.18.

butane   ester   ethylene

The alkene (sp^2) carbons (see table 3.16) should appear δ 127 (123 + 4) and δ 132 (123 + 9).

The observed chemical shifts in butyl acrylate are δ 13, 19, 30, 64, 129, 130 and 165.

Example 3.13
Question. Predict the chemical shift positions for the carbons of *p*-hydroxyacetophenone, *p*-$CH_3COC_6H_4OH$; the ^{13}C NMR spectrum is shown in figure 3.1(b).

p-hydroxyacetophenone   *ortho*   *meta*

Model answer. The CH_3 and $C=O$ carbons are listed in tables 3.18 and 3.17, respectively: the observed values are δ 26 and δ 199, respectively.

The ring carbons are considered with respect to benzene (in which all six carbons appear at δ 128) and the effect of substituents is extracted from table 3.16. Numbering the carbons as in figure 3.1(b), and setting out the data in tabular form, gives:

Carbon number	Base value	Substituent		δ_{total}	Intensity	δ_{obs}
		OH	COR			
C-1	128	− 7	+9	130	low	129
C-2/C-6	128	+ 1	+1	130	high	131
C-3/C-5	128	−13	+1	116	high	116
C-4	128	+27	+6	161	low	161

It is now possible to assign the peaks on the spectrum in figure 3.1(b).

Example 3.14
Question. Predict the chemical shift positions for the carbons of the *ortho* and *meta* isomers of hydroxyacetophenone (see example 3.13).

Model answer. The CH_3 and $C{=}O$ carbons are predicted to have the same chemical shifts as in the *para* isomer.

For the ring carbons, note that all six carbons are nonequivalent in the *ortho* and *meta* isomers. The working is as before, and is shown here only for the *ortho* isomer, and the *meta* isomer can be similarly treated. (The intensities of the substituted carbons are predicted to be lower than normal—see section 3.13.5.)

Carbon number	Base value	Substituent		δ_{total}	Intensity
		OH	COR		
C-1	128	−13	+9	124	low
C-2	128	+27	+1	156	low
C-3	128	−13	+1	116	normal
C-4	128	+ 1	+6	135	normal
C-5	128	− 7	+1	122	normal
C-6	128	+ 1	+1	130	normal

Exercise 3.20 Predict the chemical shift positions for the carbons in (a) methyl benzoate, $C_6H_5COOCH_3$, and (b) isopropyl benzoate, $C_6H_5COOCH(CH_3)_2$.

Exercise 3.21 Predict the chemical shift positions for the carbons in (a) the three isomers of dinitrobenzene (see Exercise 3.16, above); (b) the three isomeric methyl esters of chlorobenzoic acid (see Exercise 3.17, above); and (c) the three isomers of hydroxybiphenyl (see Exercise 3.17, above).

Example 3.15
Question. Predict the chemical shift positions for the carbons in 2-ethyl-1-hexanol.

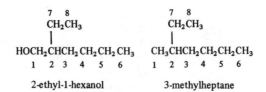

Model answer. The procedure used still demands that the δ values for the hydrocarbon are known *before* making allowance for the OH substituent. The parent, therefore, is 3-methylheptane; note the nomenclature change in the stem name. Since this hydrocarbon is not listed in table 3.11, we must predict the δ values empirically, using the data at the bottom of table 3.11. Each carbon environment is taken in turn, and its chemical shift calculated by summing together a *constant* and various *increments* caused by the adjoining carbons. (The carbons in the hydrocarbon are numbered as for the alcohol 2-ethyl-1-hexanol, to avoid confusion; for nomenclature purposes they would be numbered along the heptane chain.)

We calculate the shifts for 3-methylheptane as follows:

For C-1. This is a CH_3 carbon (constant 6.80) with a CH carbon α to it (increment 17.83). There are two CH_2 groups β to it, but their influence has already been subsumed in arriving at the value for the constant, so no further increment is applied:

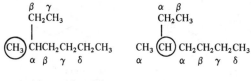

calculation of C-1 shift   calculation of C-2 shift

There are also two γ carbons (increment 2(−2.99)) and one δ carbon (increment 0.49).

The total gives the predicted value of δ 19.14.

For C-2. The constant for a CH carbon is 23.46, to which we add 2 × 6.60 (for two α CH_2 groups) and −2.07 (for one γ carbon) the δ carbon being ineffectual. Note again that no increments are taken for the α CH_3 carbon, nor for the two β carbons.

The total gives the predicted value of δ 34.59.

The other carbons are treated similarly, giving predicted values of 35.81 (C-3), 29.46 (C-4), 22.90 (C-5), 13.86 (C-6), 29.54 (C-7) and 10.87 (C-8). Since subsequent predictive ability has no better accuracy than 1 ppm, these values are shown in the formula to the nearest integer. The observed values are also shown (in parentheses).

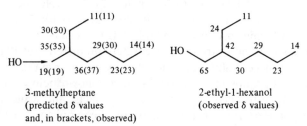

3-methylheptane
(predicted δ values
and, in brackets, observed)

2-ethyl-1-hexanol
(observed δ values)

For 2-ethyl-1-hexanol we must now add the influence of the OH group (table 3.15) giving the predicted values of C-1 through C-8 of 69, 44, 33, 29, 23, 14, 27 and 11, respectively. The observed δ values are shown on the formula, and are all within 4 ppm of predicted.

Exercise 3.22 Predict the chemical shift positions for the carbons of propane, butane, isobutane, norbornane and bicyclo(2.2.2)octane, using the empirical prediction data in table 3.11.

Example 3.16

Question. Predict the chemical shift positions for the carbons in 1-hexene.

Model answer. There are three aspects to this problem, since the sp^2 and sp^3 carbons must be treated separately, and the existence of *cis* and *trans* isomers must also be considered for all alkenes.

For the alkene carbons (sp^2) we use the empirical relationship given in table 3.12.

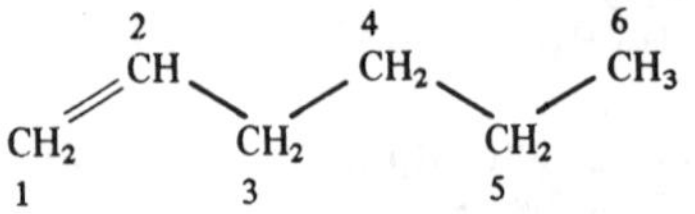

1-hexene	calculation of C−1 shift	calculation of C−2 shift

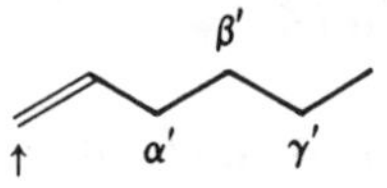

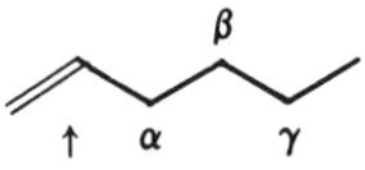

For *C-1*, we note that there is no substitution on the left-hand side of the formula (the side *near to* C-1), while there is α′, β′ and γ′ substitution (on the side *distant from* C-1). To the base value of 123 we sum the increments (−8, −2 and +2), giving the predicted value of δ 115.

For *C-2*, there is, on the side *near to* C-2, an α, a β and a γ substituent, so the predicted δ value is $(123 + 10 + 7 − 2) = \delta\ 138$. There are no substituents on the side *distant from* C-2 (α′, β′ or γ′).

For *C-3, C-4, C-5 and C-6*, since these are alkane carbons (sp^3), we take butane as base and add the appropriate increments shown in table 3.15 to account for the substituent ($—CH{=}CH_2$), as in example 3.10, etc. Predicted δ values are 35, 32, 23 and 13. All of these values are in good agreement with observed values and can be compared with, for example, the molecule shown in table 3.12.

There is no geometrical isomerism possible in this molecule, but it will be instructive to carry out the same predictive process for 2-pentene (the chemical shifts of which are shown in table 3.12). It will be found that the δ values for the sp^2 carbons are quite accurately predicted, but that the sp^3 carbons are less successfully accommodated. Now observe the δ values for the three methyl groups in 2-methyl-2-butene (shown below table 3.18): the considerable differences are due to steric effects, and while a number of cases respond to further refinement of the empirical method, the predictive accuracy is not reliable, and variations of the order shown must be tolerated.

Exercise 3.23 Predict the chemical shift positions for the alkene carbons of 2,5-dimethyl-3-hexene.

Exercise 3.24   Predict the chemical shift positions for the alkene carbons of 2-methyl-3-heptene.

Example 3.17
Question. Predict the chemical shift positions for the carbons in 1-hexyne, $C_4H_9C{\equiv}CH$.

Model answer. Table 3.13 lists the chemical shifts for the sp hybridized carbons of some alkynes; the shifts for 1-hexyne can be predicted from the empirical relationship shown at the foot of the table, operated similarly to that used for the sp^2 carbons of alkenes in example 3.16.

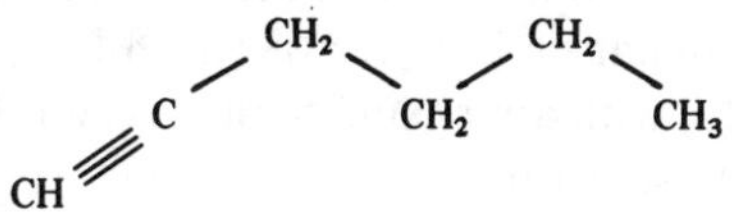

CH₂   CH₂

C   CH₂   CH₃

CH

1-hexyne

α' γ'

β' δ'

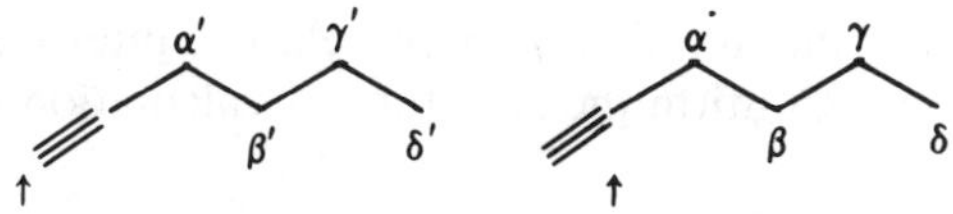

calculation of
C-1 shift

α γ

β δ

calculation of
C-2 shift

For *C-1*, the summation of base value and increments gives $(72 - 6 + 1 - 1 + 0.5)$ and a prediction of δ 67.5 for this carbon. The observed value is δ 66.

For *C-2* the summation is $(72 + 7 + 5 + 0 + 0.5)$ and a predicted shift of δ 84.5, compared with the observed δ 83.

The remaining carbons in the molecule are sp^3 hybridized, and are treated as usual from the base of butane.

It is interesting to draw attention to the way in which carbon substituents alter the chemical shift in sp carbons, in contrast to sp^2 carbons—where the effects are almost uniformly opposite in sign. Disappointingly, there is no adequate rationale for this, and it must be taken at its empirical face value.

In the following exercises ^{13}C NMR chemical shift data are given with (in parentheses) multiplicity information using the abbreviations s, d, t and q: see sections 3.13.7 and 3S.3.1. The relative intensities have been taken from real spectra: recall that such intensities are usually distorted (see section 3.13.6). In most cases it will be an advantage to construct a stick-diagram sketch of the spectrum on graph paper as an aid to familiarization with the appearance of the real spectra.

Exercise 3.25 Deduce the structure of a compound of molecular formula $C_7H_6O_2$ which was slightly acidic and gave a precipitate with 2,4-dinitrophenylhydrazine. Its ^{13}C NMR spectrum showed the following signals: δ 117 (d), δ 130 (s), δ 133 (d), δ 164 (s) and δ 191 (d). The relative intensities observed in a real spectrum were approximately 8:1:8:1:2. (Note that there are seven carbons in the molecule, but only five signals: use the intensities to tell you those signals which correspond to more than one carbon atom.)

Exercise 3.26 A neutral compound of molecular formula $C_{10}H_{12}O$ gave the following ^{13}C NMR signals: δ 22 (q), δ 68 (d), δ 128 (d), δ 129 (d), δ 131 (s), δ 132 (d) and δ 166 (s) with approximate observed relative intensities of 8:3:6:8:1:3:1. Deduce its structure.

Exercise 3.27 The ^{13}C NMR spectrum of an unknown compound ($C_6H_{10}O$) consisted of the following signals: δ 20 (q), δ 27 (q), δ 31 (q), δ 124 (d), δ 154 (d) and δ 197 (s), with relative intensities 3:4:3:4:2:1. Deduce its structure. Confirm the structure by calculation of the expected shifts.

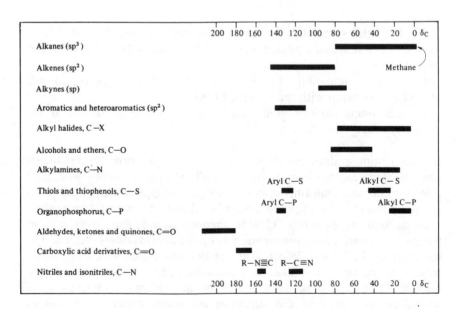

Figure 3.35 ^{13}C *chemical shift summary chart (δ values).*

3.16   TABLES OF DATA FOR ^{13}C NMR SPECTRA

Table 3.11   δ values for the carbons in alkane groups (sp^3)

REPRESENTATIVE ALKANES AND CYCLOALKANES

Methane   CH$_4$ δ −2.3   ethane   CH$_3$CH$_3$ δ 5.7

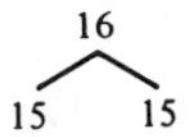

propane 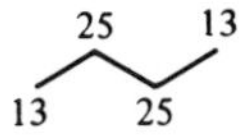   butane 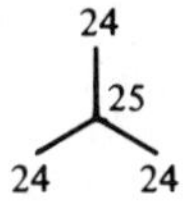   isobutane 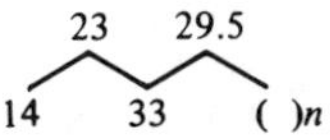   long alkyl chain

---

Cycloalkanes

Ring size	3	4	5	6	7	8	9	10
Shift (δ)	−2.9	23.3	26.5	27.3	29	28	27	26

Bicycloalkanes

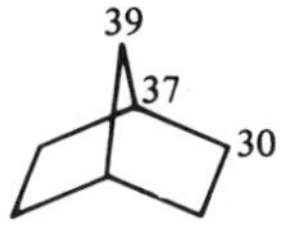

norbornane 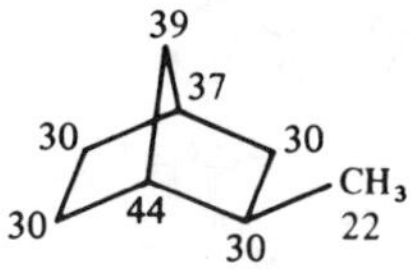 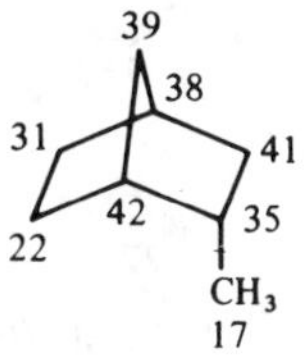 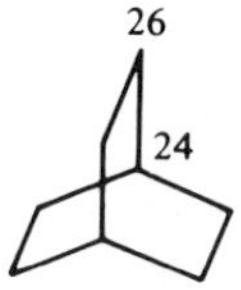   bicyclo[2,2,2] octane

---

EMPIRICAL PREDICTIONS FOR OTHER ALKANE CARBONS

$$\text{Chemical shift } (\delta) = \text{constant} + \Sigma\alpha + \Sigma\gamma + \Sigma\delta$$

α carbons − the increment varies for CH$_2$, CH and C carbons

α CH$_3$ groups ⎫
                ⎬ do not count these (the constant includes them)
β carbons   ⎭

	Constant	Increment for each carbon substituent α carbons			γ carbons	δ carbons
		—CH$_2$—	—CH—	—C—		
—CH$_3$ carbons	6.80	9.56	17.83	25.48	−2.99	0.49
—CH$_2$—carbons	15.34	9.75	16.70	21.43	−2.69	0.25
—CH—carbons	23.46	6.60	11.14	14.70	−2.07	0
—C—carbons	27.77	2.26	3.96	7.35	0.68	0

Table 3.12 δ values for the carbons in alkenes and cycloalkenes (both sp^2 and sp^3 carbons listed)

REPRESENTATIVE ALKENES AND CYCLOALKENES

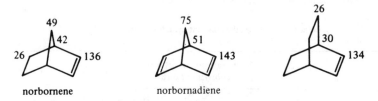

ethylene $CH_2\!=\!CH_2$

(ethene)   123

propylene $\overset{135}{CH_3CH\!=\!CH_2}$

(propene)   19   115

1–butene $\overset{28}{CH_3}\overset{}{CH_2}\overset{112}{CH\!=\!CH_2}$

            13   139

2–butene $\overset{123}{CH_3}\overset{11}{CH\!=\!CHCH_3}$

(*cis* and *trans*)   11   123

BICYCLOALKENES

norbornene

norbornadiene

EMPIRICAL PREDICTIONS FOR OTHER ALKENE CARBONS

Chemical shift (δ) = 123 + Σ(increments for carbon atoms)

$$C\!-\!C\!-\!C\!-\!C\!=\!C\!-\!C\!-\!C\!-\!C$$

	γ	β	α		α'	β'	γ'
increments	−2	+7	+10		−8	−2	+2

base value 123

Table 3.13   δ values for the carbons in alkynes (sp)

REPRESENTATIVE ALKYNES

acetylene (ethyne) $CH \equiv CH$ δ 72

R—C≡CH   R—C≡C—R' C_6H_5—C≡CH
  83   66   82   83   78

C_6H_5—C≡C—R C_6H_5—C≡C—C_6H_5
    86   90

EMPIRICAL PREDICTIONS FOR OTHER ALKYNE CARBONS

Chemical shift (δ) = 72 + Σ(increments for carbon atoms)

	C — C — C — C — C≡C — C — C — C — C							
	δ	γ	β	α	α'	β'	γ'	δ'
increments	+0.5	0	+5	+7	−6	+2	−1	+0.5

base value 72

Table 3.14 δ values for the carbons in aromatic and heterocyclic molecules (sp^2 and sp^3 carbons listed)

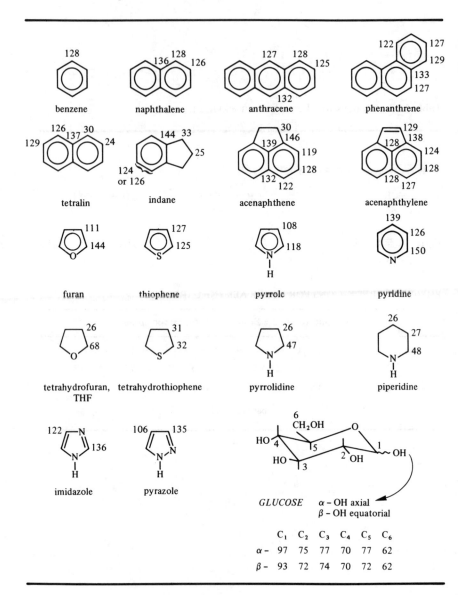

Table 3.15 Influence of functional group X on the chemical shift position (δ) of nearby carbons in alkane chains

$$X-C-C-C-C$$
$$\alpha \quad \beta \quad \gamma$$

X	α-shift			β-shift	γ-shift
	$X-CH_2-$	$X-\underset{R}{\overset{}{CH}}-$	$X-\underset{R}{\overset{R}{C}}-$		
	1° or	2° or	3°		
$-CH_3$	9	6	3	9	-3
$-R$: see table 3.11					
$\lbrace$ axial $-CH_3$	1	–	–	5	-6
equatorial $-CH_3$	6	–	–	9	0
(in cyclohexanes)					
$-CH=CH_2$	22	16	12	7	-2
$-C\equiv CH$	4	–	–	3	-3
$-C_6H_5$, $-Ar$	23	17	11	10	-3
$-F$	70	–	–	8	-7
$-Cl$	31	35	42	10	-5
$-Br$	19	28	37	11	-4
$-I$	-7 to 20	–	–	11	-2
$-NH_2$, $-NHR$, $-NR_2$	29	24	18	11	-4
$-NO_2$	62	–	–	3	-5
$-NHCOR$, $-NRCOR$	10	–	–	0	0
$-NH_3{}^+$	25	–	–	7	-3
$-CN$	3	4	–	2	-3
$-SH$	2	–	–	2	-2
$-OH$	50	45	40	9	-3
$-OR$	50	24	17	10	-6
$-OCOR$	52	50	45	7	-6
$-COOH$, $-COOR$, $-CON\langle$	20	16	13	2	-3
$-COR$, $-CHO$	30	24	17	2	-3
$-SO_3H$, $-SO_2N\langle$	50	–	–	3	0

Table 3.16   Influence of functional group X on the chemical shift positions (δ) of nearby carbons in alkene groups and benzene rings

Base values: ethylene (δ 123)   and   benzene (δ 128)

	Alkenes		Benzenes			
	C—1	C—2	C—1 (ipso)	ortho	meta	para
—CH$_3$	10	−8	9	0	0	−2
R,	16	−8	15	0	0	−2
R,	23	−8	21	0	0	−2
—CH=CH$_2$	15	−6	9	0	0	−2
—CH≡CH	−	−	−6	4	0	0
—C$_6$H$_5$, —Ar	13	−11	13	−1	1	−1
—F	25	−34	35	−14	1	−5
—Cl	3	−6	6	0	1	−2
—Br	−8	−1	−5	3	2	−2
—I	−38	7	−32	10	3	−1
—NH$_2$	−	−	18	−13	1	−10
—NHR	−	−	20	−14	1	−10
—NR$_2$	−	−	22	−16	1	−10
—NO$_2$	22	−1	20	−5	1	6
—NHCOR, —NRCOR	−	−	10	−7	1	−4
—CN	−15	15	−16	4	1	6
—SH	−	−	4	1	1	−3
—OH	−	−	27	−13	1	−7
—OR	29	−39	30	−15	1	−8
—OCOR	18	−27	23	−6	1	−2
—COOH, —COOR, —CON<	4	9	2	2	0	5
—COR, —CHO	14	13	9	1	1	6
—SO$_3$H, —SO$_2$N<	−	−	16	0	0	4
—PMe$_2$	−	−	14	1.6	0	−1
—PAr$_2$	−	−	9	5	0	0

Table 3.17 δ values for the carbons in carbonyl groups and some other multiple-bonded environments (solvent shifts, ± 2 ppm, are commonly observed for C=O)

R—CHO	aliphatic aldehydes	200–205
Ar—CHO, ⎤ CHO	aryl and conjugated aldehydes*	190–194
R—CO—R¹	dialkyl ketones	205–218
Ar—CO—R, Ar—CO—Ar, –CO–	aryl and conjugated ketones	196–199
	cyclohexanone derivatives	209–213
	cyclopentanone derivatives	214–220
	bicyclic ketones	215–219
	simple quinones	180–187
R—COOH, R—COO⊖	carboxylic acids and salts	166–181
R—COO—R′	aliphatic esters	169–176
Ar—COOAr, –COO–	esters with conjugation in the acid *or* alcohol moiety	164–169
	lactones	170–178
—CO—O—CO—	anhydrides, all classes	163–175
—CON<	amides, all classes including lactams	162–179

(*Continued on p. 200*)

Table 3.17 (*continued*)

—CO—NH—CO—	imides	168–184
—COCl	acyl chlorides, all classes	167–172
—NH—CO—NH—	ureas	153–163
—O—CO—O—	carbonates	152–156
R—C≡N	nitriles	114–124
R—N≡C	isonitriles	156–158
R—N=C=O	isocyanates	120–130
$\diagdown$C=NOH	oximes	148–158
$\diagdown$C=NNHCONH$_2$	semicarbazones	158–160
$\diagdown$C=N—NH—	hydrazones	145–149
$\diagdown$C=N—R	imines	157–175

*Note:

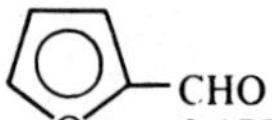

δ 177

Table 3.18 δ values for the carbons in methyl groups in common environments

CH₃ — Ar	side-chain	20–21
CH₃ — OAr	aryl ethers	56
CH₃ — OR	alkyl ethers	59

$$\underset{\displaystyle CH_3-\overset{\displaystyle O}{\overset{\displaystyle \|}{C}}-R}{}$$

	Me ketones	26–31

$$CH_3-\overset{O}{\overset{\|}{C}}-O-R$$

acetates   20–22

$$CH_3-O-\overset{O}{\overset{\|}{C}}-R$$

Me esters   51–52

$$CH_3-N\!\!<$$

2° and 3° amines   30–45

$$CH_3-\overset{O}{\overset{\|}{C}}-N\!\!<$$

acetamides   24

$$CH_3-\overset{\diagdown}{N}-\overset{O}{\overset{\|}{C}}-R$$

2° and 3° amides   31–39

(*Note*: CH₃ —NH—COOEt   15)

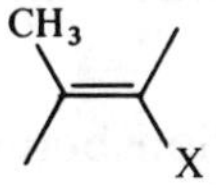

$$\overset{\displaystyle CH_3}{\diagdown}\!\!=\!\!\diagup \\ \diagup \qquad \diagdown X$$

where X is halogen, CN, CO   13–19

Note:
CH₃   H
26 \=/
17 CH₃   CH₃ 13

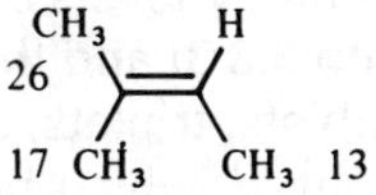

steric effects difficult to predict.

Table 3.19   Solvents used in NMR work

Solvent	Approximate δ for ^1H equivalent (as contaminant)	^{13}C δ value(s)	bp/°C	fp/°C
acetic acid-d₄	13 and 2	21, 177	118	16.6
acetone-d₆	2	30, 205	56	−95
acetonitrile-d₃	2	0.3, 117	82	−44
benzene-d₆	7.3	128	80	5.5
carbon disulfide	—	1931	46	−108.5
carbon tetrachloride	—	97	77	−23
chloroform-d	7.3	77	61	−63
deuterium oxide	4.7–5	—	101.5	3.8
dimethylsulfoxide-d₆	2	43	189	18
methanol-d₄(CD₃OD)	3.4	49	65	−98
hexachloroacetone	—	124, 126	203	−2
pyridine-d₅	7.5	124–150	115	−42
toluene-d₈	7.3 and 2.4	21, 125–138	110	−95
trifluoroacetic acid-d	13	115, 163 (quartets)	72	−15
dioxane	3.7	67	101	11.8

SUPPLEMENT 3

3S.1   SPIN–SPIN COUPLING AND DOUBLE IRRADIATION—MORE ADVANCED THEORY

Spin–spin coupling has been presented as an interaction between nuclei, transmitted via the electrons of the intervening bonds; the effect of these interactions was to alter the energy levels available to the coupling nuclei (see, for example, section 3.8.3) and this, in turn, led to splitting of the spectral lines into doublets, triplets, and so on, with spacings equal to the coupling constant, J, which had sign in addition to magnitude.

Double irradiation was introduced in sections 3.11.2 and 3.13.3, in the context of spin decoupling, to simplify proton and ^{13}C NMR spectra, respectively.

We shall now look at these mechanisms in greater detail, to see how they are interconnected: the explanations offered will also allow us to explain certain other double irradiation techniques used in the practice of NMR spectroscopy.

3S.1.1 *Electron-coupled interactions through bonds*

When molecules are arranged in rigidly close proximity in the solid state, as in a crystal lattice, their magnetic nuclei interact powerfully through space, and special techniques are necessary to minimize these nuclear dipolar effects. In nonviscous solutions, however, because of rapid tumbling, these internuclear effects are averaged out to zero and the only remaining interactions among nuclei are transmitted through the interconnecting bonds.

This is the reason for recording most NMR spectra in solution in nonviscous solvents: unless the neat sample is itself nonviscous, the slower tumbling in a viscous medium does *not* eliminate the through-space coupling, and severe line broadening is the result. Direct (through-space) coupling is about one thousand times stronger than the electron-coupled interactions.

The coupling of two magnetic nuclei via a single bond directly linking them is illustrated schematically in figure 3.36(a) and (b). We can suppose that A is a carbon-13 nucleus and X is a proton, but the argument holds for any other spin $\frac{1}{2}$ nuclei. The two electrons within the bond *must* have antiparallel spins (Pauli exclusion principle). At (a) we can show that nucleus A has its nuclear spin antiparallel to the nearest electron of the bond, as has nucleus X: this arrangement, with all four spins (both electron and nuclear) mutually antiparallel, is

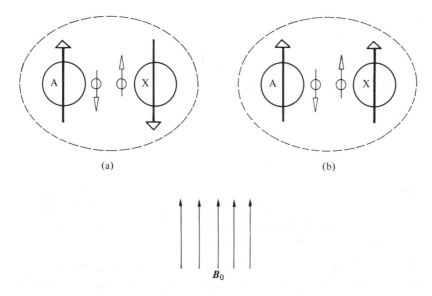

(a)   (b)

B_0

Figure 3.36 *Through-bond spin coupling for two directly bonded atoms, A and X. The electrons in the bond must retain antiparallel spins, but the nuclear spins may be* (a) *antiparallel or* (b) *parallel. Usually* (a) *is the more stable arrangement.*

usually the more stable, and the total energy of the system of spins is lowered as a result of the interaction. At (b) the electron spins are also opposed (the Pauli principle dictates this) but the nuclear spins are parallel; this is less stable and thus of higher energy than the former arrangement.

In summary, the coupling of these directly bonded nuclei leads to increased stabilization when the nuclear spins are antiparallel.

The amount of stabilization is, as we shall see, related to the coupling constant, but is of an exceedingly small order: coupling constants in organic chemistry are often less than 15 Hz, and this corresponds to about 10–13 J (from $E = h\nu$).

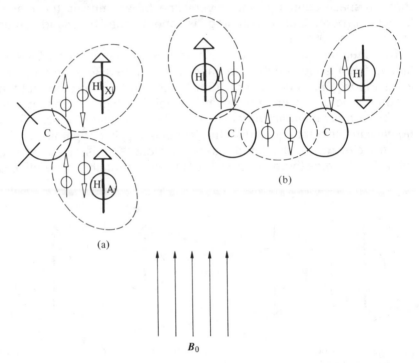

Figure 3.37 *Through-bond spin coupling for two protons joined (a) geminally or (b) vicinally through carbon. Usually two-bond coupling systems are more stable with the nuclear spins parallel to each other, and for three-bond couplings the system is usually more stable when the two coupling spins are antiparallel.*

Figure 3.37 sets out the preferred arrangements of nuclear and electron spins for nuclei coupled via two bonds (such as in the geminal coupling between the protons of CH_2 groups at (a)) and for three-bond interaction (such as the vicinal proton coupling at (b)). In each case the arrangement of electron and nuclear spins is always mutually antiparallel along the sequence from one hydrogen nucleus to the other, and these represent the most stable arrangement in each case. Note, therefore, that for geminal coupling the nuclear spins prefer the parallel orientation (antiparallel nuclear spins being less favored) and for vicinal coupling they prefer the antiparallel orientation (with parallel nuclear spins being less favored). Thus, the number of intervening bonds between coupling nuclei dictates whether a system of parallel or antiparallel nuclear spins is the more stable combination.

3S.1.2 Energy levels—the sign of J

To show how these factors lead to spin–spin splitting in the NMR experiment, we can refer to figure 3.38, which is an extended version of figure 3.16. It refers to the simple circumstance of two protons, A and X, coupling to give a system of two doublets—AX coupling—but similar arguments apply to (say) a carbon-13 nucleus coupling with a proton.

In terms of chemical shift, proton A may occupy either of two energy levels, corresponding to its being aligned with the applied field or opposed to it: likewise, there are two energy levels for proton X.

When the two nuclei interact, their four energy levels combine, creating four new *shared* levels; these are shown at (a) in the figure (not to scale).

The lowest energy level corresponds to the circumstance where both A and X are aligned with B_0—that is, (A↑)(X↑). The highest energy level has both spins opposed to B_0—that is, (A↓)(X↓). Of intermediate energy are the combinations (A↑)(X↓) and (A↓)(X↑). When proton A absorbs RF radiation and is at resonance in the NMR experiment, its aligned orientation is inverted to an opposed orientation, and this occurs both through the transition from level 1 to level 3 and through the transition from level 2 to level 4. When X absorbs RF radiation at resonance, it, respectively, undergoes both transitions, from level 1 to level 2 and from level 3 to level 4. (In all NMR transitions only one spin may invert at a time, so that transitions from (A↑)(X↑) to (A↓)(X↓) and from (A↑)(X↓) to (A↓)(X↑) are forbidden).

At (a) in the figure the energy gap from level 1 to level 3 is equal to that from level 2 to level 4; this means that proton A only absorbs RF at one frequency and thus shows only one absorption line on the spectrum. In the same way (since the energy differences in transitions 1 to 2 and 3 to 4 are equal), proton X shows only one absorption line on the spectrum.

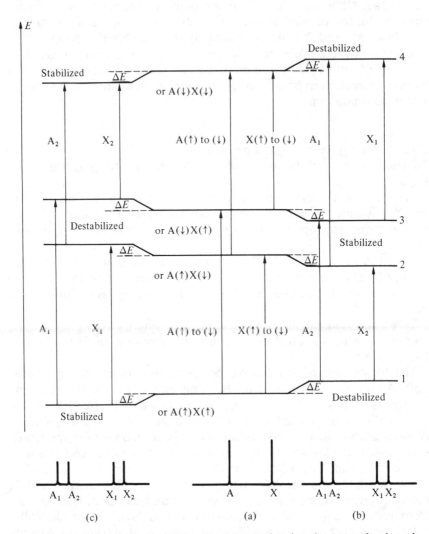

Figure 3.38 *The coupling interactions among nuclei alter the energy levels and transition frequencies: the coupling constant, J, is a measure of this energy of interaction.*

This system represents a coupling constant J = 0 Hz

Now, in addition to considering the large energy transitions of resonance, which correspond to frequencies of the order of hundreds of MHz, we must consider the effect of electron-coupled interactions. At (b) in figure 3.38 we can stipulate that the favored orientation of the nuclear spins of A and X be antiparallel, as for the vicinal protons discussed in section 3S1.1. This leads to slight changes in the energy levels: in levels 2 and 3, where the A and X spins are antiparallel, this stabilization *lowers* these levels by ΔE, while levels 1 and 4 (spins parallel) are *raised* by ΔE.

Now we can see that the four possible transitions are all different in dimension; the largest energy gap is transition 2 to 4, followed in diminishing order by 1 to 3, 3 to 4 and 1 to 2. Four frequencies are now involved, and four lines appear on the spectrum, corresponding to the A doublet (lines A_1 and A_2) and the X doublet (lines X_1 and X_2).

This is arbitrarily defined as a system in which J is positive.

If nuclear spins are more stable when parallel than when antiparallel (as in the geminal protons in section 3S.1.1), then the electron-coupled interactions will lead to slight additional stabilization—lowering—of levels 1 and 4 with destabilization—raising—of levels 2 and 3, as shown at (c) in figure 3.38. Once again four new transitions become available, and four lines appear on the spectrum, but in this case the highest-energy transition (highest-frequency absorption) is transition 1 to 3 (line A_1), then 2 to 4 (line A_2), followed by 1 to 2 and 3 to 4 (lines X_1 and X_2, respectively).

This is defined as a system in which J is negative.

Although different transitions are involved with each of the lines in the case of *J* positive compared with *J* negative, the appearance of the spectrum is identical. Conversely, it is not possible by examination of an AX spectrum to determine the sign of the coupling constant, *J*.

The dimensions of figure 3.38 are not to scale, but it is instructive to put some real values on the various energy gaps involved where A and X are both protons. A typical resonance transition for protons corresponds to 100 MHz, or 100 000 000 Hz (for a 2.35 T magnet), and a typical difference in chemical shift between A and X might be 2 ppm—or 200 Hz: transitions A_1, A_2, X_1 and X_2 are all enormous in scale (100 MHz) compared with the *difference* in dimensions between A_1 and X_1 (200 Hz).

The dimension *J* is even smaller, being typically of the order of 10 Hz. When A and X couple, one line of each doublet moves $2\Delta E$ to higher frequency and the other line moves $2\Delta E$ to lower frequency, so that the final separation within each doublet is $4\Delta E$, which is equal to the coupling constant, J. Thus, the electron-coupled interaction between these nuclei changes the energy levels by a mere 2–3 Hz. Figure 3.38 is out of scale by a factor of tens of millions.

3S1.3   Internuclear double resonance (INDOR) and selective population inversion (SPI)

In a real sample of molecules the energy levels in figure 3.38 will be populated with millions of nuclei, and at equilibrium their relative numbers can be calculated from Boltzmann distribution theory. These equilibrium populations will be distorted if the system is doubly irradiated with radiofrequency corresponding in energy to that of any of the transitions, and this, in turn, will affect the probability of other transitions arising.

We can illustrate this by considering the effect of double irradiation of the X_1 line in the AX spectrum at (b) in figure 3.38. Radiofrequency corresponding to this energy transition (from level 3 to level 4) will cause nuclei to pass upward from 3 to 4: as a consequence, the population of level 3 will be depleted, while that of level 4 will be augmented.

This will have the effect of increasing the probability of nuclei undergoing the transition from 1 to 3, since it will exaggerate the *difference* in their populations: as a result, the intensity of this 1 to 3 transition (line A_2) will increase. Conversely, since level 4 becomes overpopulated, transition 2 to 4 will be less probable, and the intensity of line A_1 will decrease. This is the basis of *selective population inversion* (SPI) and *internuclear double resonance* (INDOR). To understand the INDOR and SPI experiments, it is convenient to draw out the shared energy levels of figure 3.38(b) slightly differently, as shown in figures 3.39 and 3.41.

In the INDOR experiment applied to simple AX systems, we monitor the *line intensities* of the A signal and sweep a weak radiofrequency source through the X frequencies; as the X frequencies are stimulated, line intensities in the A signal will alter, and we can plot this change in intensity onto a recorder. The perturbing irradiation used in INDOR is about one-twentieth of that required for complete decoupling. The perturbed signals and the monitored signals can, therefore, be very much closer on the spectrum than is true in decoupling. In INDOR only the *populations* of spin states are altered—no change in *energy levels* is induced.

A simple INDOR experiment is illustrated in figure 3.39 for a four-line AX spectrum. We continuously monitor the line intensity (peak height) of line A_1, and sweep the perturbing irradiation through

X_1 and X_2; at X_1 the perturbation causes a *decrease* in the line intensity of A_1, while at X_2 it causes an *increase*. If we monitor line A_2 and repeat the irradiation of X lines, we get a second INDOR spectrum showing that the line intensity of A_2 *increases* for irradiation at X_1 and *decreases* for irradiation at X_2.

The energy diagram in figure 3.39 (compare figure 3.38) helps to explain why line intensities change in the A signal during double irradiation of the X signal. In this diagram we represent each line on the spectrum as a transition between spin states 1, 2, 3 and 4 (noting that energetically $A_1 > A_2 > X_1 > X_2$).

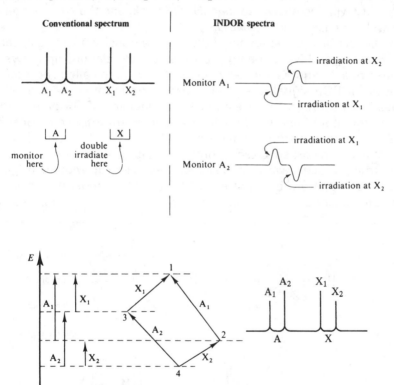

Figure 3.39   *Energy diagrams for* INDOR *study of a typical* AX *coupling system. Successive irradiation at each X line alters the intensities of the A lines. Of the alternative presentations, the second diagram emphasizes the shared nature of the E levels.*

Irradiation at X_1 will only produce an INDOR signal from line A_1 if they have an energy level in common; from the diagram we see this to be so (level 1). Irradiation at X_1 induces the transition $3 \rightarrow 1$, and the population of spin state 1 is increased. Because of this, the probability of transition $2 \rightarrow 1$ (the A_1 line) becomes less favored, and the line intensity of A_1 is reduced (negative INDOR signal).

Conversely, because the transition $3 \rightarrow 1$ is stimulated, the population of spin state 3 is reduced, and therefore transition $4 \rightarrow 3$ (the A_2 line) becomes more probable; the A_2 line intensity increases (positive INDOR signal).

We can follow this by considering irradiation at line X_2, which stimulates transition $4 \rightarrow 2$; the populations of spin states 4 and 2 are perturbed, state 4 being depleted and state 2 augmented. The transitions that share these energy levels are also affected: the A_2 line intensity will decrease (negative INDOR peak) and the A_1 line intensity will increase (positive INDOR peak).

We describe the relationship between the transition X_2 and A_1, and between A_2 and X_1, as being *progressive*; the relationship between transitions A_2 and X_2 and between X_1 and A_1, are called *regressive*. *Positive* INDOR signals are produced when monitored and irradiated lines have a *progressive* relationship; *negative* INDOR peaks arise when monitored and irradiated lines have a *regressive* relationship.

By ingenious instrumentation it can be arranged that the INDOR spectra are recorded directly onto the same trace as the normal spectrum in such a way that the INDOR peaks line up with the original transitions that are being affected. This factor is most important in complex spectra, an example of which is reproduced in figure 3.40.

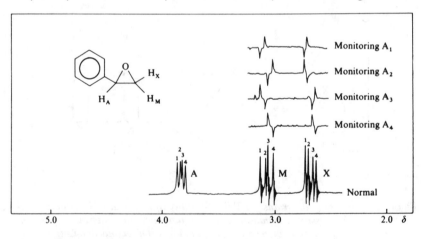

Figure 3.40   *Normal and INDOR NMR spectra of styrene oxide (epoxide ring protons only). (Recorded at 90 MHz.)*

Signs of the coupling constants. In figure 3.40 (styrene oxide) monitoring of line A_1 shows INDOR signals from lines M_1, M_3, X_1 and

X_2, which means that M_1 and M_3 correspond to the same spin orientation of X, and that X_1 and X_2 correspond to the same spin orientation of M. (The INDOR method is being used to detect those lines having common energy levels.) Since, therefore, lines A_1 and M_1 have the same spin orientation of X in common, J_{AX} and J_{MX} have the same sign. By the same argument, lines A_1 and X_1 correspond to the same spin orientation of M, so that J_{AM} and J_{MX} have the same sign.

Selective population inversion, SPI, can be understood as an extreme example of the population changes exhibited in the INDOR experiment.

In reference to figure 3.41, we may consider that the energy system relates to an AX coupling pair, in which A is a carbon-13 nucleus and X is a proton.

If a *massive* pulse of RF is applied at the frequency of line X_2, the intensity of line A_1 increases, but the population of level 1 is so depleted that nuclei are induced to drop from level 3 to level 1—giving rise to an *emission* peak for line A_2: this is *population inversion*.

The quantum theoretical treatment of the phenomenon shows that the changes in line intensity follow the ratio between the magneto-gyric ratios of the two nuclei A and X, which, in turn, follows the ratio between their NMR precession frequencies (see section 3.2). For carbon and proton this ratio is approximately 4: assuming arbitrarily that the original line intensities for the A signals were 1, then the new intensity of line A_1 is found to have a maximum intensity $(1 + 4)$ and that of line A_2 is $(1 - 4)$, thus being -3 below the baseline: see (b) in figure 3.41. It is a simple software matter to invert this negative signal, by shifting it 180° in phase, so that the spectrum has both lines 'positive', but with enhanced line intensities of 5 and 3, compared with 1 and 1 in the original spectrum: see (c) in the figure.

Having outlined the principles of *selective* population inversion and the advantages it brings to signal intensity, it can be said that the arguments largely apply to other *nonselective* procedures which are more widely used; sensitivity is a constant problem in low-abundance nuclei, such as carbon-13 or nitrogen-15, and these methods work by transfer of polarization from the strong proton system to the weaker carbon-13 or nitrogen-15 system. Since the magnetogyric ratio for proton is ten times that of nitrogen-15 (the same factor separates their precessional frequencies), then this *polarization transfer* (PT) can lead to tenfold increases in the line intensities for nitrogen-15 signals.

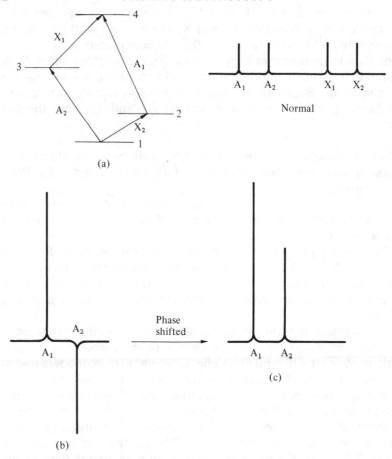

(a)

Normal

(b)

Phase shifted

(c)

Figure 3.41 *Selective population inversion, like polarization transfer, increases line intensities by double irradiation.*

3S.1.4 Nuclear Overhauser effect (NOE)

The nuclear Overhauser effect can be used to demonstrate that two protons (or groups of protons) are in close proximity within the molecule, and is therefore of considerable value in the study of molecular geometry.

The basic observation of NOE can be described by reference to the hypothetical molecule I. The two protons, H_a and H_b, we must imagine to be close enough to allow through-space interaction of their fluctuating magnetic vectors; each can contribute to the other's spin–lattice relaxation process (T_1). The number of intervening bonds between H_a and H_b is too large to allow normal coupling between them, but this is not a prerequisite for the operation of NOE.

If we double irradiate at the H_b signal, we shall stimulate absorption and emission processes for H_b, and this stimulation will be transferred through space to the relaxation mechanism of H_a. The spin–lattice relaxation of H_a will be speeded up, leading to a nett increase in the NMR absorption signal of H_a. H_a and H_b must be within 3.5 Å of each other.

In summary, provided that H_b makes a significant contribution to the spin–lattice relaxation process of H_a, then double irradiation of H_b causes an increase in the intensity of the H_a signal (by 1–50% for protons).

Since molecular oxygen is paramagnetic, and can contribute to nuclear spin–lattice relaxation processes, the NOE experiment should be carried out on deoxygenated samples, to avoid interference with the NOE observation.

A simple example of NOE is found in isovanillin. If we first record the NMR spectrum for isovanillin normally, and then while irradiating at the CH_3O frequency, the integral for the *ortho* proton (which appears as a doublet) is markedly increased.

I   isovanillin

A practical consequence of NOE is that in spin decoupling experiments the line intensities observed on a decoupled spectrum may not be the same as in the normal (nondecoupled) spectrum, and these intensities may not always correspond to integral numbers of protons. This same observation applies in the proton-decoupled spectra of ^{13}C resonances (see page 179).

The NOE can be more carefully assessed with difference spectra, in which a spectrum is first recorded without double irradiation, and then with double irradiation of one proton, or group of protons: subtraction of the former from the latter leaves residues of signals only where that signal from a nearby proton has been NOE enhanced, since all other signals will cancel out. The method has been used to establish the stereochemistry of many complex biological molecules, and also some simple *E* and *Z* alkene isomers. The most reliable results are obtained where both *E* and *Z* isomers are available; the NOE difference spectrum showing the greatest enhancement can be taken to be the isomer in which the groups are closer together in space.

3S.2 VARIABLE-TEMPERATURE NMR

Valuable structural information can be obtained by recording the NMR spectra of organic compounds at high and low temperatures, when rotational and intermolecular forces can be grossly changed. The practical difficulties in introducing a heated or cooled sample into the NMR magnet are considerable, since magnetic field changes with magnet temperature, but insulated probes allow variable-temperature NMR to be recorded routinely.

3S.2.1 The variable-temperature probe

Two main systems are in use to produce high- or low-temperature environments for the sample in the NMR spectrometer. In both systems high temperatures are achieved by passing a gas (for example, air or nitrogen) through a heater (with suitable thermostating) and then around the sample tube. Low temperatures are produced in one design by evaporation of a controlled quantity of coolant (for example, liquid nitrogen); the other design utilizes the Joule–Thomson effect, the gas being passed directly from the storage cylinder through a small orifice within the probe.

3S.2.2 Applications

A number of observations throughout this chapter have indicated the practical value of variable-temperature NMR (see pages 130, 132).

It is not difficult to see its application to studies in restricted rotation or ring inversions, etc., and there is a widely recognized advantage in the ability to record the NMR spectra of thermolabile materials, which are too transient for normal-temperature studies. Many materials (for example, polyethylene, isotactic polypropylene) are much more easily studied at high temperature and may give relatively sharp spectra, because of the attendant decrease in viscosity.

A few of the molecules whose rotational and inversional processes have been studied are shown on page 215. By noting the temperature at which such interconversions begin to occur (by gradually heating a sample in which these processes have been 'frozen'), it is possible to calculate the energies of activation for the rotations, etc.

Note that the protons of a methylene group adjacent to a chiral center will never show equivalence by increasing the temperature of the sample. No matter what their positions are, H_a and H_b will always be in different chemical environments, and are termed *diastereo-*

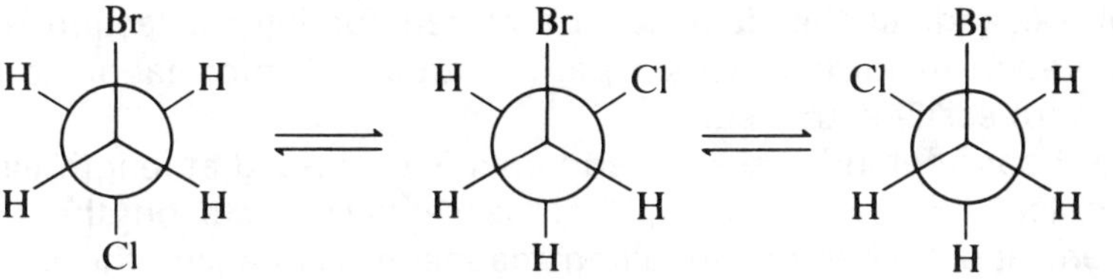

halogenoethanes

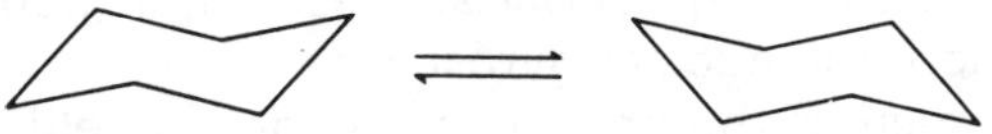

cyclohexanes

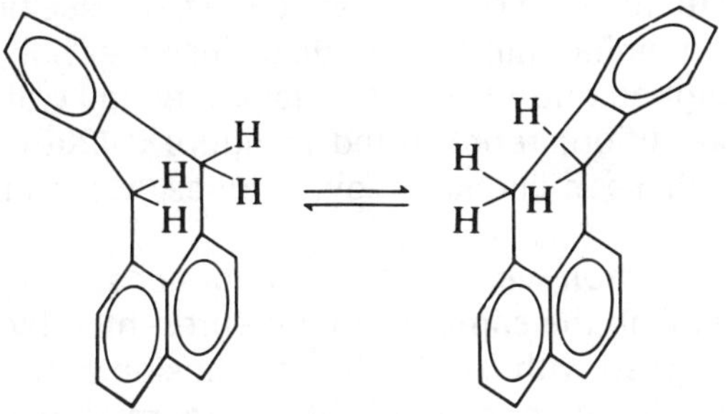

pleiadenes

topic. Chiral molecules (enantiomers) normally give rise to the same spectra, even though their absolute stereochemistries are different, but these absolute differences may be studied by using chiral solvents or chiral lanthanide-shift reagents (for example, derived from camphor) to induce different magnetic environments into the otherwise mirror image forms.

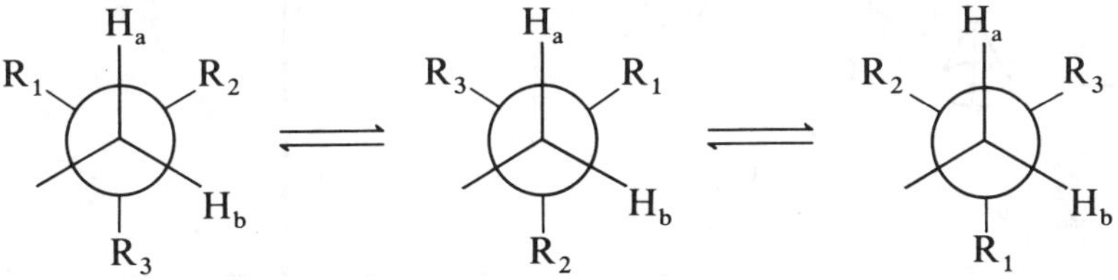

H_a and H_b are diastereotopic

3S.3 MULTIPULSE TECHNIQUES IN NMR—NETT MAGNETIZATION VECTORS AND ROTATING FRAMES

Computer control of instrumentation allows the timing and manipulation of radiofrequency pulses with accuracies of the order of nanoseconds, and many fast nuclear magnetic phenomena can be investigated by controlling sequences of pulses at predetermined intervals. In conjunction with large computer memories, vast series of individual experiments can be conducted, each with sets of variables, and all the data can be stored for later selection and presentation in formats which permit structural information to be ever more easily extracted.

The theory behind these experiments is discussed at length elsewhere (see 'Further Reading'), but essentially rests on the two different approaches to modelling magnetic resonance—the *quantum* approach, which is concerned with the energy jumps among nuclear spin states, and the *classical mechanical* model based on nuclear precession (the nuclear gyroscope).

One nucleus, such as the proton, in an applied field may precess either aligned with the field or opposed to the field: see figure 3.2. For a real sample containing millions of nuclei, just over half of them will precess in the aligned orientation and just under half will be opposed (Boltzmann distribution; represented in figure 3.42(a)). To simplify the physics of these millions of gyroscopes, we can adopt two devices.

The nett magnetization vector. The algebraic sum of all of the magnetic moments in the sample can be represented by a small nett magnetization aligned with B_0: this has dimension and direction and is a vector quantity with symbol M. At thermal equilibrium, in particular, we represent it by M_0. See figure 3.42(b): note that by convention B_0 is always taken to act along the z axis. Note also that at equilibrium the nett magnetization vector acts only along the z axis, with no component along either the x or y axis.

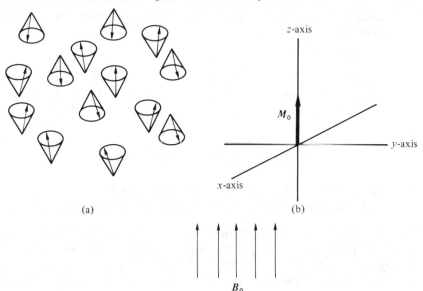

(a)   (b)

Figure 3.42 *The nett magnetization from an ensemble of nuclei at equilibrium can be represented by a vector arrow aligned along the field (defined as the z axis).*

Rotating frames of reference. Instead of considering that we observe the precessing nuclei from a viewpoint somewhere in the laboratory (a *laboratory frame of reference*), we shall imagine that we can become a part of the sample, and that we are ourselves spinning at the same frequency as the nuclear precession (say 100 MHz for protons and 25 MHz for carbon-13 nuclei). As we look around at the nuclei, we shall now see them to be static, and a description of their movements will be considerably simplified. We may select different *rotating frames of reference* for our purposes, such as the proton rotating frame or the carbon rotating frame.

When a weak pulse of radiofrequency is supplied to such an ensemble of nuclei, we now can describe the result in relation to the rotating frame of reference and the nett magnetization vector. The radiofrequency pulse is polarized, and delivered along the x axis in such a direction that it creates a magnetic effect which (with respect to the rotating frame of reference) causes the nuclear spins to focus, rather in the manner indicated in figure 3.43(a). After such a pulse, the nett magnetization vector no longer lies entirely along the x axis, but can be represented as in figure 3.43(b); we note that the vector has been rotated on the yz plane, and has components on both the z and y axes. It is this component on the y axis which is detected in the NMR experiment: it arises by the process of magnetic induction (first observed between metal conductors by Michael Faraday). For the duration of the pulse of RF, this induction is referred to as *forced induction*; when the pulse is switched off, the relaxing nuclei continue to emit energy, which is then referred to as *free induction*, and the routine pulsed NMR experiment records the *free induction decay* interferogram. See section 3.3.2 on Felix Bloch and FTNMR.

Continuing the argument, if now we deliver a more powerful RF pulse, we can rotate the nett magnetization vector through 90°, producing the situation in figure 3.43(c). *This has the maximum possible component on the y axis*: a single pulse of this power, therefore, gives the strongest signal in the NMR experiment. After such a 90° pulse, there is no residual magnetization along the z axis.

The circumstances represented by figure 3.43(d) are achieved with an even more powerful pulse—the 180° pulse. In this condition no y axis component exists (and thus no signal from the y axis detection); the nett magnetization is equal in dimension to M_0, but of opposite sign.

Lastly, a pulse of power equivalent to approximately 130° (figure 3.43(e)) gives rise to a negative y axis component and a negative signal in the NMR experiment.

Longitudinal and transverse relaxation. It is now possible to define more precisely two relaxation processes which were qualitatively described earlier, in terms of *spin–lattice relaxation* and *spin–spin relaxation.* Figure 3.42(a) represents a collection of precessing nuclei at equilibrium, and figure 3.43(a) represents them after RF excitation: in time, the nuclei in figure 3.43 will be restored to the condition of figure 3.42 by relaxation. If we now consider this in terms of the nett magnetization vector, we must assess how the x axis, y axis and z axis components of this vector behave. At equilibrium there is nett magnetization along the z axis, but within a large sample of nuclei the random orientation of magnets with respect to the x and y axes averages out to zero. After the pulse, this is no longer true (figure 3.43(a) and (b)): the z axis component has been reduced and a y axis component has been created.

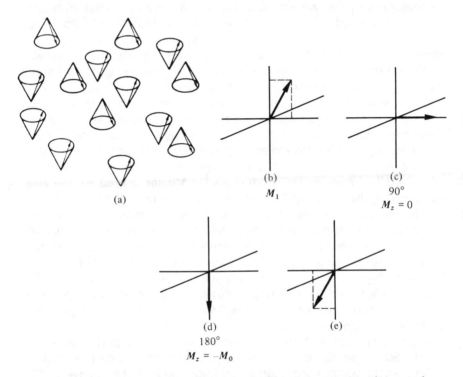

Figure 3.43   *A pulse of RF focuses the nuclei toward the y axis, and rotates the nett magnetization vector by an angle which is dictated by the pulse power.*

When the system of nuclear spins relaxes, we can identify two quite different process: (1) the reduced z axis component eventually increases back to M_0; (2) the y axis component reduces to zero.

Longitudinal relaxation is defined as that component of the whole which is resolved along the z axis; it involves the system in a loss of *enthalpy*, since it is associated with some of the nuclei dropping back down to the lower energy levels. The energy lost is transferred to other vectors in the surrounding solvent molecules and the like, and for this reason is named *spin–lattice relaxation*.

Transverse relaxation is associated with the decline in magnetization *across B_0*—that is, on the xy plane. The nuclei pictured in figure 3.43(a) are wholly in phase with the rotating frame of reference, but in time they pass out of phase. It involves a restoration of the randomness with which the nuclei are orientated at equilibrium, and is thus an *entropy* phenomenon. This can be understood by contrasting figure 3.42(a) and figure 3.43(a); an alternative description of the xy plane components, and their return to randomness, is shown in figure 3.44.

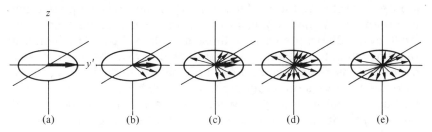

Figure 3.44 *Decline of the y axis magnetization; after a 90° pulse (see (a)), individual nuclei exchange spins and go out of phase with the rotating frame. Eventually their y axis components are randomized (e).*

Nuclei continuously and spontaneously gain and lose energy at one another's expense, moving from one energy level to another, and in the process randomizing their spin orientations with respect to one another. The alternative name *spin–spin relaxation* takes cognizance of this process.

Both longitudinal and transverse relaxation are exponential processes, and methods for measuring their separate rates are discussed in the specialized texts listed under 'Further Reading'. The terms T_1 and T_2 have been discussed earlier, in section 3.2.

3S.3.1 CH₃, CH₂ and CH *subspectra—spectrum editing—DEPT spectra*

If we consider a carbon-13 nucleus, without any attached proton, then its precessional frequency is quite easily correlated with the radiofrequency pulse, since resonance occurs when the two frequencies are matched.

A CH resonance is split into a doublet by the attached proton (with line separation equal to J_{CH}) and therefore the relationship with the radiofrequency is more complicated, since the two carbon signals have different frequencies (for clarity, suppose that $J_{CH} = 100$ Hz; then the two carbon resonances differ in frequency by 100 Hz).

For a CH_2 carbon, again if we take $J_{CH} = 100$ Hz, then in this case there will be three lines (triplet) differing by 100 Hz from one another in frequency.

Lastly, for a CH_3 carbon, four lines would appear (quartet) separated from one another by J_{CH}, again typically 100 Hz.

We should now consider this with respect to appropriate rotating frames of reference: for a 2.3 T magnet, the carbon rotating frame will have a frequency of 25 MHz and the proton rotating frame will have a frequency of 100 MHz. If an uncoupled carbon-13 nucleus has a resonance frequency of exactly 25 MHz, then we can say that the signal from this nucleus will be constantly in phase with the appropriate rotating frame. If, however, the carbon is coupled to one hydrogen, the signal is split in two, and if we assume a coupling constant of 100 Hz, then one of these signals will be 50 Hz higher in frequency than the rotating frame and the other will be 50 Hz slower: in other words, these signals are out of phase with the rotating frame. If we are observers on a rotating frame at 25 MHz, we shall observe the 'fast' signal as though it were still precessing around us with frequency 50 Hz; we would observe the 'slow' signal going in reverse at 50 Hz. It is simpler to observe from a reference frame rotating at (25 MHz + 50 Hz), in which case the other signal is out of phase by 100 Hz—the coupling constant. Every one-hundredth of a second (the reciprocal of J, the coupling constant) the other signal passes our observation frame (that is, it is in phase with the reference frame). After one-quarter of this time (one four-hundredth of a second, the reciprocal of $4J$) it has made one-quarter of a rotation (that is, it is 90° out of phase): after $1/(2J)$ it is 180° out of phase. These relationships are shown in figure 3.45.

Having discussed the case of a CH group and the way in which its signals, separated by J Hz, pass in and out of phase with the rotating frame, we can extend the principles to the signals from CH_2 and CH_3 groups.

All of these signals interact differently with the exciting radiation, since all are out of synchronization (out of phase) to different degrees, and at different times. The extent to which they are out of step is very small, since a process operating with a frequency of 100 Hz occurs every one-hundredth of a second, or every 10 ms. Fast accurate pulse control allows these small differences to be exploited, and it is possible to set up separate experiments to distinguish selectively the carbon-13 resonances from CH_3, CH_2 and CH environments.

There are several variants to the method, but we can take as an example of the results the ^{13}C NMR spectrum for menthol shown in figure 3.46: ten lines are shown in the full spectrum, each corresponding to one carbon environment. A separate subspectrum can be generated for the CH_3 carbons alone (showing three signals); likewise for the CH_2 carbons (showing three signals) and the CH carbons (showing four signals). This separation of subspectra according to the number of *attached protons* on the carbons (*spectrum editing*) constitutes a more certain method of identifying these sites than the off-resonance decoupling technique (section 3.13.7).

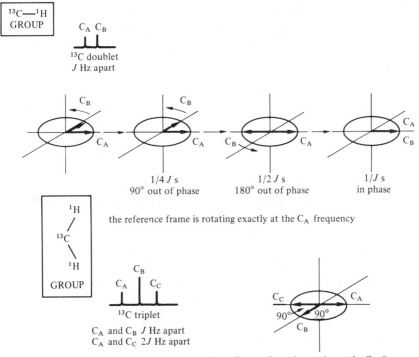

Figure 3.45 *Phase relationships in the rotating frame. Viewed from the C_A reference frame, other signals in the multiplets (C_B, C_C) move in and out of phase in a manner which is dependent on the coupling constant, J.*

The experimental method used to elicit this information involves a complex system of pulses, and the utilization of several separate phenomena. In the first place, the phase relationships outlined above, when suitably manipulated, give rise to periodic fluctuations in the signal intensity, described as *spin echoes*. Second, the decoupling pulses applied to the proton spins cause distortions in the populations of the energy levels in the carbon nuclei (to which they are attached, and with which they therefore spin-couple). These, in

222

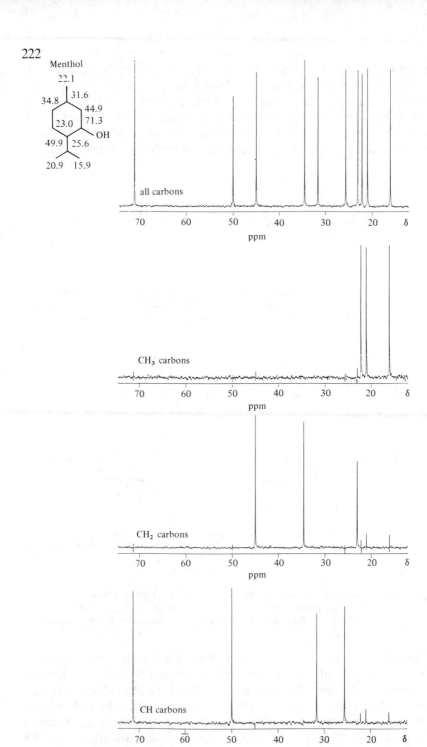

Figure 3.46 *DEPT edited spectra for the carbons in menthol: the complete spectrum is shown at the top, followed in succession by the subspectra for the CH₃ carbons, the CH₂ carbons and the CH carbons.*

turn, lead to increases in the intensities of the carbon signals, dictated by the same theory that we met in the INDOR experiment and in selective population inversion, SPI. (The case of carbon–proton coupling is only different in degree from the AX proton–proton system.)

The ability to increase the carbon-13 signal intensity in this way is referred to as *polarization transfer*; it is a general phenomenon, permitting increases in the intensity of other nuclei attached to proton (such as nitrogen-15). Spin echoes and polarization transfer can be combined in a number of different ways, among which is the above pulse technique, leading eventually to edited subspectra, which has been given the unrevealing name Distortionless Enhancement by Polarization Transfer (DEPT). The two principal features of DEPT spectra are increased sensitivity (by a factor of 4 for carbon-13 and of 10 for nitrogen-15), together with the spectrum editing facility for carbon-13 spectra.

Exercise 3.28 (a) Use the information implicit in the DEPT subspectra of figure 3.46 to confirm the assignments of carbon-13 chemical shift values shown in figure 3.33. (b) Work through the proton-coupled spectrum in figure 3.33, identifying the four doublets, three triplets and three quartets shown there.

3S.3.2   *Gated decoupling and the nuclear Overhauser effect*

We saw in section 3.13.5 and 3S.1.3 that the nuclear Overhauser effect increases the carbon-13 line intensities in routine broad-band decoupled spectra, and that it gives a welcome bonus to sensitivity in so doing. In nondecoupled spectra, because signal intensity is spread over the several lines in each multiplet, sensitivity is very low indeed, and yet the information in nondecoupled spectra is valuable.

Gated decoupling permits the recording of nondecoupled spectra, while simultaneously harnessing the increase in line intensities (and, hence, sensitivity) which the NOE generates. It can be understood by acknowledging two experimental facts: (1) coupling and decoupling among nuclei is rapidly established and broken, while (2) the NOE is only established slowly. The technique involves irradiating the proton resonances, say at 100 MHz (2.3 T), for some time (which decouples them from carbon, and distorts the population levels, inherent in the NOE); the decoupling irradiation is stopped and *immediately* the carbon resonances are pulsed (say at 25 MHz). All the C—H coupling is recorded, but with increased line intensities because of the distorted populations brought about by the NOE.

Inverse gated decoupling permits the more accurate measurement of line intensities in decoupled carbon-13 resonances, while at the same time avoiding the distortions of the NOE which invariably accompany decoupling. The technique here switches on the decoupler (100 MHz for protons at 2.3 T) *simultaneously* with the recording pulse (25 MHz for carbons) and the signal is generated so rapidly that, while the couplings are removed (rapid process), the NOE has insufficient time to develop (slow process) and thus the spectra are recorded *NOE-suppressed*.

3S.3.3   2D NMR—shift correlation spectra—COSY

The ability to present computed data in 'three dimensions' rests very much with graphics software, permitting such stack-plots as are shown (in a different context) in figure 4.3. Alternatively, the same information can be presented in cross-section, effectively a contour map of the peaks.

A conventional NMR spectrum is a plot of intensity against frequency, but for coupling nuclei (H—H or C—H, etc.) their interactions are also time-variable, as discussed above; by sampling these interactions as a function of time (and, hence, frequency) it is possible to separate out the interactions among (for example) the carbons and hydrogens of organic functions in such a way as to establish which protons couple with which carbons, or else which protons couple with which other protons. Since two frequency axes are involved, the method is called *two-dimensional NMR*, but the information is plotted in pseudo-three-dimensional form, with intensity as the third dimension as in figures 3.47 and 3.48.

There have been very many variations published using 2-D NMR, and these are dealt with in the specialist texts listed in 'Further Reading', but two of the most important are examples of *correlation spectroscopy*, either homonuclear or heteronuclear. The original use of the acronym COSY referred to the homonuclear proton–proton case; the important heteronuclear carbon–proton correlation case is not usually referred to as COSY, but this deserves to be so defined.

Figures 3.47 and 3.48 illustrate these for the molecule of menthol; it will be instructive to compare these frequently with figures 3.30 and 3.46.

Proton–proton correlation spectroscopy (homonuclear COSY, figure 3.47) sets out the proton NMR spectrum of an organic molecule such as menthol along the *x* axis, and repeats it along the *y* axis, with the signals repeated yet again in the contours of the *diagonal peaks*; wherever a proton couples with another proton (that is, wherever correlation is established), this is indicated by the contour of an off-diagonal *cross-peak*. The interpretation of the COSY spectrum is best carried out with the help of the DEPT spectrum in figure 3.46 and in conjunction with an analysis of the C—H correlation spectrum in figure 3.48.

Figure 3.48 is the C—H *shift correlation spectrum* (heteronuclear correlation) for menthol. In this, with the proton spectrum on the *y*

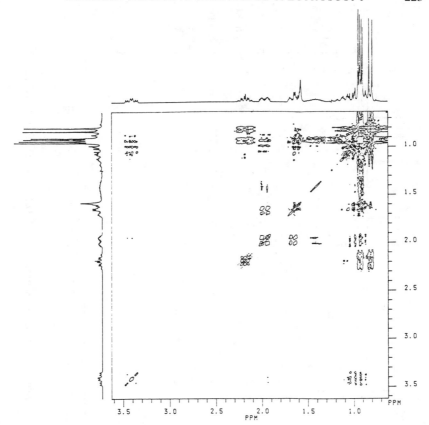

Figure 3.47 *Homonuclear proton–proton shift correlation spectrum, COSY, for menthol (200 MHz in CDCl₃). A cross-peak establishes correlation with a diagonal peak.*

axis and the carbon-13 spectrum on the *x* axis, wherever correlation exists (that is, wherever a carbon and a proton are attached to each other in menthol), cross-peaks appear in the correlation spectrum.

To simplify the interpretation, we can use the known carbon-13 chemical shift assignments shown in the figure. We already know from the DEPT spectrum in figure 3.46 how many protons are attached to each carbon atom, which allows us to identify the carbon signals from the CH_3 groups separately from the CH_2 groups, and, in turn, separately from the CH groups; it will be instructive at this stage to work through the carbon-13 spectrum, marking each signal with either s or d or t, and labeling the carbon atom it arises from (C-1, C-2, etc.).

The most easily assigned signals are those of C-1 (δ 71.3) and H-1 (δ 3.4). The multiplicity of H-1 arises from splitting with the two protons

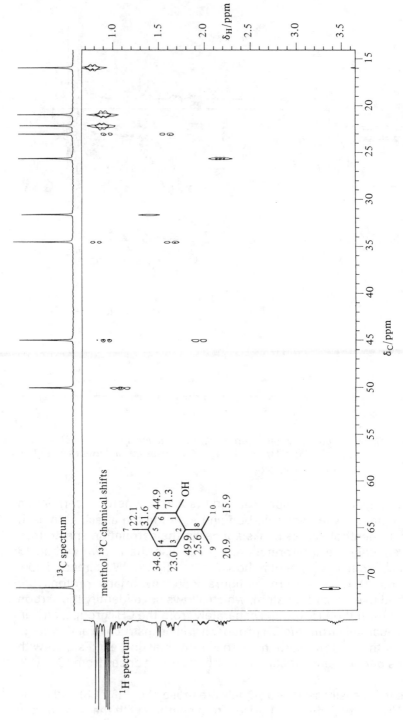

Figure 3.48 *Heteronuclear C–H correlation spectrum for menthol (200 MHz for proton, 50 MHz for carbon-13: in CDCl₃).*

on C-6 (each with a different chemical shift), one proton on C-2 and the OH proton.

The only proton signal without a carbon correlation is the OH proton, the doublet at δ 1.45.

If we continue with the CH_3 groups (carbon-13 signals at δ 15.9, δ 20.9, δ 22.1), we can identify in the proton spectrum the three doublets associated with these carbons. The protons on C-7 are split (doublet) by the proton on C-5: the protons on C-9 and C-10 are also split, within the isopropyl group, by the proton on C-8, and this allows us to distinguish the C-9 and C-10 protons from the C-7 protons. In the C—H correlation spectrum, the position of C-8 at δ 25.6 allows us to identify that H-8 appears at δ 2.16, and the signal for C-5 at δ 31.6 identifies H-5 at δ 1.4. Now in the COSY spectrum H-5 at δ 1.4 only correlates with *one* of the methyl doublets at δ 0.90, whereas H-8 at δ 2.16 correlates with *two* of the methyl doublets at δ 0.82 and δ 0.93: these methyls must be in the isopropyl group.

The next interesting point concerns the signals from the CH_2 groups, again identifiable from the DEPT spectrum. Each correlates at *two* points on the proton spectrum, because, in the chair conformation of menthol, one hydrogen of each pair is equatorial and the other is axial, so that they have different chemical shifts and appear as AX double doublets (further split by other neighbor protons). We saw at the end of section 3.4.3 that equatorial protons appear at higher frequency than do axial, so that this allows us to identify the separate equatorial/axial pairs for protons H-3, H-4 and H-6.

Continuing thus, through figures 3.47 and 3.48, and using the 600 MHz spectrum of menthol in figure 3.30, a reasonably full analysis of the menthol proton spectrum is feasible, although some uncertainties remain as a result of artefacts in the COSY spectrum associated with the overlapping of many of the proton signals. Note also that the OH proton does not appear in the same place in figures 3.47 and 3.48, since the spectra were recorded at different times, and at a different concentration.

3S.3.4 Magnetic Resonance Imaging (MRI)

Figure 3.49 is a representation of the internals of a human skull, the image being re-created from the proton NMR signals generated from the soft tissue of the brain, eyes and spinal column, etc. The ability to scan such soft tissues by NMR complements the information about the bone structure obtained from X-ray studies, and is an established medical diagnostic technique for soft tissue damage.

The experimental method has been sufficiently highly refined to produce not only very high resolution of detail, but also a whole sequence of views of the head (see small inserts at the top of the figure) which can be presented in sequence on a VDU in the form of a moving picture, so that the head can be 'rotated' to allow all aspects to be examined.

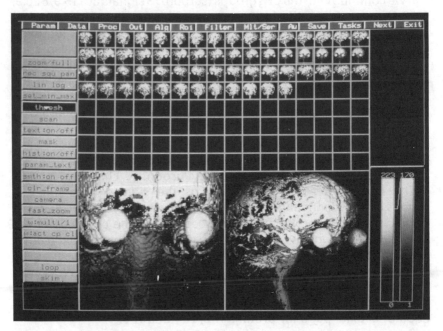

Figure 3.49   *Magnetic resonance imaging, MRI, of a human skull, showing soft tissues (eyes, brain, spinal cord) with the facility to rotate the image.*

A simple description of certain of the principles involved in NMR imaging can be obtained by recalling that protons in a head (in the water and fat) will come to sharp resonance at a given frequency only when the applied field causes them to precess at their resonance frequency. At lower and higher values of field, no NMR signal will be detected. Instead of applying a uniform field across the head, a gradient field is applied; at only one particular point along this gradient will the protons be in resonance, and therefore this distance across the head can be correlated with the field gradient. The gradient is then altered, and resonance occurs at a different point, the distance of which along the gradient is again correlated. By scanning along several gradients applied across a network of axes through a slice of the head, an image of the protons present (that is, of the soft tissue present) is constructed. The complete reconstructed image is defined by repeating the process across several slices through the head.

Two other names are applied to the method—*tomography* (derived from the Greek *tomos*, 'slice') and *zeugmatography* (from the Greek *zeugma*, 'to join'). Although referred to by chemists as NMR imaging, in medicine the adverb 'nuclear' is often omitted; hence MRI.

3S.4 CHEMICALLY INDUCED DYNAMIC NUCLEAR POLARIZATION (CIDNP)

When dipropionyl peroxide is thermolyzed in an NMR tube in the presence of thiophenol, ethyl radicals are produced, which react with the thiophenol to form thiophenetole.

$$CH_3CH_2CO-O-O-COCH_2CH_3 \xrightarrow{\Delta} 2CH_3CH_2CO-O\cdot \longrightarrow$$
dipropionyl peroxide

$$2CH_3CH_2\cdot + 2CO_2$$

$$PhSH + CH_3CH_2\cdot \longrightarrow \longrightarrow \longrightarrow PhSCH_2CH_3$$
thiophenol   thiophenetole

If the NMR spectrum of the thiophenetole is recorded *during* the course of the reaction, the ethyl signals appear as sketched in figure 3.50(a); the NMR spectrum of normal thiophenetole is also shown at (b) for comparison.

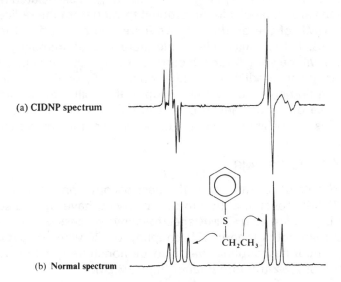

(a) **CIDNP spectrum**

(b) **Normal spectrum**

Figure 3.50 *Representations of the ^{1}H NMR spectrum of the ethyl group in thiophenetole: (a) CIDNP spectrum, recorded during the formation of thiophenetole from thiophenol and ethyl radicals; (b) normal spectrum.*

Some of the lines in spectrum (a) have increased intensity (stimulated absorption (A)), while other lines show as emissions (stimulated emission (E)).

We say that the nuclear spins in the product of this radical reaction (that is, the nuclear spins of thiophenetole) are undergoing dynamic polarization, because of the *chemical reaction* that is producing the molecule. This is an example of *chemically induced dynamic nuclear polarization* (CIDNP).

Observation of CIDNP effects during a reaction is good evidence that, at least in part, a radical mechanism is involved, and the technique is now extensively developed to study formation and decomposition of diarylmethanes, triarylmethanes, acyl peroxides, etc.

The mechanism of CIDNP is complex, and the observed effect (E or A) depends on the way in which particular nuclear spin states interact with the electron spins of the free radicals involved. Additionally, the radicals from a precursor molecule can follow several reaction paths, depending on whether the originally formed radical pair react together immediately, or diffuse out from each other, to perform other dimerization or transfer reactions. For a particular radical pair, the singlet and triplet electronic states undergo mixing processes under the influence of nearby nuclear spins; if the resultant mixed electronic state is predominantly singlet in character, the associated nuclear spin states are enhanced in the product formed from the radical pair, and the NMR of the product shows increased absorption for these spin states. If the mixed electronic state is predominantly triplet, emission will be shown for the corresponding nuclear spin states.

The ability to predict or explain whether (E) or (A) will arise in CIDNP experiments clearly follows upon the ability to calculate the importance of singlet and triplet contributions to the new energy levels: this skill will not frequently be required of the organic chemist.

3S.5 ^{19}F AND ^{31}P NMR

The concepts of ^{1}H and ^{13}C NMR spectroscopy apply equally to ^{19}F and ^{31}P NMR spectra, although the following have been discussed largely in relation to ^{1}H NMR spectroscopy: precession frequencies and field strengths (page 107); coupling of ^{19}F with ^{1}H (page 155). Proton—fluorine coupling was also demonstrated in the ^{1}H NMR spectrum of 2,2,2-trifluoroethanol (figure 3.24).

3S.5.1 ^{19}F NMR
^{19}F is the naturally occurring isotope of fluorine (100 per cent abundance). Apart from the provision of an appropriate radio-

frequency source (56.46 MHz at 1.4 T), no major instrument modification is needed to change an NMR spectrometer from ^{1}H work to ^{19}F work.

Chemical shift positions are most often measured from $CFCl_3$ or trifluoroacetic acid, and figure 3.51 shows a few of these positions for the most important organic situations. The range of chemical shift covered by even this limited number of fluorine environments is 200 ppm, compared with $\approx$ 10 ppm for the most common ^{1}H positions. In consequence, fluorine resonances tend to be well separated on the spectrum, and first-order spectra are the norm rather than the exception.

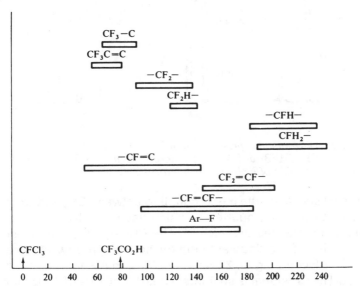

Figure 3.51 *Chemical shift ranges (ppm) for ^{19}F in common organic environments.*

Coupling constants between fluorine nuclei cover a wider range than in ^{1}H NMR. Geminal F—F coupling ranges from 43 Hz to 370 Hz, and vicinal F—F from 0 to 39 Hz; *cis* fluorines show J, 0–58 Hz, and *trans* fluorines show J, 106–148 Hz. Long-range coupling over five bonds (through F—C—C—C—C—F) is 0–18 Hz.

Coupling between ^{1}H and ^{19}F is also strong. Geminal coupling ranges from 42 Hz to 80 Hz, and vicinal ^{1}H—F from 1.2 Hz to 29 Hz; *cis* ^{1}H—F shows J, 0–22 Hz, and *trans* ^{1}H—F shows J, 11–52 Hz.

Fluorine attached to benzene also couples with protons on the ring, the $J_{H,F}$ ranges being: *ortho*, 7.4–11.8 Hz; *meta*, 4.3–8.0 Hz; *para*, 0.2–2.7 Hz.

Figure 3.52 is a diagrammatic presentation showing the ^{1}H and ^{19}F NMR spectra of 1-bromo-1-fluoroethane (CH_3CHFBr). Each spectrum is recorded using a 1.4 T magnet, so that the ^{1}H spectrum is at 60 MHz and the ^{19}F spectrum is at 56.46 MHz.

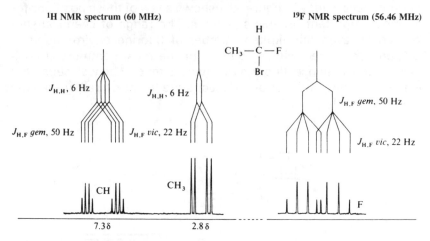

Figure 3.52 *Diagrammatic presentation of ^{1}H and ^{19}F NMR spectra of CH_3CHFBr at 1.4 T. The fluorine spectrum is lower in frequency than that of hydrogen by approximately 60 000 ppm.*

In the ^{1}H spectrum the H—H coupling between CH_3 and CH should give rise to an AX_3 quartet and doublet (*J*, 6 Hz), but the H—F coupling complicates this picture. The CH_3 signal is further split by ^{19}F, each line of the doublet being further split into two by the large vicinal H—F coupling (*J*, 22 Hz); the CH_3 signal is therefore a double doublet. The CH signal is split into a quartet by CH_3 (*J*, 6 Hz) and then each line is split into two by the enormous geminal H—F coupling (*J*, 50 Hz); the CH signal is therefore a doublet of quartets.

In the ^{19}F spectrum the fluorine signal is split into a doublet by the geminal methine proton (*J*, 50 Hz), and each line is further split into four by the vicinal methyl protons (*J*, 22 Hz); the fluorine resonance is therefore an overlapping doublet of quartets.

3S.5.2 ^{31}P NMR

Like ^{1}H and ^{19}F, ^{31}P has $I = \frac{1}{2}$ and the multiplicity that it engenders in ^{1}H and ^{19}F spectra is easily predicted, using the same rules that we have seen applied to protons and fluorine. We shall restrict ourselves to studying the multiplicity in the ^{1}H and ^{19}F spectra of Me_2PCF_2Me; coupling of phosphorus with Me_2 gives rise to a doublet, and the single Me group is split by ^{31}P into a doublet, and then by the two ^{19}F nuclei, so that the Me signal consists of overlapping triplets (see figure 3.53). The fluorine resonance is split into two by the phosphorus, each line being further split into four by the CH_3 protons. If we could observe the ^{31}P NMR spectrum (24.3 MHz at 1.4 T), what multiplicity would we predict? The signal would be split into three by the two ^{19}F nuclei (J, 70 Hz); each line would then be split into four by the single Me protons (J, 8 Hz); finally, each of these twelve lines would be split into seven by the Me_2 protons (J, 4 Hz). Considerable overlap would arise, but because of the high multiplicity the signal would be difficult to analyze.

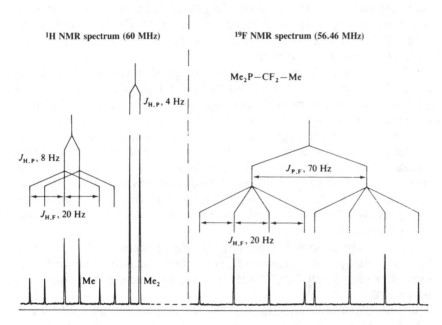

Figure 3.53 *Diagrammatic presentation of ^1H and ^{19}F NMR spectra of*
MeP—CF_2Me (1.4 T).

3S.6 ^{14}N, ^{15}N AND ^{17}O NMR

Nitrogen and oxygen are the next most important elements to carbon and hydrogen in organic chemistry.

While nitrogen-14 NMR has been studied for some time, the nucleus of nitrogen-14 has an electric quadrupole moment (see section 3.2), so that its signals are often several hundred Hz in width; virtually all current interest lies in the sharper spectra of the ^{15}N isotope. The main problem with both ^{15}N and ^{17}O is their natural abundances, which are 0.37 per cent and 0.037 per cent, respectively. With highly sensitive instruments, and making maximum use of various signal enhancement methods (including spectra summation, DEPT and NOE), natural-abundance spectra can be recorded on these nuclei.

3S.6.1 ^{15}N NMR

Since the magnetic shielding of a nucleus is dependent mainly on the electron density around it, chemical shifts for ^{15}N and ^{14}N are the same; these shifts are usually quoted in ppm differences from an external sample of liquid ammonia (which is inconvenient) or measured initially against the shift for nitromethane, CH_3NO_2, used as an internal primary standard, with a resonance 380 ppm to higher frequency than liquid ammonia.

Nitrogen-15 NMR spectra are recorded at 10 MHz in a field strength of 2.3 T; since ^{15}N has $I = \frac{1}{2}$, its behavior is similar to that of the proton, with simple multiplicity rules in NH groups, etc. Most routine spectra are proton-decoupled (^{15}N-$\{^1$H$\}$) to give signal enhancement, and in these each nitrogen environment gives one signal.

The approximate shift ranges for a few environments are: *aliphatic amines* (δ 0–100), *amino acids* (δ 40–50), *amides* (δ 100–160), *nitro groups* (δ 360–410), *oximes* (δ 380–420), *cyanides* (*ca* δ 230), *iso-cyanides* (ca δ 200), *azo groups* (δ 530–610), *diazonium salts* (two signals, *ca* δ 210 and δ 340), *pyridine derivatives* (δ 240–520) and *pyrrole derivatives* (δ 100–180).

Thus, *p*-nitroaniline has a ^{15}N-$\{^1$H$\}$ NMR spectrum showing only two signals, widely separated, and easily assigned; that for *p*-cyanobenzamide, $NCC_6H_4CONH_2$ likewise has two signals, near δ 120 and δ 220.

In non-decoupled spectra, the ^{15}N resonance will be split by any attached protons, so that the NH_2 signal in amines will appear as a triplet. In primary amides (such as $HCONH_2$) restricted rotation (see section 3.5.2) means that the two NH_2 protons have different environments and do not couple equally with the nitrogen atom, so that the signal does not appear with simple triplet multiplicity. For example, in $HCONH_2$ the three protons all split the nitrogen signal, but with different coupling constants (14 Hz, 88 Hz and 92 Hz; the smallest J value is associated with the formyl proton).

One odd feature of the ^{15}N nucleus is revealed in the nuclear Overhauser effect during proton decoupling: instead of a positive increase in intensity (as in ^{13}C-$\{^1$H$\}$ spectra), the effect is negative; this is associated with the fact that the ^{15}N nucleus behaves in a manner opposite to that of the proton (and most other nuclei) in relation to its magnetic moment and direction of spin. Suffice it to say that, although the NOE is opposite in sign to that of the proton, signal enhancement can still be achieved. This produces signals which are negative (below the baseline), and spectra are often plotted in this way; since it is a trivial problem to invert this subsequently, many ^{15}N NMR spectra are also plotted conventionally, with the peaks above the baseline.

The role of ^{15}N NMR is paramount in biological and heterocyclic chemistry and the subject is treated extensively in the texts listed in 'Further Reading'.

3S.6.2 ^{17}O NMR

In addition to its low abundance (0.037 per cent), [17]O has $I = \frac{5}{2}$ and possesses an electric quadrupole moment, giving fast relaxation and thus broad lines (10–1000 Hz wide).

Recording oxygen-17 NMR spectra (13.5 MHz at 2.3 T) is far from trivial, but oxygen occurs sufficiently widely in organic and inorganic environments to lend impetus to the achievement.

Chemical shifts are usually measured from water (defined at δ 0) or dioxane (δ 12); with an overall range of about 1000 ppm, many distinctions are possible. *Aldehydes* (ca δ 600), *ketones* (ca δ 560), *carboxylic acids* (ca δ 250) and *amides* (ca δ 300) all show well-separated signals, but these functions are more easily detected by other means. *Esters* are also easily otherwise identified, but it is worth a comment that the acyl and alkyl oxygens have different shifts (ca δ 350 and δ 120, respectively); thus, the ^{17}O-{^{1}H} NMR spectrum of ethyl acetate consists of two single lines, etc.

Primary, secondary and tertiary alcohols all have different oxygen-17 chemical shifts (δ 30, δ 50 and δ 70, respectively), and this is a useful distinction.

In carbohydrates, too, there are subtle differences in δ values which can be correlated with stereochemistry and reactivity: in glucose the six oxygens have δ values which distinguish O-1 (δ 36–47) from O-5 (δ 50–65), from O-6 (δ −7 to −12), from O-2, O-3 and O-4 (δ 6–11).

Non-decoupled oxygen-17 NMR spectra are much more difficult to record, not only because of the loss of sensitivity, but also because the signals tend to be broad, and this makes small couplings impossible to observe. The ^{17}O NMR spectra of both H_2O and H_3O^+ have been recorded—the former appears as a triplet and the latter as a quartet, because of coupling to protons, with $n + 1$ multiplicity.

The application to inorganic chemistry is considerable, since there are wide chemical shift differences among the various oxidation states of oxyacids; some approximate δ values are phosphite (δ 105), phosphate (δ 90); sulfite (δ 240), sulfate (δ 180); nitrite (δ 650), nitrate (δ 410).

3S.7   ELECTRON SPIN RESONANCE SPECTROSCOPY (ESR)

Just as the ^1H nucleus has spin and therefore a magnetic moment, so the electron, with its spin, is paramagnetic and also possesses a magnetic moment. Like the proton, too, the electron will precess in an applied magnetic field with a precise precessional frequency and will undergo transitions between spin states (spin orientations) if energy of the correct frequency is applied. I (or, more usually, S) for the electron is also $\frac{1}{2}$.

Measurement of electron spin transitions is the basis of *electron spin resonance* spectroscopy (ESR), also called *electron paramagnetic resonance* (EPR) spectroscopy.

The magnetic moment of an unpaired electron is about 700 times that of the proton, so that the sensitivity of ESR detection is very much higher than in NMR (which is fortunate, since the concentrations of unpaired electrons are often correspondingly lower): ESR spectra can be recorded for radical concentrations down to 10^{-4} mol dm^{-3}, irrespective of the number of nonradical species present.

The ESR experiment can only detect species having unpaired electron spin—for example, organic free radicals. Paired electrons cannot be detected, since the spins of an electron pair within an orbital must remain antiparallel; if one of the spins is reversed, the Pauli exclusion principle will be violated, and if both spins could be reversed, there would be no net absorption of energy.

For the electron, the energy necessary to induce spin transitions in modern ESR instruments is in the microwave region of the electromagnetic spectrum, with very much higher frequencies than radiofrequency. (At 1.4 T v is 3.95×10^4 MHz.) The field strength commonly used for ESR work is 0.34 T, and for this field the precessional frequency of the electron is $\approx$ 9.5 GHz (9.5 gigahertz, 9.5×10^9 Hz).

We must now discuss a few features in the language of ESR which differ from the corresponding features in NMR.

3S.7.1   Derivative curves

To begin with, ESR spectra look different from NMR spectra, since ESR spectra are virtually always plotted as first-derivative spectra. Conventional NMR signals are a plot of absorption against field strength, while ESR signals are a plot of the *rate of change of absorption* against field strength; see figures 3.54, 3.55 and 3.56. In general, ESR absorptions are broad compared with NMR absorptions, and first-derivative spectra enable more accurate measurement of the spacings to be made: conversely, first-derivative spectra of NMR signals would be hopelessly overlapped with one another in complex spin-coupling systems.

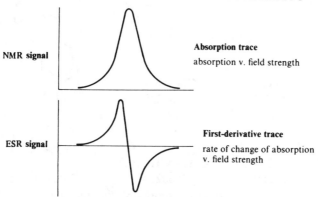

Figure 3.54 *A NMR absorption trace and an ESR first-derivative trace.*

Another difference between ESR and NMR is that while a molecule may contain several ^1H nuclei and give rise to several ^1H NMR signals (usually split by coupling) a radical contains only *one* unpaired electron, and therefore gives rise to only *one* signal (split by coupling).

3S.7.2 g values

In ESR, as in NMR, the position of the resonance signal has to be specified as a function of both field strength and frequency. In NMR, chemical shift positions (for example, 60 MHz at 1.4 T) are expressed as field-independent units (δ). In ESR the resonance positions are expressed as a g value (also a function of field strength and frequency). The energy of the ESR transition is given by

$$E = h\nu = gB_0 \frac{eh}{4\pi m_e c}$$

where B_0 is the external magnetic field, m_e is the electron mass, e is the electronic charge and c is the speed of light: g is a proportionality constant, and has the value 2.002 319 for an unbound electron. When the unpaired electron is present in an organic substrate, magnetic interactions will shift g from this value. The g values for some common organic radicals are shown below.

Note that the g values for carbon radicals are not substantially shifted from g for the unbound electron, but that oxygen and nitrogen radicals have much higher g values: this constitutes an important application to organic free-radical chemistry.

$$CH_3^{\bullet} \qquad CH_2{=}CH^{\bullet}$$
$$2.00255 \qquad 2.00220 \qquad 2.0155$$

$$C_2H_5^{\bullet} \qquad CH_2{=}CHCH_2^{\bullet}$$
$$2.00260 \qquad 2.00254 \qquad 2.00585$$

g values for representative radicals
(free electron $g = 2.00232$)

3S.7.3 Hyperfine splitting

The resonance lines in ESR spectroscopy are frequently split in a way reminiscent of spin–spin splitting in NMR, but the mechanism operating in ESR has no strict counterpart in NMR. The σ electrons in an organic molecule are normally represented as lying substantially between the atoms they bind, but electrons are not constrained to individual bonds. In particular, an unpaired electron can be associated with several atoms within the molecule in differing degrees, and if these atoms have magnetic nuclei (for example, ^1H, ^{13}C), the interaction of this magnetic moment with the electron will cause splitting of the ESR signal. The benzene radical anion ($C_6H_6^{-}$) gives rise to a seven-line signal, since the six protons cause splitting analogous to nuclear coupling: the same ($n + 1$) rule holds here for nuclei with $I = \frac{1}{2}$ (see figure 3.55).

This splitting in ESR is called *hyperfine splitting*, with symbol a: the dimensions of the splitting are measured in Hz, but in older literature ESR splitting is usually measured in gauss ($G \equiv mT$).

The appearance of seven equally spaced lines in the ESR spectrum of $C_6H_6^{-}$ is proof that all six hydrogens are equivalent, and therefore that the unpaired electron is coupling equally with all six hydrogens, rather than localized at one particular position (which would have given rise to *ortho*, *meta* and *para* splittings). Since ^{12}C is nonmagnetic, no further hyperfine splitting is produced.

The electron spin density on each carbon atom of $C_6H_6^{-}$ is directly proportional to the observed hyperfine splitting constant a_H (3.75 G).

A second example is the ESR spectrum of $CH_3\overset{\bullet}{C}HOCH_2CH_3$, shown in figure 3.56. The various hyperfine splittings correspond to coupling with H_3, H and H_2, in order of decreasing a_H values. The γ-CH$_3$ group is too distant to contribute further splitting.

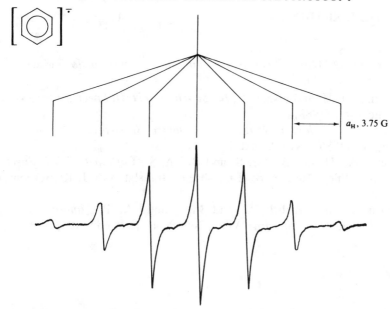

Figure 3.55 *ESR spectrum of $C_6H_6^{-}\cdot$ showing the hyperfine splitting through equal coupling with six protons.*

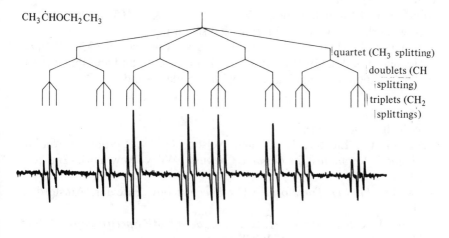

Figure 3.56 *ESR spectrum of $CH_3\dot{C}HOCH_2CH_3$. (Radical is produced when $\dot{H}$ is abstracted from diethyl ether.)*

FURTHER READING

MAIN TEXTS

Kemp, W., *NMR in Chemistry: A Multinuclear Introduction*, Macmillan, London (1986).

Abraham, R. J. and Fisher, J., *Introduction to NMR Spectroscopy*, Wiley, Chichester (1988).

Williams, D., *Nuclear Magnetic Resonance Spectroscopy*, Wiley, Chichester (1986). ACOL text.

Kalinowski, H. O., Berger, S. and Braun, S., *Carbon-13 NMR Spectroscopy*, Wiley, New York (English edn, translated by J. K. Becconsall, 1988).

Wehrli, F. W., Wehrli, S. and Marchand, A. P., *Interpretation of Carbon-13 NMR Spectra*, Wiley, New York (2nd edn, 1988).

SPECTRA CATALOGS

Proton Spectra

NMR Spectra Catalog, Volumes 1 and 2, Varian Associates, Palo Alto, California. 700 spectra.

Sadtler Handbook of Proton NMR Spectra, Sadtler, Pennsylvania, and Heyden, London (1978). 3000 spectra.

Aldrich Library of NMR Spectra, Aldrich Chemical Co., Milwaukee, 2nd edn, 2 volumes. 8500 spectra.

Carbon-13 Spectra

Johnson, L. F. and Jankowski, W. C., *Carbon-13 NMR Spectra*, Wiley, New York, and Krieger, Florida. 500 spectra.

Breitmaier, E., Haas, G. and Voelter, W., *Atlas of Carbon-13 NMR Data*, Heyden, London (1979). Carbon-13 chemical shifts and multiplicities for 3000 compounds; ongoing series with cumulative indexes.

SUPPLEMENTARY TEXTS

Levy, G. C., Lichter, R. L. and Nelson, G. L., *Carbon-13 Nuclear Magnetic Resonance for Organic Chemists*, Wiley, New York (2nd edn, 1980).

Shaw, D., *Fourier Transform NMR Spectroscopy*, Elsevier, Amsterdam (1987).

Sanders, J. K. M. and Hunter, B. K., *Modern NMR Spectroscopy*, Oxford University Press, Oxford (1987).

Derome, A. E., *Modern NMR Techniques for Chemistry Research*, Pergamon, Oxford (1987).

Freeman, R., *A Handbook of Nuclear Magnetic Resonance*, Longman, London (1988).

*Becker, E. D., *High Resolution NMR*, Academic Press, New York (2nd edn, 1980).

*Gunther, H., *NMR Spectroscopy*, Wiley, New York (1980).

*Harris, R. K., *Nuclear Magnetic Resonance Spectroscopy: A Physicochemical View*, Pitman, London (1983).

Pine, S. H., *J. Chem. Educ.* (1972), **664**. Good review of CIDNP.

Emsley, J. W. and Phillips, L., Fluorine chemical shifts, *Prog. NMR Spectr.* (1971), **7**, 1.

Crutchfield, M. M. *et al.*, Phosphorus-31 NMR, in *Topics in Phosphorus Chemistry*, Volume 5, Wiley, New York (1967).

Sorenstein, D. G., *Phosphorus-31 NMR,* Academic Press, London (1984).

Levy, G. C. and Lichter, R. L., *Nitrogen-15 NMR Spectroscopy*, Wiley, New York (1979).

Harris, R. K. and Mann, B. E. (Eds.), *NMR and the Periodic Table*, Academic Press, London (1978).

Neuhaus, D. and Williamson, M., *The Nuclear Overhauser Effect in Structural and Conformational Analysis*, VCH Publishers Inc., New York (1989).

Farrar, T. C. and Becker, E. D., *Pulse and Fourier Transform NMR*, Academic Press, New York (1971).

* These three texts treat theory very extensively, but do not include much organic stuctural interpretation.

Ultraviolet and Visible Spectroscopy

Routine UV/VIS spectrometer in the middle, with associated printer and PC for data storage and manipulation.

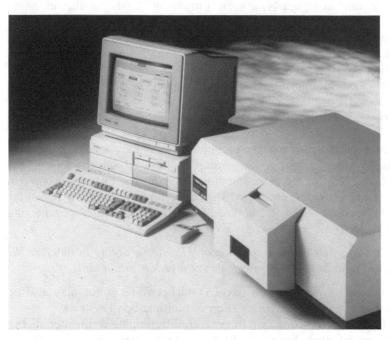

Luminescence spectrometer capable of measuring fluorescence, phosphorescence and bioluminescence.

The absorption of ultraviolet/visible radiation by a molecule leads to transitions among the electronic energy levels of the molecule, and for this reason the alternative title *electronic spectroscopy* is often preferred.

A typical electronic absorption spectrum is shown in figure 1.4, and we can see that it consists of a series of *absorption bands*: each of these bands corresponds to an electronic transition for which $\Delta E \approx 5 \times 10^5$ J mol^{-1} or 500 kJ mol^{-1}—somewhere around the bonding energies involved in organic compounds (see sections 1.2 and 1.3).

All organic compounds absorb ultraviolet light, albeit, in some instances, of very short wavelength. For practical reasons, we shall be concerned with absorptions above 200 nm (see section 4.3.4), and even with this restriction we find that most organic compounds show some ultraviolet absorption.

Historically, routine ultraviolet spectrometers were developed before infrared, NMR or mass spectrometers, and we find now that some of the ultraviolet correlations that were previously useful have long been super-seded by the later-developed techniques, so that the application of ultraviolet spectroscopy to structural investigation may appear disappoint-ingly restricted. To take a few examples, infrared spectroscopy is the method of choice for the detection of nitrile groups or carbonyl groups (in ketones, acids, amides, etc.); NMR spectroscopy will normally reveal far more information about the nature of substituents on the benzene ring than will electronic spectroscopy, etc.

The strength of electronic spectroscopy lies in its ability to measure the extent of multiple bond or aromatic conjugation within molecules.

The nonbonding electrons on oxygen, nitrogen and sulfur may also be involved in extending the conjugation of multiple-bond systems.

Electronic spectroscopy can, in general, differentiate conjugated dienes from nonconjugated dienes, conjugated dienes from conjugated trienes, αβ-unsaturated ketones from their βγ-analogs, etc. Since the degree of conjugation may suffer in strained molecules (examples are the loss of π-orbital overlap in 2-substituted biphenyls or acetophenones), electronic spectroscopy may be used to study such strain by correlating the change in spectrum with angular distortion. The position of absorption may also be influenced in a systematic way by substituents, and a particularly successful application has been the correlation of substituent shifts in conjugated dienes and carbonyl compounds.

The method rests heavily on empiricism, particularly in the study of aromatic and heteroaromatic systems, but here it can provide information completely unobtainable from any other spectroscopic technique.

Figure 4.1 introduces some of the qualitative aspects of electronic absorption spectroscopy: the spectra of series of isomers are presented, and the ease of distinction is noteworthy. These distinctions are achieved largely on the basis that the longer the conjugation the longer the maximum wavelength of the absorption spectrum.

Example 4.1

Question. Which of the isomers of pentadiene, (a) or (b), will show the longest wavelength of UV absorption? (a) CH_2=$CHCH$=$CHCH_3$; (b) CH_2=$CHCH_2CH$=CH_2.

Model answer. Isomer (a), since conjugated dienes have longer wavelengths of absorptions than do nonconjugated.

Exercise 4.1   The following are pairs of isomers, and one member of each pair absorbs ultraviolet/visible light at longer wavelength than the other: in each case state whether it is the first or second member. (a) 1,3-Hexadiene and 1,4-hexadiene; (b) 4-hept-1-enone and 4-hept-2-enone; (c) ethyl phenyl ketone (propiophenone) and benzyl methyl ketone (phenylacetone); (d) 1,2-dihydronaphthalene and 1,4-dihydronaphthalene.

Exercise 4.2   Sun-tan protective creams protect our skins from over-exposure to the sun's ultraviolet rays, and boat sails are often left rolled up and covered with a 'sacrificial' strip of material to protect the sails from deterioration in the sun. For the former, esters of glycerol and *p*-aminobenzoic acid can be used, and for the latter acrylic fibre (polyacrylonitrile) is common. Explain this.

4.1   COLOR AND LIGHT ABSORPTION—THE CHROMOPHORE CONCEPT

Compounds that absorb light of wavelength between 400 and 800 nm (visible light) appear colored to the human eye, the precise color being a complicated function of which wavelengths the compounds subtract from

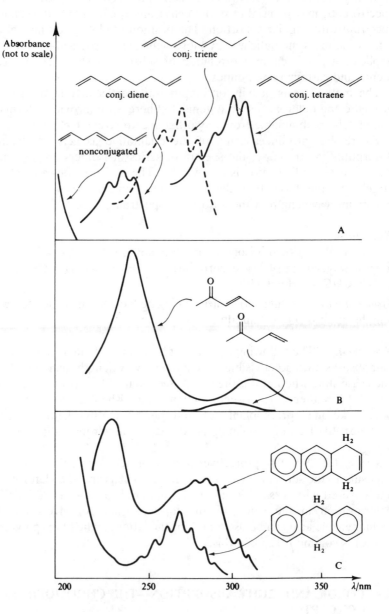

Figure 4.1 *Qualitative applications of electronic spectroscopy. (A) number of double bonds in conjugation can be determined, (B) conjugated carbonyl compounds can be distinguished from nonconjugated, (C) extent of aromatic π system can be distinguished.*

white light. Very many compounds have strong ultraviolet absorption bands, the shoulders of which may tail into the visible spectrum—absorbing the violet end of the white-light spectrum. Subtraction of violet from white light leaves the complementary colors, which appear yellow/orange to the human eye, and for these reasons yellow and orange are the most common colors among organic compounds. Progressive absorption from 400 nm upward leads to progressive darkening through yellow, orange, red, green, blue, violet and ultimately black.

Chromophores. Originally, the term *chromophore* was applied to the system responsible for imparting color to a compound. (The derivation is from the Greek *chromophoros*, or *color carrier*.) Thus, in azo dyes the aryl conjugated azo group (Ar—N=N—Ar) is clearly the principal chromophore; in nitro compounds the yellow color is carried by —NO_2; etc.

The term has been retained within an extended interpretation to imply *any functional group that absorbs electromagnetic radiation*, whether or not a 'color' is thereby produced. Thus, the carbonyl group is a chromophore in both ultraviolet and infrared terms, even though one isolated C=O group is insufficiently 'powerful' to impart color to a compound. (An isolated carbonyl group, as in acetone, absorbs ultraviolet light around 280 nm.)

Important examples of organic chromophores are listed in table 4.1, and it should be stressed again that the organic application of electronic spectroscopy is mainly concerned with the conjugated chromophores at the bottom of table 4.1, to a lesser extent with those others that give rise to absorption above 200 nm, and hardly at all with those absorbing below 200 nm.

In table 4.1 the position of the maximum point of the absorption band (λ_{max}) is given. The intensity of the band at this maximum (ϵ_{max}) is defined in section 4.2.2.

An *auxochrome* was an earlier-defined term for a group that could enhance the color-imparting properties of a chromophore without being itself a chromophore, examples being —OR, —NH_2, —NR_2, etc. As in the case of chromophores, the definition of auxochromes has been modified in the light of modern theory. The synergist effect of auxochromes is coupled with their ability to extend the conjugation of a chromophore by sharing of the nonbonding electrons: in a very real sense, the auxochrome then becomes part of a new, extended chromophore, and their action is only different in degree from the effect of extending the conjugation by adding a chromophore to a chromophore. There is no need to retain the term, and it will not be further used in this book.

Table 4.1 Some simple organic chromophores and the approximate wavelengths at which they absorb. The intensity of absorption is discussed in section 4.2.2

Chromophore	Wavelength, λ_{max}/nm (typical)	Intensity, $\epsilon_{max}/10^{-2}m^2\,mol^{-1}$ (typical)
C=C	175	14 000
C≡C	175	10 000
	195	2 000
	223	150
C=O	160	18 000
	185	5 000
	280	15
R—NO$_2$	200	5 000
	274	15
C≡N	165	5
C=C—C=C	217	20 000
C=C—C=O	220	10 000
	315	30
C=C—C≡C	220	7 500
	230	7 500
benzene	184	60 000
	204	7 400
	255	204

Example 4.2
Question. From table 4.1 state the increase in the wavelength of λ_{max} when (a) CH_2=CH_2 is joined in conjugation with C=C, to give CH_2=CH—CH=CH_2, and (b) CH_2=CH_2 is joined in conjugation with C=O, to give CH_2=CH—C=O.

Model answer. (a) λ_{max} for CH_2=CH_2 is 175 nm, and for C=C—C=C it is 217 nm, so the absorption band moves 42 nm to longer wavelength. (b) λ_{max} for the high-intensity band of C=C—C=O is 220 nm, so the shift on conjugation is 45 nm.

Exercise 4.3 Given that extending a chromophore involves additional conjugation or the involvement of additional nonbonding electrons, which of the following, as a substituent in benzene, will substantially increase the maximum wavelength of absorption in the electronic spectrum: (a) —CH=CH_2; (b) —CO—Me; (c) —CH_2COMe; (d) —$CH_2CH_2CH_3$; (e) —OH; (f) —N=NMe?

4.2 THEORY OF ELECTRONIC SPECTROSCOPY

4.2.1 ORBITALS INVOLVED IN ELECTRONIC TRANSITIONS

When a molecule absorbs ultraviolet/visible light of a particular energy, we assume as a first approximation that only one electron is promoted to a higher energy level and that all other electrons are unaffected. The *excited state* thus produced is formed in a very short time (of the order of 10^{-15} s) and a consequence is that during electronic excitation the atoms of the molecule do not move (the Franck–Condon principle).

The most probable ΔE transition would appear to involve the promotion of one electron from the highest occupied molecular orbital to the lowest available unfilled orbital, but in many cases several transitions can be observed, giving several absorption bands in the spectrum. Not all transitions from filled to unfilled orbitals are allowed, the symmetry relationship between the two orbitals being important, and this aspect is discussed further in supplement 4. Where a transition is 'forbidden', the probability of that transition occurring is low, and correspondingly the intensity of the associated absorption band is also low.

The relative energies of the molecular orbitals that most concern us are illustrated in figure 4.2. From this we can extract several generalizations.

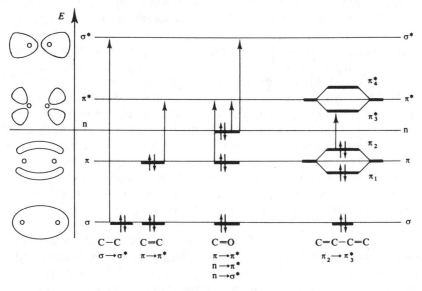

Figure 4.2 *Relative energies of orbitals most involved in electronic spectroscopy of organic compounds.*

In *alkanes* the only transition available is the promotion of an electron from a low-lying σ orbital to a high-energy σ* antibonding orbital: this is a high-energy process, and requires very-short-wavelength ultraviolet light (around 150 nm). This type of transition is classed as $\sigma \to \sigma^*$ ('sigma to sigma star').

In simple *alkenes* several transitions are available, but the lowest-energy transition is the most important: this is the $\pi \to \pi^*$ transition, which is responsible for the absorption band around 170–190 nm in unconjugated alkenes (see table 4.1).

In *saturated aliphatic ketones* the lowest-energy transition involves the nonbonding electrons on oxygen, one of which can be promoted to the relatively low-lying π^* orbital: this $n \to \pi^*$ transition is 'forbidden' in symmetry terms (see supplement 4) and therefore the intensity is low, although the wavelength is long (around 280 nm). Two other transitions available are $n \to \sigma^*$ and $\pi \to \sigma^*$: these are both 'allowed' transitions and give rise to strong absorption bands, but the energy involved is higher than for $n \to \pi^*$; therefore, the wavelength of the absorption is shorter (around 185 for $n \to \sigma^*$ and around 160 for $\pi \to \pi^*$). The most intense band for these compounds is always due to the $\pi \to \pi^*$ transition (see table 4.1).

Many factors influence the relative energies of molecular orbitals and a knowledge of these factors is the essence of an understanding of electronic spectroscopy.

The influence of solvents and substituents will be discussed later, but the most dramatic effect is brought about by conjugation: the simplest case is the conjugation of two alkene groups.

In *conjugated dienes* the π orbitals of the separate alkene groups combine to form new orbitals—two bonding orbitals named π_1 and π_2 (or Ψ_1 and Ψ_2) and two antibonding orbitals named π_3^* and π_4^* (or Ψ_3^* and Ψ_4^*). The relative energies of these new orbitals are shown in figure 4.2 and it is easily apparent that a new $\pi \to \pi^*$ transition of very low energy ($\pi_2 \to \pi_3^*$) is now possible as a result of the conjugation. Conjugated dienes, therefore, show absorption at much longer wavelength than do isolated alkene groups, a typical value being around 217 nm. Table 4.1 also shows the wavelength of absorption for a typical conjugated ketone: again the strong absorption at 220 nm is the $\pi \to \pi^*$ transition, the weaker absorption being the forbidden $n \to \pi^*$ transition.

Exercise 4.4   Assuming that figure 4.2 is reasonably to scale, assess in the case of conjugated dienes the relative energies of the following possible transitions: (a) $\pi_1 \to \pi_3^*$; (b) $\pi_2 \to \pi_4^*$; (c) $\pi_2 \to \pi_3^*$. Compare them also with (d) the $\pi \to \pi^*$ transition for unconjugated alkenes.

Absorption bands appear rather than *absorption lines*, because vibrational and rotational effects are superimposed on the electronic transitions, so that an envelope of transitions arises.

The fine structure in the absorption spectra of aromatic compounds (for example, benzene) is also associated with vibrational and rotational effects: in the vapor phase this fine structure is very marked, but in the liquid phase, or especially in polar solvents, the energies of the transitions are blurred out and the fine structure is less apparent.

4.2.2 LAWS OF LIGHT ABSORPTION—BEER'S AND LAMBERT'S LAWS

These two early empirical laws govern the absorption of light by molecules: Beer's law relates the absorption to the concentration of absorbing solute, and Lambert's law relates the total absorption to the optical path length.

They are most conveniently used as the combined Beer–Lambert law:

$$\log (I_0/I) = \epsilon c l \qquad \text{or} \qquad \epsilon = A/cl$$

where I_0 is the intensity of the incident light (or the light intensity passing through a reference cell),

I is the light transmitted through the sample solution,

$\log (I_0/I)$ is the *absorbance* (A) of the solution (formerly called the *optical density*, OD),

c is the concentration of solute (in mol l^{-1}, mol dm^{-3}),

l is the path length of the sample (in cm, m × 10^{-2}) and

ϵ is the *molar absorptivity* (formerly called the *molecular extinction coefficient*, in $l\ mol^{-1}\ cm^{-1}$, $m^2\ mol^{-1}$ × 10^{-2}).

The molar absorptivity, ϵ, is constant for a particular compound at a given wavelength, and is most commonly expressed as ϵ_{max}—the molar absorptivity at an absorption band maximum: typical values for ϵ_{max} are listed in table 4.1. Note that ϵ is not dimensionless, but is correctly expressed in units of $10^{-2}\ m^2\ mol^{-1}$. It is customary practice among organic chemists to omit the units (and tacitly to redefine ϵ) and the practice will be perpetuated in this book. Since values for ϵ_{max} can be very large, an alternative convention is to quote its logarithm (to the base 10), $\log_{10} \epsilon_{max}$.

The absorbance, A, defined as $\log (I_0/I)$, is recorded directly by all modern double-beam instruments (see section 4.3, 'Instrumentation and Sampling'.

If the relative molecular mass, M_r (molecular weight), of a compound is unknown, and consequently ϵ cannot be calculated, a convenient expression is $A = acl$; here, a is the *absorptivity* (contrast ϵ) with c being the concentration in g l^{-1}. A somewhat older expression is $E_{1\ cm}^{1\%}$ ('E one centimeter one per cent'), defined as $E_{1\ cm}^{1\%} = A/cl$, where c in this case is the concentration in g/100 cm^3. It follows that $\epsilon = 10^{-1} E_{1\ cm}^{1\%} \times M_r$.

Example 4.3

Question. An αβ-unsaturated ketone of relative molecular mass 110 has an absorption band with λ_{max} at 215 nm and ϵ 10 000. A solution of this ketone showed absorbance $A = 2.0$ with a 1 cm cell. Calculate the concentration of the ketone in this solution, expressed in grams per liter, $g \, l^{-1}$.

Model answer. Since $A = \epsilon cl$, where ϵ is the molar absorptivity, then c (the concentration in moles per liter) is given by $A/\epsilon l = 2.0/(10\,000 \times 1) = 2 \times 10^{-4}$ mol l^{-1}. We know that the relative molecular mass is 110, so the concentration can be expressed as $2 \times 110 \times 10^{-4}$ or 2.2×10^{-2} g l^{-1}.

Exercise 4.5 (a) Calculate the molar absorptivity, ϵ, for a solution containing 1.0 mmol dm^{-3} (1.0×10^{-3} moles per liter) of solute, when the absorbance of a 1 cm cell was 1.5. (b) What would be the value of A for a solution of double this concentration?

Exercise 4.6 An unknown compound (0.01 g) was dissolved in ethanol in a 250 ml standard flask, and its absorption spectrum was measured using 1 cm cells: the absorbance at the maximum of one absorption band was found to be 2.2. The compound was known to have a relative molecular mass of 186. Calculate ϵ at this λ_{max}. At another λ_{max}, A was found to be 1.2 for the same solution: calculate ϵ for this absorption band too.

Exercise 4.7 (a) A solution of benzene in ethanol showed $A = 1.4$ at 255 nm (path length 1 cm): use the data in table 4.1 to calculate the concentration of benzene in the solution (in g l^{-1}). (b) For this same solution, state the value of A at 255 nm for a path length of 2 cm.

4.2.3 CONVENTIONS

Absorption spectra in organic work are most often plotted with increasing ϵ (or log ϵ or $E^{1\%}_{1\,cm}$) on the ordinate, and with wavelength (λ) on the abscissa increasing from left to right. We must acknowledge that these are mere conventions: a few spectrometers plot absorbance 'upside-down' to the normal convention, and, especially in physical chemistry, frequency units are sometimes used on the abscissa, corresponding to λ being plotted reciprocally and from right to left.

When a change in solvent or a change in a substituent causes λ_{max} for a band to shift, we can unambiguously state whether the shift occurs *to longer wavelength* or *to shorter wavelength*. (Thus, conjugation of an alkene group causes a shift to longer wavelength of the $\pi \rightarrow \pi^*$ band.)

Some older-established terms should be mentioned here, although their use should be progressively discouraged. *Bathochromic shift* (or *red shift*) is a shift to longer wavelength (that is, toward the red end of the spectrum). *Hypsochromic shift* (or *blue shift*) is a shift to shorter wavelength. (The expression *blue shift* is most confusing, since it could be applied to a shift from 300 nm → 250 nm, a direction that is receding from the wavelength of blue light.) A *hyperchromic effect* is one that leads to increased intensity of absorption: a *hypochromic effect* is the opposite.

4.3 INSTRUMENTATION AND SAMPLING

4.3.1 THE ULTRAVIOLET–VISIBLE SPECTROMETER—DISPERSIVE, PHOTO-DIODE ARRAY AND FOURIER TRANSFORM INSTRUMENTS

Dispersive instruments
In principle, dispersive ultraviolet–visible spectrometers are similar to the infrared spectrometers discussed in section 2.4 (page 39), consisting of a light source, double beams (reference and sample beams), a mono-chromator, a detector, and amplification and recording devices. Several of the details differ, however: the source usually incorporates a tungsten filament lamp for wavelengths greater than 375 nm, a deuterium discharge lamp for values below that and a solenoid-operated mirror, which automatically deflects light from either one as the machine scans through the wavelengths. All UV/VIS spectrometers are *ratio recording*; the ratio between reference beam and sample beam intensities (I_0/I) is fed to a pen recorder (the optical null technique is not used). The recorder trace is invariably absorbance (A) against wavelength (λ), and most instruments record linearly over the range 0–3 absorbance units.

Photomultiplier detectors
The detector is commonly a photomultiplier tube. Several designs of photomultiplier are used, and specific details depend on the wavelength range to be detected, but the principle of operation depends on the *photoemissive* properties of many compounds, a simple example being beryllium oxide (BeO).

Photons of UV/VIS light, impacting on the BeO, expel several electrons for each photon absorbed. If the BeO is made the cathode in an evacuated tube, with an anode nearby, and with a voltage of 50–100 V between them, then the current (electrons) flowing from the cathode to the anode will markedly increase if the cathode is irradiated: this increase in current is proportional to the intensity of the incident light. A series of such units (called *dynodes*) can be assembled in series, such that the electrons expelled from the first cathode do not meet the anode, but are directed electrostatically to a second photoemissive cathode, from which an ever-greater stream of electrons is ejected. Repetition over up to ten dynodes generates a cascade effect, so that the final electron flow reaching the anode is amplified manyfold over the series.

The geometry within the photomultiplier which permits the cascade consists of a series of approximately ten curved plates, arranged in two sets facing each other, with each partially overlapping its neighbors, so that the electrons bounce down the line, as in an arcade game of bagatelle. A similar device, the electron multiplier, is shown in figure 5.3.

Rapid-scan ultraviolet–visible spectrometers
The spectrometer described above uses a grating monochromator to scan the wavelengths (frequencies) in succession over a detector, recording the absorbance at each wavelength as a plotted absorption spectrum.

Many analytical operations benefit from the ability to record an entire spectrum (190–900 nm) in one second, rather than in several minutes; kinetic processes, flow-cell reactions and HPLC (see section 1S.3) are examples. Two different approaches to this are available.

Linear photodiode arrays
Microelectronics allows the construction of a series of some hundreds of photodiode elements, each a few nanometers in width, arranged in a linear array approximately 1 cm long; this is the linear photodiode array (PDA or LDA).

In the spectrometer the light beam is diffracted and directed to this array, such that each of the diodes receives a discrete wavelength. The output from each diode gives the absorbance at that wavelength, and the set of diodes together allows the construction of the whole absorption spectrum. The wavelength resolution of a PDA is effectively the dimension of each photodiode, and this can range from 5 nm (for qualitative detection purposes) to 0.25 nm for high-accuracy analytical work.

Discounting the infinitesimally small delay required to compute these data in a microprocessor, the information can be examined in real time on a video display unit, or dumped in memory for subsequent reprocessing. With scan times of a second or so, spectra summation can quickly improve the sensitivity for weak absorbers or dilute solution.

One useful presentation of the data is in the form of a stack plot showing the variation in absorption spectrum with time during, for example, an HPLC separation of dyestuffs, or drugs or polycyclic aromatic hydro-carbons. Such a plot (or 'chromascan') is shown in figure 4.3 for a chromatogram of benzene, naphthalene, xylenes, biphenyl, fluorene, anthracene, fluoranthene, pyrene, chrysene, *m*-diphenylbenzene and its *para* isomer. Three dimensions can be recognized: wavelength, absorbance and time. Graphics software allows the stack plot to be rotated and viewed from any angle, or amplified along any axis of choice.

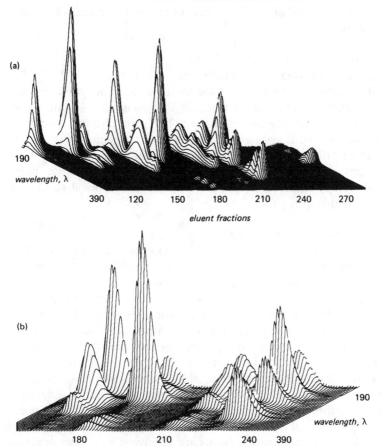

Figure 4.3 *Chromascan recorded on an HPLC separation of polycyclic aromatic hydrocarbons. Wavelength range 190–390 nm. (a) Complete chromascan and (b) part chromatogram, rotated and expanded.*

Fourier Transform ultraviolet–visible spectrometers

We saw in section 2.4.5 the use of interferometry (the Michelson interferometer) to give FTIR spectra, and similar operation can be applied to the ultraviolet–visible region of the spectrum. The differences between FTIR and FTUV spectrometers are in degree rather than in kind.

The efficiency and resolving power of the interferometer rests on the accuracy with which the mirrors can be aligned and (for the moving mirror) transported. Since the wavelengths in the ultraviolet are much smaller than those of the infrared region, the mirror geometry must be defined with that much greater precision.

The advantages of the method over dispersive instruments have already been expounded in section 2.4.5.

4.3.2 SAMPLE AND REFERENCE CELLS

For most organic work, cells of path length 1 cm are used, although 0.1 cm and 10 cm cells are commercially available: for accurate work, the sample and reference cells should be matched in optical path length. Synthetic silica and natural silica are both used, but glass absorbs strongly from 300 nm down, and is not particularly useful for organic work.

Matched silica cells are expensive and fragile, and they must be thoroughly cleaned after each use and wiped with soft tissue dipped in methanol: they must be properly stored, and the optical surfaces must never be handled.

4.3.3 SOLVENTS AND SOLUTIONS

Electronic spectra are usually measured on very dilute solutions, and the solvent must be transparent within the wavelength range being examined. Even with double-beam operation, we cannot argue that 'solvent absorption will cancel out between the sample and reference beams', since a strongly absorbing solvent will allow so little light to pass through the cells that the photomultiplier will be effectively 'blind': this causes the amplifier to produce a very noisy background on the recorded spectrum. The lower-wavelength limits for common spectroscopic-grade solvents are shown in table 4.2: below these limits, the solvents show excessive absorbance and *sample* absorbance will not be recorded linearly. The solvents are listed in table 4.2 in an approximate order of preference, and, indeed, a choice of water, ethanol and hexane will meet most requirements. Spectroscopic-grade solvents are expensive and may not always be justified: for example, absolute ethanol is now obtainable from ethylene hydration in the petrochemicals industry in sufficiently high purity for most spectroscopic use (the older benzene-azeotrope method always left traces of benzene in the ethanol, whereas the newer benzene–heptane azeotrope method produces ethanol with less than 2 ppm of residual benzene).

Solvent shifts may occur, and comparisions between spectra should only be made with this realization (see section 4.4).

Solution preparation should always be done with great care: standard solutions are prepared in volumetric flasks, the concentration usually being around 0.1 g l^{-1}. This concentration can be used to record the absorbances (and ϵ) of the weak bands, but the solution may have to be diluted to bring the intense bands on scale: if dilution is necessary, it should always be done by a known factor (for example, dilute 2 cm^3 to 20 cm^3), so that ϵ for strong bands can also be calculated. Some compounds do not obey Beer's law exactly, so that the spectra recorded at differing concentrations do not match. This is usually a minor effect, unless the compound ionizes to an increased extent at high dilutions.

Table 4.2 Solvents for electronic spectroscopy

Solvent	Lower wavelength limit/nm
water	205
ethanol (95 per cent or absolute)	210
hexane	210
cyclohexane	210
methanol	210
diethyl ether	210
acetonitrile	210
tetrahydrofuran	220
dichloromethane	235
chloroform	245
carbon tetrachloride	265
benzene	280

It is worth stressing that modern ultraviolet–visible spectrometers are extremely accurate in the measurement of absorbance, and a more significant source of error is in our ability to prepare standard solutions at low concentrations. For accurate work, all standard flasks, etc., should be of the highest analytical quality, and if dilution of the original solution is necessary, it should be carried out on a volume that can be measured with the required accuracy; too small a volume will certainly lead to dilution errors.

Replotting of the spectrum may be desirable to convert the absorbance spectrum (from the spectrometer) into the ϵ or log ϵ form: this involves calculating ϵ from concentration and absorbance figures and subsequently redrawing the spectrum on graph paper. Microprocessor programs can perform these operations routinely.

Exercise 4.8 An $\alpha\beta$-unsaturated ketone, relative molecular mass 84, has an absorption band at λ_{max} 220 nm, with ϵ_{max} 10 000 (see table 4.1 and figure 4.1). As a practice exercise, draw out four properly scaled spectra (on graph paper) as follows: (a) a plot of ϵ against λ; (b) a plot of log ϵ against λ; (c) a plot of A against λ for a solution with A_{max} 3.0 (path length 1 cm); and (d) a plot of A against λ for a solution with A_{max} 1.5 (path length 1 cm). What are the concentrations, expressed in grams per liter, needed to give rise to plots (c) and (d)?

4.3.4 VACUUM ULTRAVIOLET

The conventional electronic spectroscopic techniques outlined above cannot be used below 200 nm, since oxygen (in the air) begins to absorb strongly there. To study the higher-energy transitions below 200 nm (see table 4.1), the entire path length must be evacuated, and for this reason the region below 200 nm is usually referred to as the *vacuum ultraviolet*. The equipment for this work is more expensive, and a more exacting technique is demanded to operate it: it has mainly been exploited in studying bond energies, etc., and is not usually very helpful in organic structural determinations. The availability now of FT vacuum UV instruments may lead to greater utilization of the region.

4.4 SOLVENT EFFECTS

The position and intensity of an absorption band may shift when the spectrum is recorded in different solvents. For changes to solvents of *increased polarity* we can summarize the normal pattern of shifts as follows.

Conjugated dienes and *aromatic hydrocarbons* experience very little solvent shift.

$\alpha\beta$-*Unsaturated carbonyl compounds* show two different shifts: (1) the $\pi \to \pi^*$ band moves to longer wavelength (red shift), while (2) the $n \to \pi^*$ band moves to shorter wavelength (blue shift).

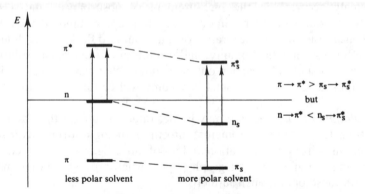

Figure 4.4 *Effect of solvation on the relative energies of orbitals and transitions in $\alpha\beta$-unsaturated carbonyl compounds. In more polar solvents $\pi \to \pi^*$ absorptions move to longer λ, $n \to \pi^*$ move to shorter λ.*

We can express this general picture in the form of an energy diagram, as in figure 4.4; solvation by a polar solvent stabilizes π, π^* and n orbitals: the stabilization of nonbonding orbitals is particularly pronounced with hydrogen-bonding solvents (such as water or ethanol), and π^* orbitals are more stabilized by solvation than are π orbitals, presumably because π^* orbitals are the more polar. The net result is as shown in figure 4.4: the energy of transition $\pi \rightarrow \pi^*$ becomes less with solvation (red shift), while the energy of transition $n \rightarrow \pi^*$ becomes greater (blue shift).

The dimensions of the shift in $\alpha\beta$-unsaturated ketones are listed in table 4.4: the maximum shift (19 nm) occurs in the $\pi \rightarrow \pi^*$ band on changing the solvent from hexane or cyclohexane to water.

4.5 APPLICATIONS OF ELECTRONIC SPECTROSCOPY— CONJUGATED DIENES, TRIENES AND POLYENES

Figure 4.1 shows the characteristic progression in the electronic spectra of dienes, trienes and tetraenes: the progression continues with increased conjugation to a limit of 550–600 nm (more than 20 double bonds in conjugation) by which stage the polyenes are strongly yellow in color. The red color of tomatoes and carrots arises from conjugated molecules of this type.

Table 4.3 Conjugated dienes and trienes (in ethanol);
λ_{max} for $\pi \rightarrow \pi^*$ transitions;
ϵ_{max} 6000–35 000 ($\times$ 10^{-2} m^2 mol^{-1})

acyclic and heteroannular dienes	215 nm
homoannular dienes	253 nm
acyclic trienes	245 nm
Addition for each substituent	
—R alkyl (including part of a carbocyclic ring)	5 nm
—OR alkoxy	6 nm
—SR thioether	30 nm
—Cl, —Br	5 nm
—OCOR acyloxy	0 nm
—CH=CH— additional conjugation	30 nm
if one double bond is exocyclic to one ring	5 nm
if exocyclic to two rings simultaneously	10 nm
solvent shifts minimal	

For dienes and trienes the position of the most intense band can be correlated in most instances with the substituents present: table 4.3 summarizes these empirical relationships (usually called the Woodward rules—since they were first enunciated by Nobel Laureate R. B. Woodward in 1941). To use the rules, it is simply a matter of choosing the correct base value and summing the contributions made by each substituent feature.

Two worked examples will illustrate the method.

I   II   III

IV

Example 4.4

Question. Predict the wavelength of an absorption band in the UV spectrum of compound I.

Model answer. For compound I the base value is 215 nm, since the two double bonds are heteroannular: there are 4 alkyl substituents (the ring residues a, b, c and the methyl group d), adding 4×5 nm: the double bond in ring A is exocyclic to ring B, adding 5 nm: the total is $215 + 20 + 5 = 240$ nm. This is within 2 nm of the observed value.

Example 4.5

Question. Predict λ_{max} of a UV absorption band for compound II.

Model answer. For compound II the base value is 253 nm, to which we add 2×5 (two ring residues a and c but not b and d) and 2×5 (both double bonds are exocyclic, to rings A and C, respectively), giving a total of 273 nm. This is within 2 nm of the observed value.

Exercise 4.9   Use the Woodward rules to calculate a λ_{max} for each of compounds III and IV.

Further examples of the application of Woodward's rules are given in chapter 6.

It is worth restating that these rules are empirically based on analogous model compounds, and will only hold good for other compounds whose structures are close to those of the models. The influence of contradictory factors is discussed in section 4.11.

It is not possible to predict ϵ_{max} other than within the range given in table 4.3: the only general rule is that longer conjugation leads to higher intensity, the *approximate* relationship in many cases being $\epsilon \propto$ (length of chromophore)^2.

4.6 APPLICATIONS OF ELECTRONIC SPECTROSCOPY —CONJUGATED POLY-YNES AND ENEYNES

Interest in these compounds arose from their being discovered in plant material, and electronic spectroscopy has been of paramount importance in structure elucidation: the number of possible permutations between triple and double bonds in polyeneynes is very large indeed, but two typical groupings are shown below:

The absorption spectrum of a typical poly-yne and a typical polyeneyne are shown in figure 4.5: as expected, the intensity and λ_{max} values increase with increasing conjugation and substitution, so that assistance in proof of structure of a new compound can be achieved by comparison of the new compound's spectrum with that of a closely related model compound.

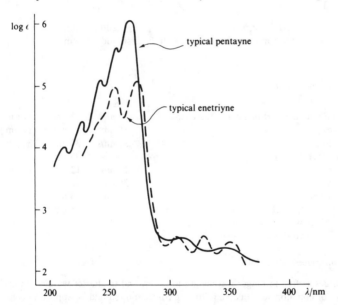

Figure 4.5 *Electronic absorption spectra typical of poly-yne and polyeneyne chromophores.*

4.7 APPLICATIONS OF ELECTRONIC SPECTROSCOPY —αβ-UNSATURATED CARBONYL COMPOUNDS

Table 4.4 shows the correlations that can apply to αβ-unsaturated carbonyl compounds: the original data were compiled by Woodward, with many extensions and verifications being added over the years.

The method of use is similar to that for conjugated dienes (table 4.3), the only two variants being (1) the distinction between α and β substituents and (2) the importance of solvent shifts (see also section 4.4).

The typical appearance of an $\alpha\beta$-unsaturated ketone spectrum is shown in figure 4.1: note that these Woodward rules calculate the position of the

Table 4.4 $\alpha\beta$-Unsaturated carbonyl compounds (in ethanol); λ_{max} for $\pi \rightarrow \pi^*$ transitions; ϵ_{max} 4500–20 000 ($\times$ 10^{-2} m^2 mol^{-1})

$$\overset{\beta}{|}\quad\overset{\alpha}{|}$$

ketones —C=C—CO— acyclic or 6-ring cyclic	215 nm
5-ring cyclic	202 nm
aldehydes —C=C—CHO	207 nm
acids and esters —C=C—CO$_2$H(R)	197 nm

Additional conjugation

$$\overset{\delta}{|}\quad\overset{\gamma}{|}\quad\overset{\beta}{|}\quad\overset{\alpha}{|}$$
—C=C—C=C—CO— etc.   add   30 nm
(If the second double bond is homoannular with the first, add   39 nm)

Addition for each substituent

	α	β	γ	δ
—R alkyl (including part of a carbocyclic ring)	10 nm	12 nm	17 nm	17 nm
—OR alkoxy	35 nm	30 nm	17 nm	31 nm
—OH hydroxy	35 nm	30 nm	30 nm	50 nm
—SR thioether	—	80 nm	—	—
—Cl chloro	15 nm	12 nm	12 nm	12 nm
—Br bromo	25 nm	30 nm	25 nm	25 nm
—OCOR acyloxy	6 nm	6 nm	6 nm	6 nm
—NH$_2$, —NHR, —NR$_2$ amino	—	95 nm	—	—
if one double bond is exocyclic to one ring		5 nm		
If exocyclic to two rings simultaneously		10 nm		

Solvent shifts (see section 4.4)

Above values shifted to longer wavelength in water, and to shorter wavelength in 'less polar' solvents. For common solvents the following corrections should be applied in computing λ_{max}

water	+8 nm
methanol	0
chloroform	−1 nm
dioxan	−5 nm
diethyl ether	−7 nm
hexane	−11 nm
cyclohexane	−11 nm

intense $\pi \to \pi^*$ transition, not the weak $n \to \pi^*$ transition. As for conjugated dienes, ϵ_{max} cannot be predicted with accuracy.

Two worked examples will illustrate the method.

3-methylpent-3-en-2-one

I   II   III   IV

Example 4.6

Question. Predict λ_{max} for a $\pi \to \pi^*$ absorption band in the UV spectrum of compound I.

Model answer. For compound I the base value is 215 nm: for one α-alkyl group add 10 nm, for one β-alkyl group add 12 nm; the total is $215 + 10 + 12 = 237$ nm. This is within 1 nm of the observed value (for which ϵ_{max} is 4600). Had the spectrum been recorded in water, λ_{max} would have moved to 245 nm.

Example 4.7

Question. Predict the λ_{max} for an absorption band in the UV spectrum of compound II.

Model answer. For compound II the base value is again 215 nm: for one α-alkyl group add 10 nm, for two β-alkyl groups add 2×12 nm, for a double bond exocyclic to two rings add 10 nm: the total is $215 + 10 + 24 + 10 = 259$ nm. This is within 3 nm of the observed value (for which ϵ_{max} is 6500). Had the spectrum been recorded in hexane, λ_{max} would have moved to 248 nm.

Exercise 4.10 Use the Woodward rules to calculate a λ_{max} for each of compounds III and IV.

In general, the same reservations apply to the predictions of $\alpha\beta$-unsaturated ketone absorptions as apply to conjugated dienes (see section 4.5).

4.8 APPLICATIONS OF ELECTRONIC SPECTROSCOPY —BENZENE AND ITS SUBSTITUTION DERIVATIVES

It has already been stated (in section 4.2.1) that the electronic absorption spectrum of benzene shows a great deal of fine structure in the vapor phase; less is seen in hexane solution, and less still in ethanol solution.

A similar trend takes place when benzene is substituted—simple alkyl substituents shift the absorptions slightly to longer wavelength, but do not destroy the fine structure, while nonbonding pair substituents (—OH, —OR, —NH$_2$, etc.) shift the absorptions more substantially to longer wavelength and seriously diminish (or wholly eliminate) the fine structure.

It is only very occasionally necessary to rely on the electronic spectrum to infer structural relationships in substituted benzenes: additionally, the empirical rules that have been compiled for calculating their λ_{max} values are often ambiguous, and do not permit predictions concerning the number of bands present or their intensity. Given the enormous number of possible substitution patterns, the only consistently valid approach to predicting the electronic spectrum of anything but the simplest benzene derivative is to record the electronic spectrum of a suitable model compound or to find such a spectrum in the literature (see 'Further Reading'). It would be unwise to extrapolate substantially from the model, but the following indicators are worth recording.

1.   The most 'influential' combination of substituents is a −M group *para* to a +M group, so that nonbonding pair donation (by the +M group) is effectively complementary to the electron-withdrawing −M group: the shift in λ_{max} is greater than the sum of the individual shifts. (Examples are *p*-nitroaniline and *p*-nitrophenol.)
2.   Usually a −M group *ortho* or *meta* to a +M group produces merely a small shift from that of the isolated chromophores.
3.   Altering the nonbonding pair 'availability' will alter the position of λ_{max}.

Exercise 4.11   When *p*-nitrophenol is dissolved in water, the color is yellow, but when NaOH is added, the color deepens in intensity and moves to longer wavelength. Explain this.

Exercise 4.12   When *p*-aminophenol is dissolved in water, λ_{max} is at longer wavelength than in acid solution. Explain this.

4.9   APPLICATIONS OF ELECTRONIC SPECTROSCOPY —AROMATIC HYDROCARBONS OTHER THAN BENZENE

Figure 4.6 shows the characteristic development of ultraviolet–visible absorption as benzene rings are fused together in the linear series benzene–naphthalene–anthracene or angularly in the series benzene–naphthalene–phenanthrene. The fusion of additional rings leads to more and more complex spectra (see that of pyrene in figure 4.6), *but these spectra are uniquely characteristic of each aromatic chromophore*, so that it is a very simple matter to compile a spectra catalog of aromatic hydrocarbons for identification purposes. The introduction of alkyl groups has little influence on λ_{max} or ϵ_{max}, so that (for example) methylanthracenes are readily identified as possessing the anthracene chromophore from their electronic spectra, etc.

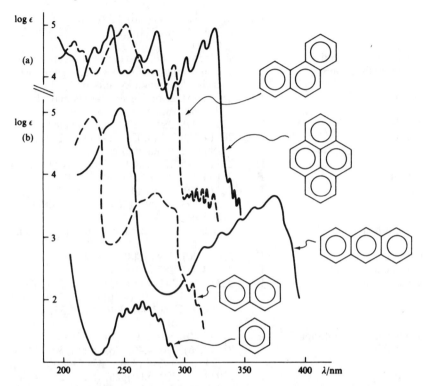

Figure 4.6 *Electronic absorption spectra of typical polynuclear aromatic hydrocarbon chromophores. All spectra recorded in hexane. For clarity the upper set* (a) *(phenanthrene and pyrene) has been displaced upward on the ordinate from set* (b) *(benzene, naphthalene and anthracene). The wavelength scale for both sets is identical.*

Table 4.5 lists the λ_{max} and ϵ_{max} values for many common hydrocarbon systems.

Exercise 4.13 The electronic spectra of a series of polycyclic aromatic hydrocarbons (PAHs) had λ_{max} values as follows: identify these hydrocarbons unequivocally from table 4.5. (Note that simple alkyl substituents would not interfere with the identification of the hydrocarbon chromophore; look up any structures which are unknown to you.) (a) 231, 241, 251, 262, 272, 292, 305, 318, 334, 352, 362, 372; (b) 236, 276, 287, 309, 323, 342, 359; (c) 229, 264, 274, 311, 323, 334, 340, 440, 468.

Table 4.5   Principal maxima in the electronic spectra of aromatic hydrocarbons[a]

Hydrocarbon	Solvent	Principal maxima
benzene	E	229 (1.21), 234 (1.46), 239 (1.76), 243 (2.00), 249 (2.30), 254 (2.36), 260 (2.30), 268 (1.04)
toluene, xylenes and trimethylbenzenes	E	similar appearance to benzene spectrum; but peaks move to longer λ and higher ϵ with each additional alkyl group
biphenyl	E	250 (4.15); 2- or 2'-substituents may change the spectrum considerably
binaphthyls	H	220 (5.00), 280 (4.20)[b]; see biphenyl
indene	H	209 (4.34), 221 (4.03), 249 (3.99), 280 (2.69), 286 (2.35); many inflections
styrene	H	247 (4.18), 273 (2.88), 283 (2.87), 291 (2.76)
fluorene	E	261 (4.23), 289 (3.75), 301 (3.99)
naphthalene	E	221 (5.00), 248 (3.40), 266 (3.75), 275 (3.82), 285 (3.66), 297 (2.66), 311 (2.48), 319 (1.36)
acenaphthene	E	similar in appearance to naphthalene spectrum
trans-stilbene	E	230 (4.20), 299 (4.48), 312 (4.47)
cis-stilbene	E	225 (4.34), 282 (4.10)
phenanthrene	E	223 (4.25), 242 (4.68), 251 (4.78), 274 (4.18), 281 (4.14), 293 (4.30), 309 (2.40), 314 (2.48), 323 (2.54), 330 (2.52), 337 (3.40), 345 (3.46)
phenalene (perinaphthene)	E	234 (4.40), 320 (3.90), 348 (3.70)
fluoranthene	E	236 (4.66), 276 (4.40), 287 (4.66), 309 (3.56), 323 (3.76), 342 (3.90), 359 (3.95)
chrysene	E	220 (4.56), 259 (5.00), 267 (5.20), 283 (4.14), 295 (4.13), 306 (4.19), 319 (4.19), 344 (2.88), 351 (2.62), 360 (3.00)
pyrene	E	231 (4.62), 241 (4.90), 251 (4.04), 262 (4.40), 272 (4.67), 292 (3.62), 305 (4.06), 318 (4.47), 334 (4.71), 352 (2.82), 362 (2.60), 372 (2.40)
anthracene	E	252 (5.29), 308 (3.15), 323 (3.47), 338 (3.75), 355 (3.86), 375 (3.87)
perylene	E	245 (4.44), 251 (4.70), 387 (4.08), 406 (4.42), 434 (4.56)
acenaphthylene	H	229 (4.72), 264 (3.46), 274 (3.43), 311 (3.93), 323 (4.03), 334 (3.70), 340 (3.70), 440 (2.00), 468 (1.56)

[a] Values quoted are for λ_{max} in nm, with log ϵ_{max} in parentheses. Solvents used were either hexane (H) or ethanol (E).
[b] 2,2'-Binaphthyls are different: 255 (4.9), 320 (4.4).

4.10 APPLICATIONS OF ELECTRONIC SPECTROSCOPY —HETEROCYCLIC SYSTEMS

Once again the most successful approach to the electronic spectra of heterocyclic systems has been an empirical one, coupled with a few guidelines on substituent effects.

Table 4.6 lists the principal λ_{max} and ϵ_{max} values for common heterocyclic systems.

Table 4.6 Principal maxima in the electronic spectra of some common heterocyclic systems[a]

Compound	Solvent	Principal maxima
pyrrole	E	235 (2.7, shoulder); no sharp maxima
furan	H	207 (3.96)
thiophene	H	227 (3.82), 231 (3.85), 237 (3.82), 243 (3.58)
indole	H	220 (4.42), 262 (3.80), 280 (3.75), 288 (3.61)
carbazole	E	234 (4.63), 244 (4.38), 257 (4.29), 293 (4.24), 324 (3.55), 337 (3.50)
pyridine	H	251 (3.30), 256 (3.28), 264 (3.17)
quinoline	E	226 (4.53), 230 (4.47), 281 (3.56), 301 (3.52), 308 (3.59)
isoquinoline	H	216 (4.91), 266 (3.62), 306 (3.35), 318 (3.56)
acridine	E	249 (5.22), 351 (4.00)
pyridazine	H	241 (3.02), 246 (3.15), 251 (3.15), 340 (2.56)
pyrimidine	H	242 (3.31), 293 (2.51), 307 (2.40), 313 (2.18), 317 (2.04), 324 (1.73)
pyrazine	H	254 (3.73), 260 (3.78), 267 (3.57), 315 (2.93), 322 (2.99), 328 (3.02)
barbituric acids	water	256–257 ($\approx$ 4.4, concentration-dependent)

[a] Values quoted are for λ_{max} in nm, with log ϵ_{max} in parentheses. Solvents used were usually hexane (H) or ethanol (E); change of solvent may affect the spectrum considerably.

Simple alkyl substituents, as usual, have little effect on the spectra, but polar groups (electron donors or attractors) can have profound effects, which are usually highly dependent on substitution position in relation to the heteroatom(s). The possibility that tautomeric systems may be generated should also be borne in mind, the classic case here being the 2-hydroxypyridines, which tautomerize almost entirely to 2-pyridones, with substantial changes in the electronic spectra.

4.11   STEREOCHEMICAL FACTORS IN ELECTRONIC SPECTROSCOPY

The empirical rules met until now are much honored in the breach, a variety of reasons being responsible for the failures: consistently, stereochemical reasons are to be blamed, and it is worth taking a few examples to illustrate this. In all cases it can be shown that angular strain or steric overcrowding has distorted the geometry of the chromophore, so that, for example, conjugation is reduced by reducing the π-orbital overlap, etc.

4.11.1   BIPHENYLS AND BINAPHTHYLS

Biphenyl (I) is not completely planar (the two rings being at an angle of approximately 45°) and in 2-substituted biphenyls (II) the two rings are pushed even further out of coplanarity: the result is that diminished π-orbital overlap in the 2-substituted derivatives leads to shifts to shorter λ and diminished intensity in their electronic spectra.

Thus, biphenyl (I) has λ_{max} 250 (ϵ, 19 000), while 2-methylbiphenyl (II) has λ_{max} 237 (ϵ, 10 250). Adding more methyl groups is complicated by the effect of the methyl groups themselves (shifts to longer λ), but an interesting comparison can be made between the hexamethylbiphenyl (III) and mesitylene (1,3,5-trimethylbenzene, IV): these two exhibit the same λ_{max} (266 nm) and their ϵ_{max} values are 545 and 260, respectively (in ethanol).

The 1,1'-binaphthyls cannot be coplanar (and indeed are enantiomeric) and exhibit electronic absorption at much shorter wavelength than for naphthalene itself (see table 4.5): in 2,2'-binaphthyls, with much less overcrowding, the λ_{max} values are nearer those expected from the benzene-biphenyl analogy.

I   II   III   IV

4.11.2   cis and trans ISOMERS

Where an alkene chromophore is capable of geometrical isomerism, it is normally found that the *trans* isomer exhibits longer-wavelength absorption (and higher intensity) than the *cis* isomer: this can be rationalized in terms of the more effective π-orbital overlap possible in the *trans* isomer, and is illustrated well in the case of *cis*- and *trans*-stilbenes (see table 4.5).

4.11.3 ANGULAR DISTORTION AND CROSS-CONJUGATION. STERIC
        INHIBITION OF RESONANCE

The Woodward rules for conjugated dienes and carbonyl compounds give reliable results only where there is an absence of strain around the chromophore: thus, the rules are successful for acyclic and (most) six-membered ring systems. Well-authenticated violations are abundant, whether the chromophore distortion is engendered by ring strain or by the introduction of additional conjugation other than at the end of the chromophore (cross-conjugation). Examples of systems whose electronic spectra could not satisfactorily be predicted by the Woodward rules are shown below:

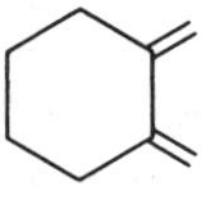

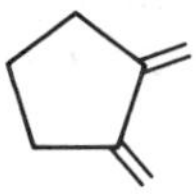

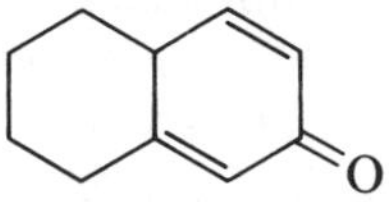

So sensitive is electronic spectroscopy to distortion of the chromophore that one can turn this to advantage in *demonstrating* that distortion is present in the molecule. In molecules IV, V and VI, the observed λ_{max} values are lower than calculated, thus demonstrating the influence of steric inhibition of resonance in the conjugation; further examples of steric inhibition of resonance arise in substituted biphenyls and binaphthyls (see section 4.11.1).

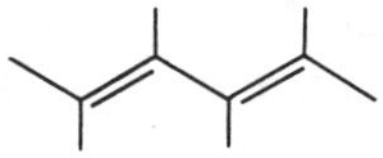

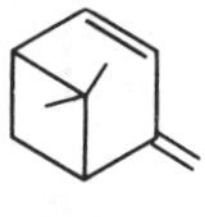

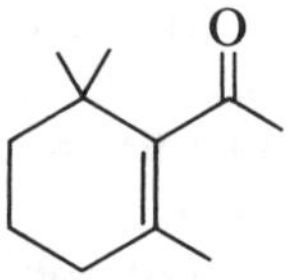

SUPPLEMENT 4

4S.1 QUANTITATIVE ELECTRONIC SPECTROSCOPY

The high sensitivity of electronic spectroscopy, and the ease and accuracy with which quantitative work can be carried out, make it a valuable analytical method—limited principally by the need for a chromophore in the system under analysis.

From the Beer–Lambert law (see section 4.2.2), expressed in the form $\epsilon = A/cl$, we can see that the concentration of a chromophoric molecule can be measured in solution, provided that we know the value of ϵ: the path length (l) is the cell dimension, and absorbance (A) is obtained as log (I_0/I) from the spectrometer.

A simple example of an analytical problem would be the estimation of a mixture of anthracene and naphthalene. Figure 4.6 shows that in ethanol anthracene absorbs at λ_{max} 375 nm (log ϵ, 3.87), well clear of any naphthalene absorption. We can prepare a standard solution of the anthracene–naphthalene mixture and measure the absorbance at 375 nm: from the Beer–Lambert law the concentration of pure anthracene can be calculated and, hence, the proportion of anthracene in the mixture.

Direct measurement of the naphthalene concentration is complicated by the fact that, at all λ_{max} for naphthalene, part of the absorbance will be due to a contribution from anthracene. The procedure would be to choose a suitable λ_{max} for naphthalene (for example, λ_{max} 285 nm), and measure *on the anthracene spectrum the value of ϵ for anthracene* at this wavelength. (From figure 4.6, log ϵ at 285 nm for anthracene is 2.2.) Since we already know the concentration of anthracene, and now also ϵ at 285 nm, we can *calculate* the absorbance due to anthracene (A_a) at 285 nm (for a given mixture solution). We can *measure* the total absorbance at 285 nm for the same solution (A_t) and hence calculate the absorbance due to naphthalene at 285 nm (A_n), since $A_n = A_t - A_a$. From A_n we can then calculate the concentration of naphthalene in the solution and, hence, the proportion of naphthalene in the mixture.

This latter technique must be used if (say) anthracene and naphthalene are present in a mixture with other nonabsorbing constituents, so that the proportion of naphthalene cannot be calculated simply by difference.

Simple extensions of the two basic analyses outlined above, with the ever-desirable compiling of calibration curves for more complex cases, have been used to meet the vast majority of quantitative applications.

Exercise 4.14 Modern spectrometers can construct a *working curve* from a series of solutions of known concentration. On graph paper manually prepare a working curve (a plot of absorbance at a fixed wavelength against known concentration of standards) from the following data points: (1) $A = 0$ at $c = 0$ g/l; (2) $A = 0.5$ at $c = 0.01$ g/l; (3) $A = 1.0$ at $c = 0.02$ g/l; (4) $A = 1.5$ at $c = 0.03$ g/l; (5) $A = 2.0$ at $c = 0.04$ g/l. From the working curve, calculate the concentrations of two solutions whose measured absorbances at the same wavelength were (a) 1.8 and (b) 0.7.

4S.2 FLUORESCENCE AND PHOSPHORESCENCE

While anthracene is 'colorless', in the sense that its electronic absorption spectrum lies wholly within the ultraviolet region (see table 4.5), pure samples of anthracene viewed in ultraviolet light give off a blue visible light, which we call fluorescence.

Fluorescence is light that is emitted from a molecule after the molecule has absorbed light of a different (and shorter) wavelength.

One characteristic feature of fluorescent radiation is that the fluorescence stops whenever the irradiating light is removed. A related phenomenon, phosphorescence, arises when molecules continue to emit the longer-wavelength radiation even after the exciting radiation has been removed.

The explanation of fluorescence and phosphorescence can best be viewed from a study of figure 4.7: this represents the one-electron excitation process discussed earlier (in section 4.2) but introduces additional nomenclature and two additional considerations.

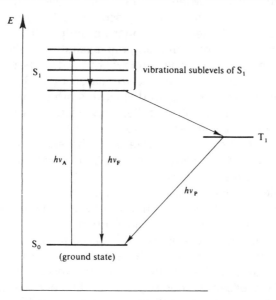

Figure 4.7 *Mechanisms of fluorescence and phosphorescence. The electronic transitions $h\nu_A$, $h\nu_F$ and $h\nu_P$, are, respectively, absorption, fluorescence and phosphorescence.*

The absorption of light, $h\nu_A$, by the ground-state molecule leads to promotion of one electron from a ground-state MO to one of the vibrational sublevels of the first excited-state MO, S_1 (or to some higher electronic excited state S_2, S_3, etc.). This rapidly deactivates in solution, by a radiationless process, to the *lowest vibrational sublevel of* S_1.

For the electron to return from S_1 to the ground state, it must emit radiation ($h\nu_F$), which is clearly of lower energy and thus of longer wavelength than $h\nu_A$: the *fluorescent radiation* corresponds to $h\nu_F$.

Exercise 4.15 In Example 1.3 we saw that UV light of 200 nm wavelength corresponds to energy absorbed by a molecule of 600 kJ mol^{-1}, while visible light must have a wavelength above 400 nm. Calculate for a hypothetical molecule the corresponding energy emitted in fluorescence, assuming a fluorescent excitation of 200 nm and a fluorescent emission of 450 nm, and hence calculate how much energy is 'lost' by deactivation in the vibrational sublevels.

An alternative path involves loss of energy from S_1 to T_1: the principal difference between S_1 and T_1 is the electron spin orientation. For the two electrons originally occupying the ground-state MO under consideration, their spins must be antiparallel, following the Pauli exclusion principle: the original excitation, followed by the decay to the lowest vibrational sublevel of S_1, does not alter the spin of the promoted electron, but the transition S_1 to T_1 does. Energy states containing only spin-paired electrons are called *singlet states* (S_0, S_1, S_2), while those with parallel spin electrons are called *triplet states* (T_1, etc.). Triplet states are more stable than singlet states, and are longer-lived: they may survive after the exciting radiation has been removed, and *thereafter* decay to S_0 by emitting the radiation $h\nu_P$, the *phosphorescent radiation*.

For complex molecules such as anthracene, several fluorescent bands are emitted, constituting the fluorescence spectrum, which is therefore, in origin, not unlike the Raman spectrum discussed in section 2S.3: anthracene will emit this same fluorescence spectrum even though the wavelength of the excitation is varied within quite wide limits (although the intensity of the fluorescence spectrum will vary). The fluorescence spectrum for anthracene consists of four maxima, with λ_{max} 380 nm, 400 nm, 420 nm and 450 nm: this brings the emission into the blue region of the visible spectrum, and therefore anthracene fluoresces blue.

Important uses of the fluorescence phenomenon are represented by the molecules of fluorescein and the 'optical brightener' derivative of 4, 4'-diaminostilbene (also called a 'fluorescent brightening agent' or 'optical bleach').

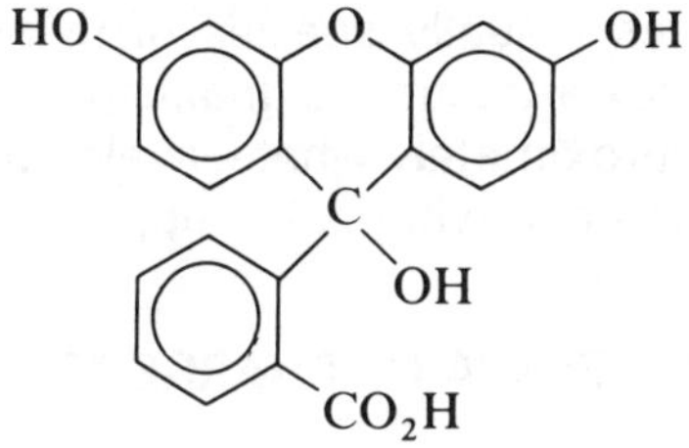

fluorescein
(acid form)

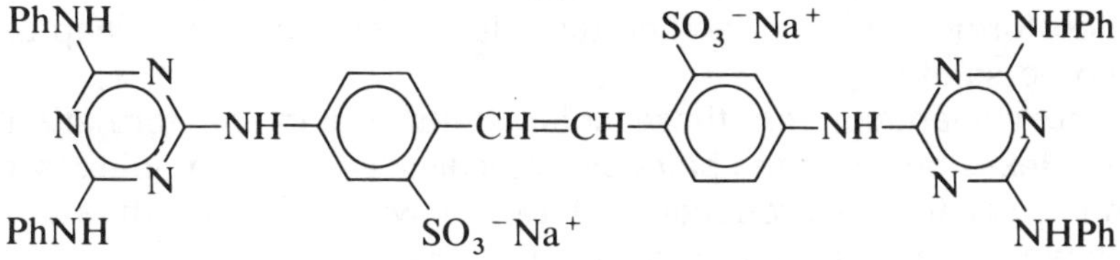

an optical brightener

The intense green fluorescence of aqueous fluorescein solutions makes it an excellent material to add to water systems for leak detection, and an excellent 'marker' for sea rescue operations, etc.

Minute quantities of optical brighteners are added to detergents, and are retained on the fabrics after washing: in sunlight (or other ultraviolet source) they fluoresce blue and add brightness (and whiteness?) to the fabric.

Other applications of fluorescence depend on its extremely low detection limit: it is used in polymer chemistry to detect and identify plasticizers, and in the study of impurities (for example, oxidation impurities in polyethylene and polypropylene). Fluorescent material dissolved in solution or in solid plastic bases can also detect radioactive decay: this is the foundation of *scintillation counting* of β-emitters, etc. Biological applications include the study of the three-dimensional tertiary structure of proteins by measuring the proximity of known fluorescent groups within the protein: these fluorescent groups can be either aromatic amino acids or specifically added fluorescent labels.

(It might be interesting to note in contrast that the 'phosphorescence' seen at night in the sea is due to several species of marine micro-organism, which, when agitated by a bow-wave or an oar-splash, undergo an enzymatic alarm reaction which liberates energy in the form of green light. This phenomenon is more correctly called *bioluminescence*.)

Exercise 4.16 Fireflies (usually species of flying beetles) use bio-luminescence as a sex-attracting signal, and commonly greenish light is emitted: at approximately what wavelength is this? If red light is used in a similar context, what is the approximate wavelength?

4S.3   ABSORPTION SPECTRA OF CHARGE-TRANSFER COMPLEXES

Solutions of iodine in hexane are violet in color, while in benzene they are brown: if tetracyanoethylene (TCNE; colorless) is added to a chloroform solution of aniline (colorless), the result is a deep-blue-colored solution.

The explanation for these color shifts lies in the formation of complexes between the pairs of molecules, and this complex forma-tion leads to the production of two new molecular orbitals and, consequently, to a new electronic transition.

Perhaps the best-known of such complexes are the picrates of aromatic hydrocarbons, ethers and amines; these picrates are usually sufficiently stable to be isolable as crystalline material, although some picrates are only stable in solution.

Since the formation of these complexes involves transfer of elec-tronic charge from an 'electron-rich' molecule (a Lewis-base donor) to an 'electron-deficient' molecule (a Lewis-acid acceptor) they are called *charge-transfer complexes*: common donors and acceptors are shown below:

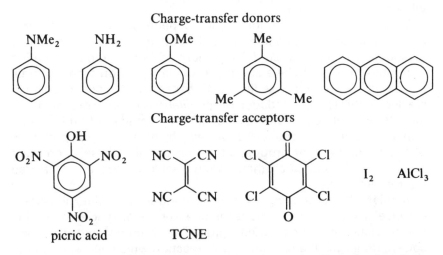

Charge-transfer donors

picric acid   TCNE

Bond formation between the molecular pairs is brought about when filled π orbitals (or nonbonding orbitals) in the donor overlap with depleted orbitals in the acceptor. The two new molecular orbitals formed are illustrated in figure 4.8: the lower-energy MO for the complex is occupied in the ground state, and transitions from this MO to the new upper MO are responsible for the new absorption bands formed.

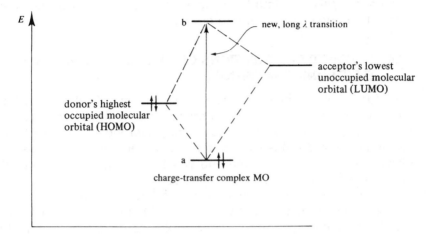

Figure 4.8   *Electronic transitions for charge-transfer complexes. Donor and acceptor orbitals combine to form two new orbitals (a and b) for the complex. New electronic transitions of long λ are then possible between a and b.*

The structure of most charge-transfer complexes can be visualized as a face-to-face association on a 1:1 donor:acceptor basis: only thus, for example, can maximum overlap of aromatic π orbitals take place. This kind of structure is difficult to draw, and most representations use one or other of the conventions shown below:

From the large number of possible donors and acceptors, a very large number of charge-transfer complexes can be formed, and it would be impossible to discuss their electronic-absorption spectra. The benzene–iodine and aniline–TCNE cases are representative; λ_{max} values for aniline and TCNE are 280 nm and 300 nm, respectively, while the deep-blue complex has λ_{max} well into the visible at 610 nm. In the benzene–iodine case λ_{max} for benzene is 255 nm, while for molecular iodine in hexane λ_{max} is in the visible around 500 nm: the charge-transfer complex has an intense additional band around 300 nm, but this tails into the visible region and modifies the violet color of the molecular iodine to brown.

4S.4 SYMMETRY RESTRICTIONS ON THE ALLOWEDNESS OF ELECTRONIC TRANSITIONS

Even for simple organic molecules, it is usually difficult to calculate the energies of the molecular orbitals that might be involved in electronic transitions and, hence, in the absorption of ultraviolet or visible light: where this can be done, it is then necessary to decide which transitions are allowed and which are forbidden, before the principal electronic absorption bands can be predicted.

The allowedness of a transition is associated with the relationship between the geometries of the lower- and higher-energy molecular orbitals and the symmetry of the molecule as a whole, but the mathematics of group theory is an essential prerequisite to a full exposition of the restrictions that apply. We can, however, state a few qualitative generalizations, which are indicators of the symmetry rules that must be observed.

Symmetrical moelcules (that is, molecules with a high degree of symmetry) have more restrictions on their electronic transitions than have less symmetrical molecules. As an example, benzene is highly symmetrical, having a large number of symmetry elements: many restrictions apply to the electronic transitions of the benzene molecule, and therefore its electronic absorption spectrum is simple.

For a *totally* unsymmetrical molecule, the only symmetry operation that can be performed is the identity operation (which changes nothing). For such a molecule, no symmetry restrictions apply to the electronic transitions, so that transitions may be observed among *all* of its molecular orbitals except, of course, among filled orbitals: a complex electronic absorption spectrum will result.

Between these two extremes lies the majority of organic compounds that absorb light in the ultraviolet–visible region. For any of these compounds, to decide whether a transition between two given molecular orbitals is allowed or forbidden, and will give rise to ultraviolet–visible absorption, we must consider (a) the geometry of the ground-state MO, (b) the geometry of the excited-state MO and (c) the orientation of the electric dipole of the incident light that might induce the transition (when referred to the same coordinates). Provided that these three have an appropriate symmetry relationship, the transition will be allowed.

4S.5 OPTICAL ROTATORY DISPERSION AND CIRCULAR DICHROISM

4S.5.1 Definitions and nomenclature

The optical activity of a compound is its ability to rotate the plane of polarized light, and we define the *specific rotation*, $[\alpha]$, as $[\alpha] = 100\alpha/lc$, where α is the observed rotation, l is the length of polarimeter tube (in dm) and c is the concentration (in g/100 cm^3). Since $[\alpha]$ may vary with temperature and with the wavelength of light used, these two must also be specified; routine measurements use the sodium D line at 589 nm, so that for data at 20 °C we would quote specific rotation as $[\alpha]_D^{20}$. The solvent used must also be quoted, and for accurate work it is better also to quote the concentration used.

The fact that $[\alpha]$ varies with wavelength is an important one: the phenomenon is named *optical rotatory dispersion* (ORD), and ORD curves play an important role in structure elucidation in optically active compounds.

Figure 4.9 shows that we can regard plane polarized light as the resultant of two equal and opposite beams of circularly polarized light, and this leads to two derivations.

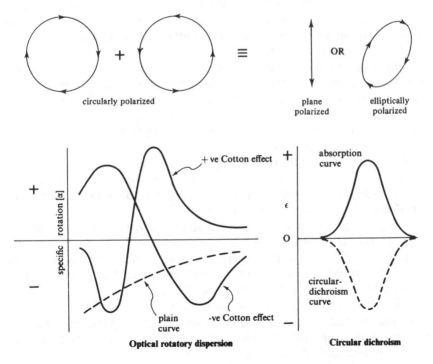

Figure 4.9 *Optical rotatory dispersion and circular dichroism curves.*

i. If one of the circularly polarized components passes through the medium more slowly than the other, *the plane of polarization will be rotated.* (For this to arise, the refractive index of the medium for one circularly polarized component must be different from that for the other, the phenomenon being named *circular birefringence.*)

ii. Whenever we find circular birefringence, we also find that the two circularly polarized components are differentially absorbed by the medium: this leads to the emergent beam having an imbalance between the strengths of the two circularly polarized beams, so that the emergent beam is not truly plane polarized but *elliptically polarized* (see figure 4.9). We name this phenomenon *circular dichroism* (CD).

The combined phenomena of circular birefringence and circular dichroism are named the *Cotton effect.*

4S.5.2   *Cotton effect and stereochemistry*

The ORD curves for many compounds show typically the *plain curve* variance indicated in figure 4.9: but if the compound has an electronic absorption band in the region under study, we obtain an anomalous ORD curve, called a *Cotton effect* curve (figure 4.9). For example, if we measure the ORD curve for an optically active ketone in the region of the $n \rightarrow \pi^*$ transition band ($\approx$ 280 nm), we shall see an anomalous curve. As shown in figure 4.9, the Cotton effect can be positive or negative, depending on whether a peak or a trough is met first in going from long to short λ.

The shape of the Cotton effect curves (positive or negative) can be correlated empirically with very many features of the stereochemistry around the chromophore, and while most work has centered on the weak $n \rightarrow \pi^*$ chromophore of carbonyl compounds, the Cotton effect can be extended to the study of any optically active molecule that possesses a chromophore.

The $\pi \rightarrow \pi^*$ transitions (of conjugated ketones, etc.) are of much higher intensity than are $n \rightarrow \pi^*$ transitions and are, consequently, more difficult to study, and the multiple Cotton effect curves observed make interpretation more complex.

Anomalous curves in ORD have their counterpart in circular dichroism curves (see figure 4.9): although it is difficult to record the variation of CD with wavelength, especially near the chromophores that absorb in the vacuum ultraviolet, equipment that renders this region accessible down to 165 nm is available.

Chromophores successfully studied include aromatic systems, heteroaromatic systems and those groups that have absorptions only in the less accessible vacuum ultraviolet (such as OH groups in alcohols and carbohydrates).

Many examples of CD have also been recorded in the infrared region.

4S.5.3 The octant rule

The sign of the Cotton effect in ORD and CD curves can be computed empirically from a knowledge of the absolute stereochemistry of substituents around the chromophore: conversely, the absolute stereochemistry of substituents can be deduced from the sign of the Cotton effect. The enormous potential in this has been fully exploited in a wide range of molecules, but we can illustrate the principle of the method by reference to the *ketone octant rule*.

If an organic molecule is placed at the origin of three-dimensional orthogonal axes, the three orthogonal planes will cut the molecule into eight parts, each lying in an 'octant' defined by the intersecting coordinate planes (see figure 4.10). For cyclohexanones the molecule is placed as shown, with the C=O group along the z axis, the xy plane bisecting the C=O bond and the yz plane bisecting C-1 and C-4. The molecule is now viewed in projection along the z axis, and except in rare cases the molecule can be seen to occupy only the four rear octants.

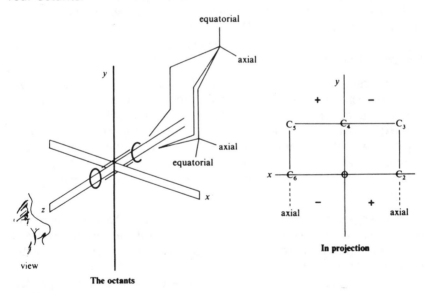

Figure 4.10 *The octant rule for ketones.*

The octant rule, applied to cyclohexanones, states that (a) substituents lying *on the coordinate planes* have no influence on ORD, (b) substituents lying in the *far upper left* and *far lower right* octants make a positive contribution to the sign of the Cotton effect: the other substituents make a negative contribution (see figure 4.10).

3-Methylcyclohexanone is an optically active ketone, and therefore can be treated by the octant rule. Write out the possible conformations of the dextrorotatory and levorotatory forms (four forms in all) and decide which will exhibit positive or negative Cotton effects. Given that $(+)(R)$-3-methylcyclohexanone exhibits a positive Cotton effect, what is the preferred conformation of this isomer?

4S.6 ELECTRON SPECTROSCOPY FOR CHEMICAL ANALYSIS (ESCA)—X-RAY PHOTOELECTRON SPECTROSCOPY (XPS)

Electronic absorption spectra in the ultraviolet–visible region arise when electrons in n and π orbitals undergo upward transitions on bombardment by light: at higher energies (in the vacuum ultraviolet) transitions from σ orbitals are also observed. At still higher energies (using X-rays for bombardment) we can induce the nonvalency *core electrons* of the atoms to undergo transitions: thus, if we bombard an organic molecule with $MgK\alpha_{1,2}$ or $AlK\alpha_{1,2}$ radiation, electrons will be ejected from the carbon 1s orbitals, and the spectrum of energies in these ejected electrons can be recorded as a direct measure of the *core binding energies* for C 1s orbitals.

The bases on which this *electron spectroscopy* is used *for chemical analyses* (ESCA) are

i. Different elements have different 1s binding energies, and we can detect the presence of each element in a molecule from the ESCA spectrum (from a knowledge of these binding energies).

ii. Atoms of the same element, which are in different environments within a molecule, will have different core binding energies: this means that a *chemical shift* phenomenon exists in ESCA, and in organic molecules the different carbon atoms will give rise to separate signals (analogous to ^{13}C NMR shifts).

As a very simple example, CH_3CHO gives rise to two signals for the two carbon atoms, separated by 2.6 eV: therefore, the difference in C 1s core binding energies in CH_3CHO is 2.6 eV. A third signal arises from the oxygen atom.

The development of ESCA, though recent, has been rapid, particularly in its application to the study of surfaces: improved X-ray monochromators can now give much enhanced resolution, although the wide natural line-width (of the order of 1.0 eV) presents an upper limit to this. The method (also called *X-ray photoelectron spectroscopy*, XPS) is in some ways complementary to NMR and to other photoelectron spectroscopies, such as Auger.

As in NMR, it is usually more important to an organic chemist to measure relative shifts than to measure the absolute core binding energies, but these latter can be measured and correlated with theoretical values. Typical C 1s binding energies are around 290 eV: within hydrocarbons the chemical shifts are small, while in carbocations major shifts can be observed. Shifts have been correlated with electronegativity of substituent, it being found that increasing electronegativity (C, Br, N, Cl, etc.) leads to increases in the core-binding energies. It becomes possible, therefore, to deduce structural information from ESCA spectra as from NMR spectra.

The time scale of ESCA is very short: the lifetimes of 'hole states' (those from which the core electron, usually 1s, has been ejected) are of the order of 10^{-4}–10^{-17} s. Therefore, the method is well suited to studying fast reactions, transient species, radicals, carbocations, etc.: in radical studies ESCA gives similar information to that given by ESR (see section 3S.7).

Dramatic insights into carbocation structures have been achieved. We can now quantitatively compare the delocalization of charge in CH_3CO^+ and $PhCO^+$, by measuring the ESCA shift between the CO carbon atom and the other carbon atoms of the ion structure. For CH_3CO^+ the shift is 6.0 eV, while for $PhCO^+$ it is 5.1 eV: the lesser shift in $PhCO^+$ implies a lesser difference in core binding energies and *therefore* greater delocalization of charge.

In the *tert*-butyl cation the charge is substantially localized on C-1 (ESCA shift between C-1 and the methyl carbons is 3.4 eV), while in the trityl and tropylium cations only a single C 1s line is observed, showing that the charge is efficiently delocalized over all carbons in these systems.

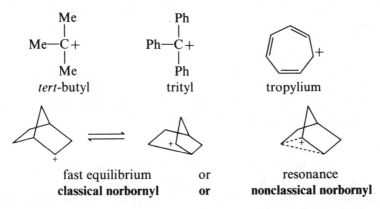

Me	Ph	
Me—C+	Ph—C+	
Me	Ph	
tert-butyl	trityl	tropylium

fast equilibrium   or   resonance
classical norbornyl   or   **nonclassical norbornyl**

In the norbornyl cation, the argument concerning its classical or nonclassical nature hinges on whether we have a fast-equilibrating system or a true resonance system. By comparing the ESCA shifts within the molecules of a series of norbornyl cations, Olah has claimed that in the parent norbornyl cation the shift (1.7 eV) is much less than would be expected in a charge-localized equilibrating ion, and concludes that this ion is nonclassical. Others dissent from this view.

FURTHER READING

Main Texts

Stern, E. S. and Timmons, C. J., *An Introduction to Electronic Absorption Spectroscopy in Organic Chemistry*, Arnold, London (3rd edn, 1970).

Friedel, R. A. and Orchin, M., *Ultraviolet Spectra of Aromatic Compounds*, Wiley, New York (1951).

These two books, although now rather old, contain a fund of data, spectra and interpretation.

Scott, A. I., *Interpretation of the Ultraviolet Spectra of Natural Products*, Pergamon, Oxford (1964). The authoritative work on Woodward rules, etc.

Hunt, R. W. C., *Measuring Colour*, Wiley, Chichester (1987). ACOL text.

Denney, R. and Sinclair, R., *Visible and Ultraviolet Spectroscopy*, Wiley, Chichester (1987). ACOL text.

Knowles, A. and Burgess, C. (Eds.), *Practical Absorption Spectrometry*, Chapman and Hall, London (1984).

Chamberlin, G. J. and Chamberlin, D. G., *Colour*, Heyden, London (1980).

Spectra Catalogs

Friedel and Orchin (1951) includes large numbers of spectra.

Clar, E., *Polycyclic Hydrocarbons*, Volumes 1 and 2, Springer, Berlin, and Academic Press, London (1965). Very many condensed aromatic hydrocarbon systems described, with electronic spectra.

DMS UV Atlas of Organic Compounds, Verlag Chemie, Weinheim, and Butterworths, London. Up-to-date series of well-presented spectra.

Sadtler Handbook of Ultraviolet Spectra, Sadtler, Pennsylvania, and Heyden, London (1979). 200 UV–VIS spectra.

Supplementary Texts

Crabbé, P., *Optical Rotatory Dispersion and Circular Dichroism in Organic Chemistry*, Holden-Day, San Francisco (1965).

Nordling, C., *Angew. Chem. Int. Ed.*, **11** (1972), 83. Review of ESCA.

Brundle, C. R., Baker, A. D. and Thompson, M., *Chem. Soc. Rev.*, **1** (1972), 355. Review of ESCA.

Nakanishi, K. and Horada, N., *Circular Dichroism Spectroscopy*, Oxford University Press, New York (1983).

Rendell, D., *Fluorescence and Phosphorescence*, Wiley, Chichester (1987).

Nowicki-Jankowska, T., Gorkzynska, K., Malik, A. and Wietecka, E., *Analytical Ultraviolet and Visible Spectroscopy*. Volume 19 of *Comprehensive Analytical Chemistry*, Elsevier, Amsterdam (1986).

Mass Spectrometry

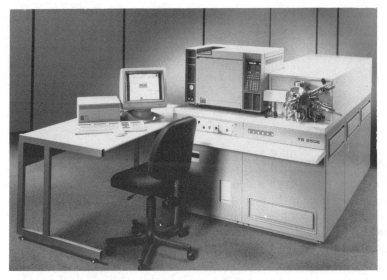

Totally computer controlled mass spectrometer. The inlet system is on the right, with a gas chromatograph in the middle (for GC-MS)

High performance organic mass spectrometer.

Organic chemists use mass spectrometry in three principal ways: (1) to measure *relative molecular masses* (molecular weights) with very high accuracy; from these can be deduced exact molecular formulae: (2) to detect within a molecule the places at which it *prefers to fragment*; from this can be deduced the presence of recognizable groupings within the molecule: and (3) as a method for identifying analytes by comparison of their mass spectra with libraries of digitized mass spectra of known compounds.

5.1 BASIC PRINCIPLES

In the simplest mass spectrometer (figure 5.1), organic molecules are bombarded with electrons and converted to highly energetic positively charged ions (*molecular ions* or *parent ions*), which can break up into smaller ions (*fragment ions*, or *daughter ions*); the loss of an electron from a molecule leads to a radical cation, and we can represent this process as $M \rightarrow M^{+}$. The molecular ion M^{+} commonly decomposes to a pair of fragments, which may be either a radical plus an ion, or a small molecule plus a radical cation. Thus,

$$M^{+} \rightarrow m_1{}^{+} + m_2{}^{\cdot} \quad \text{or} \quad m_1{}^{+} + m_2$$

The molecular ions, the fragment ions and the fragment radical ions are separated by deflection in a variable magnetic field according to their mass and charge, and generate a current (the *ion current*) at the collector in proportion to their *relative abundances*. A *mass spectrum* is a plot of relative abundance against the ratio mass/charge (the *m/z* value). For singly charged ions, the lower the mass the more easily is the ion deflected in the magnetic field. Doubly charged ions are occasionally formed: these are deflected twice as much as singly charged ions of the same mass, and they appear in the mass spectrum at the same value as do singly charged ions of half the mass, since $2m/2z = m/z$.

Neutral particles produced in the fragmentation, whether uncharged molecules (m_2) or radicals ($m_2^{\bullet}$), cannot be detected directly in the mass spectrometer.

Figure 5.1 shows a block layout for a simple mass spectrometer. Since the ions must travel a considerable distance through the magnetic field to the collector, very low pressures ($\approx 10^{-6}-10^{-7}$ mmHg $\equiv 10^{-4}$ N m^{-2}) must be maintained by the use of diffusion pumps.

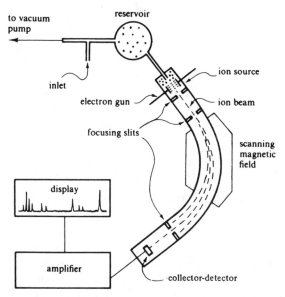

Figure 5.1 *A simple mass spectrometer. Molecules drift from the reservoir into the ion source, where they are ionized by electron bombardment. The resulting ion beam consists of molecular ions, fragment ions and neutral fragments; the ions are deflected by the magnetic field onto the collector-detector.*

Figure 5.2 shows a simplified line-diagram representation of the mass spectrum of 2-methylpentane (C_6H_{14}). The most abundant ion has an m/z value of 43 (corresponding to $C_3H_7^+$), showing that the most favored point of rupture occurs between C_x and C_y: this most abundant ion (the *base peak*) is given an arbitrary abundance of 100, and all other intensities are expressed as a percentage of this (*relative abundances*). The small peak at m/z 86 is obviously the molecular ion. The peaks at m/z 15, 29 and 71 correspond to CH_3^+, $C_2H_5^+$ and $C_5H_{11}^+$, respectively, etc.: the fragment ions arise from the rupture of the molecular ion, either directly or indirectly, and the analysis of many thousands of organic mass spectra has led to comprehensive semiempirical rules about the preferred *fragmentation modes* of every kind of organic molecule. The application of these

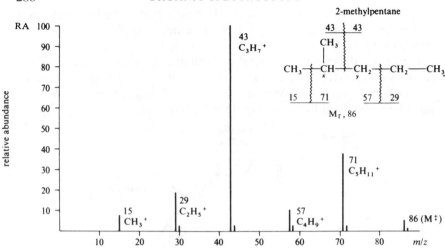

Figure 5.2 *Simplified mass spectrum of 2-methylpentane showing how the chain can break in several places; but breaks occur preferentially at the branching point, the most abundant ion being m/z 43 ($C_3H_7^+$).*

rules to organic structural elucidation will be developed later in the chapter.

The mass spectrum of a compound can be obtained on a smaller sample size (*in extremis* down to 10^{-12} g) than for any other of the main spectroscopic techniques, the principal disadvantages being the destructive nature of the process, which precludes recovery of the sample, the difficulty of introducing small enough samples into the high-vacuum system needed to handle the ionic species involved and the high cost of the instruments. Mass spectrometry is unlike the other spectroscopic techniques met in this book in that it does not measure the interaction of molecules with the spectrum of energies found in the electromagnetic spectrum, but the output from the instrument has all other spectroscopic characteristics, in showing an array of signals corresponding to a spectrum of energies; to highlight this distinction, the name *mass spectrometry* is preferred.

5.2 INSTRUMENTATION—THE MASS SPECTROMETER

We can conveniently study the design of a sophisticated mass spectrometer by considering each of its five main parts successively, with reference to figure 5.3.

5.2.1 SAMPLE INSERTION—INLET SYSTEMS
Organic compounds that have moderate vapor pressures at temperatures up to around 300 °C (including gases) can be placed in an ampoule

connected via a reservoir to the ionization chamber. Depending on volatility, it is possible to cool or heat the ampoule, etc., to control the rate at which the sample volatilizes into the reservoir, from which it will diffuse slowly through the sinter into the ionization chamber. Samples with lower vapor pressures (for example, solids) are inserted directly into the ionization chamber on the end of a probe, and their volatilization is controlled by heating the probe tip.

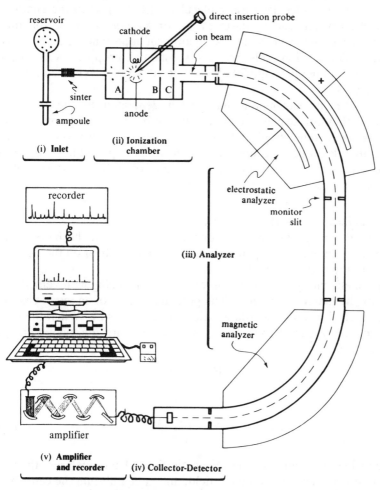

Figure 5.3 *High-resolution, double-focusing mass spectrometer. (Evacuation system not shown.)*

5.2.2 ION PRODUCTION IN THE IONIZATION CHAMBER
The several methods available for inducing the ionization of organic compounds are discussed in section 5S.1, but electron bombardment is

routinely used. Organic molecules react on electron bombardment in two ways: either an electron is captured by the molecule, giving a radical anion, or an electron is removed from the molecule, giving a radical cation:

$$M \xrightarrow{e} M^{\bar{\cdot}} \text{ or } M \xrightarrow{-e} M^{\dotplus} + 2e$$

The latter is more probable by a factor of 10^2, and positive-ion mass spectrometry is the result.

Most organic molecules form molecular ions ($M^{\dotplus}$) when the energy of the electron beam reaches 10–15 eV ($\approx 10^3$ kJ mol^{-1}). While this minimum *ionization potential* is of great theoretical importance, fragmentation of the molecular ion only reaches substantial proportions at higher bombardment energies, and 70 eV ($\approx 6 \times 10^3$ kJ mol^{-1}) is used for most organic work.

When the molecular ions have been generated in the ionization chamber, they are expelled electrostatically by means of a low positive potential on a repeller plate (A) in the chamber. Once out, they are accelerated down the ion tube by the much higher potential between the accelerating plates B and C (several thousand volts). Initial focusing of the ion beam is effected by a series of slits.

5.2.3   SEPARATION OF THE IONS IN THE ANALYZER

Theory. In a magnetic analyzer ions are separated on the basis of m/z values, and a number of equations can be brought to bear on the behavior of ions in the magnetic field.

The kinetic energy, E, of an ion of mass m travelling with velocity v is given by the familiar $E = \frac{1}{2}mv^2$. The potential energy of an ion of charge z being repelled by an electrostatic field of voltage V is zV. When the ion is repelled, the potential energy, zV, is converted into the kinetic energy, $\frac{1}{2}mv^2$, so that

$$zV = \tfrac{1}{2}mv^2$$

Therefore,

$$v^2 = \frac{2zV}{m} \tag{5.1}$$

When ions are shot into the magnetic field of the analyzer, they are drawn into circular motion by the field, and at equilibrium the centrifugal force of the ion (mv^2/r) is equalled by the centripetal force exerted on it by the magnet (zBv), where r is the radius of the circular motion and B is the field strength. Thus,

$$\frac{mv^2}{r} = zBv$$

Therefore,

$$v = \frac{zBr}{m} \qquad (5.2)$$

Combining equations (5.1) and (5.2),

$$\frac{2zV}{m} = \left[\frac{zBr}{m}\right]^2$$

Therefore,

$$\frac{m}{z} = \frac{B^2r^2}{2V} \qquad (5.3)$$

It is from equation (5.3) that we can see the inability of a mass spectrometer to distinguish between an ion m^+ and an ion $2m^{2+}$, since the ratio, m/z, between them has the same value of $(B^2r^2/2V)$, and these three parameters B, r and V dictate the path of the ions.

To change the path of the ions so that they will focus on the collector and be recorded, we can vary either V (the accelerating voltage) or B (the strength of the focusing magnet). *Voltage scans* (the former) can be effected much more rapidly than *magnetic scans* and are used where fast scan speed is desirable.

Resolution. The ability of a mass spectrometer to separate two ions (the spectrometer resolution) is acceptably defined by measuring the depth of the valleys between the peaks produced by the ions. If two ions of m/z 999 and 1000, respectively, can just be resolved into two peaks such that the recorder trace almost reaches back down to the baseline between them, leaving a valley which is 10 per cent of the peak height, we say the resolution of the spectrometer is '1 part in 1000 (10 per cent valley resolution)'. Simple magnetic-focusing instruments have resolving powers of around 1 in 7500 on this basis.

Double focusing. Ions repelled by the accelerator plates do not in practice all have identical kinetic energies, and this energy spread constitutes the principal limiting factor in improving the resolution of a magnetic analyzer. Preliminary focusing can be carried out by passing the ion beam between two curved plates, which are electrostatically charged—an *electrostatic analyzer.*

Different equations hold for the behavior of ions in an electrostatic (rather than magnetic) field. Here the centrifugal force $(mv^2/r) = zE$, and combining this with $zV = \frac{1}{2}mv^2$ gives

$$r = \frac{2V}{E} \qquad (5.4)$$

Thus, the radial path followed by ions of a given velocity is independent of m and z (and m/z values). The electrostatic analyzer focuses ions of identical kinetic energy onto the monitor slit, whatever their m/z values, and, coupled with a magnetic analyzer to resolve m/z values, this *double-focusing* spectrometer can attain resolutions of 1 in 60 000.

In a low-resolution instrument we may identify an ion of m/z 28 as possibly CO^+ or N_2^+ or $C_2H_4^+$; the accurate mass measurement possible at high resolution enables us to distinguish among several possible exact formula masses—for example, CO^+ (27.9949), N_2^+(28.0062), $C_2H_4^+$ (28.0312).

5.2.4   THE DETECTOR–RECORDER

The focused ion beam passes through the collector slit to the detector, which must convert the impact of a stream of positively charged ions into an electrical current. This must be amplified and recorded, either graphically or digitally.

Several different amplification systems are used by different manufacturers, but the most common is the *electron multiplier*, which operates in a manner similar to the photomultiplier detector described in section 4.3.1. A series of up to twenty copper–beryllium dynodes transduces the initial ion current, and the electrons emitted by the first dynode are focused magnetically from one dynode to the next; the final cascade current is thus amplified more than one million times. Figure 5.3 illustrates this schematically.

Two essential features of the recording system in a mass spectrometer are that it must (a) have a very fast response, and be able to scan several hundred peaks per second, and (b) be able to record peak intensities varying by a factor of more than 10^3. The problem of fast response can be met by using mirror galvanometers, which reflect an ultraviolet light beam onto fast-moving photographic paper. Problem (b) was formerly overcome by having a series of mirror galvanometers covering a range of sensitivities (for example, three covering the ratio 1:10:100, or five covering the ratio 1:3:10:30:100). Simultaneous scanning produces the mass spectrum trace shown in figure 5.4.

5.2.5   DATA HANDLING

The analog signal coming from the detector is first converted to digital form in an analog-to-digital convertor, or ADC, and the digitized data are stored in computer memory.

Computer-controlled instruments produce the mass spectral data in several forms, either as a list of fragment ions or plotted directly as a bar diagram such as figure 5.2. Improvements in instrumentation have largely eliminated the need for sets of mirror galvanometers. Accurate mass calibration is carried out each day (or every other day) by recording the mass spectrum of appropriate reference compounds, such as perfluoro-kerosene, PFK, or cesium iodide clusters for very high molecular masses: see also section 5.4.3.

Identification of known compounds can be carried out by searching through computer-held digitized mass spectra; many collections are available commercially containing up to 100 000 compounds on file. See also section 5S.2.

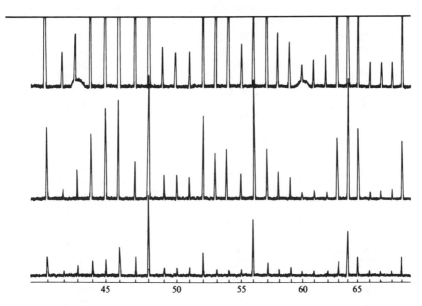

Figure 5.4 *Typical appearance of mass spectra with three mirror galvanometers scanning simultaneously. Note the metastable ions at m/z 43.4 and m/z 60.2.*

5.3 ISOTOPE ABUNDANCES

Few elements are monoisotopic, and table 5.1 gives the natural isotope abundance of the elements we might expect to encounter in organic compounds. Ions containing different isotopes appear at different m/z values.

For an ion containing n carbon atoms, there is a probability that approximately $1.1n$ per cent of these atoms will be ^{13}C, and this will give rise to an ion of mass one higher than the ion that contains only ^{12}C atoms. The molecular ion for 2-methylpentane (figure 5.2) has an associated M + 1 ion, whose intensity is approximately 6.6 per cent of that of the molecular ion; while the molecular ion has only ^{12}C atoms, the M + 1 ion contains ^{13}C atoms. (The contribution of ^{2}H atoms should not be overlooked, even though the probability of their presence is not as high as for ^{13}C, and in exact work the statistics of both ^{13}C and ^2H abundances must be calculated.)

Table 5.1   Natural isotope abundances and isotopic masses (^{12}C = 12.000 000) for common elements

Isotope	Natural abundance (%)	Isotopic mass/m_u
^1H	99.985	1.007 825
^2H	0.015	2.014 102
^{12}C	98.9	12.000 000
^{13}C	1.1	13.003 354
^{14}N	99.64	14.003 074
^{15}N	0.36	15.000 108
^{16}O	99.8	15.994 915
^{17}O	0.04	16.999 133
^{18}O	0.2	17.999 160
^{19}F	100	18.998 405
^{28}Si	92.2	27.976 927
^{29}Si	4.7	28.976 491
^{30}Si	3.1	29.973 761
^{31}P	100	30.973 763
^{32}S	95.0	31.972 074
^{33}S	0.76	32.971 461
^{34}S	4.2	33.967 865
^{35}Cl	75.8	34.968 855
^{37}Cl	24.2	36.965 896
^{79}Br	50.5	78.918 348
^{81}Br	49.5	80.916 344
^{127}I	100	126.904 352

A second associated peak can arise at m/z M + 2 if two ^{13}C atoms are present in the same ion (or if two ^2H atoms, or one ^{13}C and one ^2H, are present); these probabilities can be calculated and may be a help in deciding the formula for an ion in the absence of exact mass measurement. For example, the two ions $C_8H_{12}N_3^+$ and $C_9H_{10}O_2^+$ have the same unit mass (m/z 150), and the M + 1 relative abundances are similar (9.98 per cent and 9.96 per cent, respectively); however, the M + 2 abundances are sufficiently different to enable differentiation of the structures (0.45 per cent and 0.84 per cent, respectively). The ability to see M + 2 peaks of such low abundance depends on there being a large $M^{\ddagger}$ peak.

Ions containing one bromine atom create a dramatic effect in the mass spectrum because of the almost equal abundance of the two isotopes; pairs of peaks of roughly equal intensity appear, separated by two mass units.

Equally characteristic are the ions from chlorine compounds engendered by the ^{35}Cl and ^{37}Cl isotopes; for ions containing one Cl atom, the relative intensities of the lines, separated by two mass units, is 3:1. Ions containing one sulfur atom also have associated m + 2 peaks.

The picture becomes much more complicated when one considers the relative abundances of ions containing several polyisotopic elements; the presence of two bromine atoms in an ion gives rise to three peaks at m, m + 2 and m + 4, the relative intensities being 1:2:1, while for three bromines the peaks arise at m, m + 2, m + 4, m + 6, with relative intensities 1:3:3:1. These figures ignore any contribution from ^{13}C that may be present.

For each element in a given ion, the relative contributions to m + 1, m + 2 peaks, etc., can be calculated from the binomial expansion of $(a + b)^n$, where a and b are the relative abundances of the isotopes and n is the number of these atoms present in the ion. Thus, for three chlorine atoms in an ion, expansion gives $a^3 + 3a^2b + 3ab^2 + b^3$. Four peaks arise; the first contains three ^{35}Cl atoms and each successive peak has ^{35}Cl replaced by ^{37}Cl until the last peak contains three ^{37}Cl atoms. The m/z values are separated by two mass units, at m, m + 2, m + 4, m + 6. Since the relative abundances of ^{35}Cl and ^{37}Cl are 3:1 (that is, $a = 3$, $b = 1$), the intensities of the four peaks (ignoring contributions from other elements) are $a^3 = 27$, $3a^2b = 27$, $3ab^2 = 9$, $b^3 = 1$ (that is, 27:27:9:1).

5.4   THE MOLECULAR ION

5.4.1   STRUCTURE OF THE MOLECULAR ION

In previous sections we represented the molecular ion as $M^{\ddot{+}}$, signifying a radical cation produced when a neutral molecule loses an electron: where does the electron come from?

For electron bombardment around their ionization potentials (10–15 eV $\approx 10^3$ kJ mol^{-1}), it is meaningfully possible in organic molecules to say which are the likeliest orbitals to lose an electron. The highest occupied orbitals of aromatic systems and nonbonding orbitals on oxygen and nitrogen atoms readily lose an electron; the π electrons of double and triple bonds are also vulnerable. At the instant of ionization in a Franck–Condon process, before any structural rearrangement can occur, these ionizations can then be represented as shown below. In alkanes, all we can say is that the ionization of C—C σ bonds is easier than that of C—H bonds.

In the strictest sense, however, we must not attempt to localize the excitation produced by the loss of an electron when the electrons themselves are not localized: in the absence of specific evidence, we must be satisfied with the generalized view of an electron being expelled from the whole molecular orbital and of the excitation energy being spread through-out the molecule. For electron bombardment of complex molecules around 70 eV (6×10^3 kJ mol^{-1}), any specificity in the site of electron removal is lost entirely. These arguments apply also to the structures of fragment ions and fragment radical ions, and these are frequently written in square brackets (for example, $[C_5H_5]^{\ddagger}$ or $[C_4H_7]^{\ddagger}$), no attempt being made to speculate on the precise structures. The use of a partial square bracket (for example, $C_5H_5]^{\ddagger}$ or $C_4H_7]^{\ddagger}$, or as in structure I below) is a useful alternative, especially for larger structures. However, organic chemists have successfully adapted the postulates of resonance theory to explain the reactivities of functional groups and the mechanistic processes inherent in syntheses, rearrangements and degradations. It is not surprising that they have extended their mechanistic interpretations to include fragmentation reactions of molecular ions, and in so doing they have perhaps oversimp-lified the problems faced by theoretical workers. The organic chemist's argument is that it may be inexact to write the single structure II rather than I for the molecular ion of 2-propanol after 70 eV bombardment: but II makes it easy to *rationalize* its fragmentation to a methyl radical ($CH_3{}^{\bullet}$) and the ion $[C_2H_5O]^+$ at *m/z* 45.

At the instant of fragmentation, it is certainly true that sufficient excitation energy must be concentrated within the appropriate σ bond to exceed the energy of activation for its rupture. Provided that mechanistic notations are used with full acknowledgement of their implications and limitations, they can supply to organic chemists a familiar rational framework for the interpretation of fragmentation processes that cannot be supplied by the unadulterated precision of molecular-orbital theory.

The fragmentation of molecular ions, etc., is discussed systematically in section 5.6.

5.4.2 RECOGNITION OF THE MOLECULAR ION

The molecular ions of roughly 20 per cent of organic compounds decompose so rapidly ($< 10^{-5}$ s) that they may be very weak or undetected in a routine 70 eV spectrum. For most unknown compounds the ion cluster appearing at highest m/z value is likely to represent the molecular ion with its attendant M + 1 peaks, etc., but we must apply a number of tests to ensure that this is so.

Abundant molecular ions are given by aryl amines, nitriles, fluorides and chlorides. Aromatic hydrocarbons and heteroaromatic compounds give strong M^+ peaks, provided that no side-chain of C_2 or longer is present: indeed the M^+ peak is often the base peak, and doubly charged ions may often be observed in the mass spectra of these compounds appearing at $m/2z$ values. Aryl bromides and iodides lose halogen too readily to give strong molecular ion peaks. Other classes with weakened M^+ peaks are aryl ketones (which fragment easily to $ArCO^+$) and benzyl compounds such as side-chain hydrocarbons ($ArCH_2R$) or $ArCH_2X$ (both of which fragment at the benzylic carbon).

Absence of molecular ions (or an extremely weak M^+ peak) is characteristic of highly branched molecules whatever the functional class. Alcohols and molecules with long alkyl chains also fragment easily and lead to very weak M^+ peaks.

Isotope abundances should correlate with the appearance of the purported molecular-ion cluster, as discussed in section 5.3. The intensities of M + 1, M + 2 peaks, etc., are obviously most easily measured and of greatest value when the M^+ peak is fairly abundant.

Nitrogen-containing compounds with an *odd* number of nitrogen atoms in the molecule must have an *odd* molecular weight (relative molecular mass).

An *even* number of nitrogen atoms, or no nitrogen at all, leads to an *even* molecular weight (relative molecular mass).

Common fragment ions in the spectrum contribute positive support for the assignment of M^+; what constitutes 'common fragment ions' is the subject of section 5.6.

Unusual fragment ions should make one suspicious: for example, molecular ions can give rise to a series of weaker ions at M − 1, M − 2 and M − 3 due to successive loss of hydrogen, but a *specific* fragmentation leading directly to M − 3 (or M − 4 or M − 5) is never observed. An apparent M − 14 peak (corresponding to direct loss of a CH_2 fragment) is so rare that the purported $M^{\ddagger}$ ion should be discounted; a more likely explanation is the presence of a lower homolog as contaminant in the sample.

5.4.3 MOLECULAR FORMULA FROM THE MOLECULAR ION

For an unknown organic compound, the ability to measure its relative molecular mass (molecular weight) to within four decimal places leads immediately to an accurate molecular formula for the compound.

Accurate isotopic masses for the principal isotopes of the elements commonly met in organic chemistry are given in table 5.1. Note that the *relative atomic mass*, A_r (formerly called the *atomic weight*), of an element is the weighted mean of the masses of the naturally occurring isotopes. Thus, since the natural abundances of ^{12}C (12.000 000) and ^{13}C (13.003 354) are 98.9 per cent and 1.1 per cent, respectively, we can calculate the relative atomic mass of carbon as 12.01.

> *In the mass spectrum of an organic molecule, each peak corresponds to an ion of a particular isotopic composition, and its* m/z *value is calculated from the* isotopic masses *in table 5.1 and* not from the relative atomic masses of the elements.

Similarly, in calculating the molecular formula from the molecular ion, we must be sure that we know which isotopes are present in that ion. In the case of compounds containing only C, H and O, we can easily identify the ^{13}C isotope peak (see section 5.3), and ^2H and ^{17}O are at extremely low abundance; the main peak in the molecular ion cluster then corresponds solely to a combination of ^{12}C, ^1H and ^{16}O.

As an example, consider an unknown compound X whose relative molecular mass (to the nearest integer) is measured at low resolution to be 100. From this and other evidence the compound could be either (A) $C_6H_{12}O$ or (B) $C_4H_4O_3$. High-resolution mass measurement of the molecular ion gives *m/z* 100.088 71, proving that the correct structure is A.

Mass measurement can be carried out at its most sophisticated level by having a high-resolution instrument interfaced to a computer, which is programmed to identify reference peaks and to interpolate from their known masses to the masses of other individual ions from the sample. An additional program can convert the accurate masses to the element compositions of the ions, so that the printout contains a list of ion masses, abundances and compositions. The positions of known accurate masses can be obtained from an electronic *mass marker* or by using reference compounds such as perfluorokerosene, PFK (mixed long-chain perfluoro-alkanes, C_nF_{2n+2}): PFK fragments to a series of well-spaced fragment ions up to high *m/z* values, for example, *m/z* 169, 181, 219. In *double-beam* spectrometers PFK can be fed simultaneously into the instrument with the sample, and the ion masses of the sample displayed in parallel with those of the PFK.

An alternative method of mass measurement is to couple the output of the instrument to a cathode-ray oscilloscope, and to display alternately on the oscilloscope tube the peak whose mass is to be measured and an ion of known mass from the reference compound. The accelerator voltage of the spectrometer can then be adjusted until these two peaks overlap, and the difference in mass between them can be calculated as a function of the change in accelerator voltage. This process is known as *peak matching*.

5.5 METASTABLE IONS

5.5.1 THE NATURE OF METASTABLE IONS

In the diagrammatic mass spectrum in figure 5.4 can be seen broad peaks at non-integer masses m/z 60.2 and m/z 43.4. The ions producing these peaks are termed *metastable ions*: they have lower kinetic energy than have normal ions, and arise from fragmentations that take place during the flight down the ion tube rather than in the ionization chamber. The exact position where they are formed in the tube determines whether or not we can easily observe them.

Up until now we have tacitly assumed that molecular ions formed in the ionization chamber do one of two things: either they decompose completely and very rapidly in the ion source and never reach the collector (as in the case of highly branched molecular ions with lifetimes less than $\approx 10^{-5}$ s), or else they survive long enough to reach the collector and be recorded there (lifetimes longer than $\approx 10^{-5}$ s). We have also assumed that fragment ions are produced by decompositions of a proportion of the molecular ions in the ion source.

Depending on the inherent stability of an ion, and on the amount of excitation energy absorbed on bombardment, ion lifetimes will vary in a complex manner: a given molecular ion may possess a spread of energies and, not surprisingly, some of the molecular ions will have intermediate lifetimes ($\approx 10^{-5}$ s) and so leave the ionization chamber intact, but decompose *en route* to the collector. Some fragment ions behave similarly, decomposing between ion source and collector.

Suppose that a large number of molecules of M are converted to molecular ions M^{+}: not all of the M^{+} will possess the same excitation energy and therefore some will have longer lifetimes than others.

The M^{+} ions with shortest lifetimes may decompose in the ionization chamber to stable daughter ions A^{+} and radicals $B^{\cdot}$; the daughter ions A^{+} will be detected at the collector normally. The molecular ions that leave the ion source intact will be accelerated by the accelerator voltage and will then possess a translational energy eV. Some of these M^{+} ions may survive intact to the collector and be detected normally. If, however, others of these M^+ ions decompose to A^+ and $B^.$ immediately *after* acceleration, the translational energy of the parent M^+ (eV) will be shared *between* A^+ and $B^.$ in proportion to their masses (principle of conservation of momentum).

The translational energy of the daughter ion A^+ must then be *lower* than that of the parent ion, and this ion A^+ will arrive at the collector differently from the 'normal' A^+ ion produced in the ion source.

> *The ion A^+ with 'abnormal' translational energy is a* metastable ion. *Note that metastable A^+ ions have* the same mass *as normal A^+ ions, but simply have less translational energy.*

5.5.2 ION TUBE REGIONS

Of the metastable ions produced in the ion tube, only a fraction come to reasonable focus at the collector (unless by the use of special techniques): we must consider the successive regions of the ion tube to understand why this should be so.

The first field-free region, in a double-focusing instrument, lies between the ion source and the electrostatic analyzer. (This region has no counterpart in single-focusing instruments.) If a metastable ion is produced here, it will be focused out by the electrostatic analyzer because of its abnormal kinetic energy. Such ions, and any metastable ions formed *in the analyzer*, will appear out of focus (randomly) at the collector as background current, and will be undetected.

The second field-free region in a double-focusing instrument lies between the electrostatic and magnetic analyzers. (In single-focusing spectrometers the corresponding region is between the ion source and the magnetic analyzer.) Metastables produced in this region will be focused reasonably sharply by the magnetic analyzer on the bases of their masses and translational energies, but since a metastable A^+ ion has the same mass and lower translational energy (less momentum) than the normal A^+ ion, the metastable A^+ ion is deflected more easily in the analyzer than the normal A^+ ion: these metastable A^+ ions will appear on the spectrum among ions of lower mass, and are the only metastable ions detected routinely. *Metastable peaks* are broadened for a number of reasons, one of which is the possibility that some of the excitation energy leading to bond rupture may be converted to additional kinetic energy.

Ions produced *in the magnetic analyzer* will be focused at the collector, but there will be substantial differences in energy between those formed at the *beginning* compared with those formed at the *end* of the analyzer: this produces a continuum of low-intensity signals between the positions of normal A^+ and metastable A^+, and is usually too weak to be detected.

The *third field-free region* lies between the magnetic analyzer and the collector. Since no focusing takes place in this region, a parent ion is already immutably on path, and if it decomposes to the metastable A^+, then this metastable will continue on the same path as the parent ion. The metastable ion is detected at the same m/z value as the parent ion.

5.5.3 CALCULATION OF METASTABLE ION m/z VALUES

The apparent mass of a metastable ion A^+ (m^*) can be calculated fairly accurately from the masses of the parent ion (m_1) and the normal daughter ion A^+ (m_2) from the equation

$$m^* = \frac{(m_2)^2}{m_1}$$

This equation often gives an apparent mass 0.1—0.4 mass units lower than is in fact observed.

As an example, the mass spectrum of toluene shows strong peaks at m/z 91 and m/z 65, together with a strong broad metastable peak at m/z 46.4. Now $65^2/91 = 46.4$, so we can interpret that the ion m/z 91 is decomposing by loss of 26 mass units to the daughter ion m/z 65, and that some of this fragmentation takes place in the second field-free region, leading to a metastable ion peak of m/z 46.4. (The value 46.4 is the *apparent* mass of the metastable ion, the *real* mass being the same as that of the normal daughter—that is, 65.)

It is worth restating that some of the m/z 91 ions will decompose before and after the second field-free region, producing m/z 65 ions, which are *not* focused at m/z 46.4.

Tables and computer programs are widely used to relate metastable ions to the corresponding parent and daughter ions; the use of the nomogram in figure 5.5 for this purpose is sufficiently accurate, but computer programs are now universally used.

Example 5.1

Question. Calculate the expected apparent mass of the metastable ion produced when m/z 77 decomposes by loss of $CH{\equiv}CH$ to m/z 51.

Model answer. The equation $m^* = (m_2)^2/m_1$ becomes $51^2/77 = 33.4$.

Exercise 5.1   (a) Using the nomogram on figure 5.5, predict the m/z value for the normal daughter ion generated from the decomposition of $PhCO^+$ (m/z 105), given that a metastable ion (near m/z 56.5) is associated with this decomposition. Verify the value by calculation. What neutral molecule is being extruded in this process?

(b) Similarly predict the approximate apparent mass of the metastable ion associated with the decomposition of m/z 129 to m/z 91, and verify the answer by calculation. What is a possible formula for the neutral fragment being extruded?

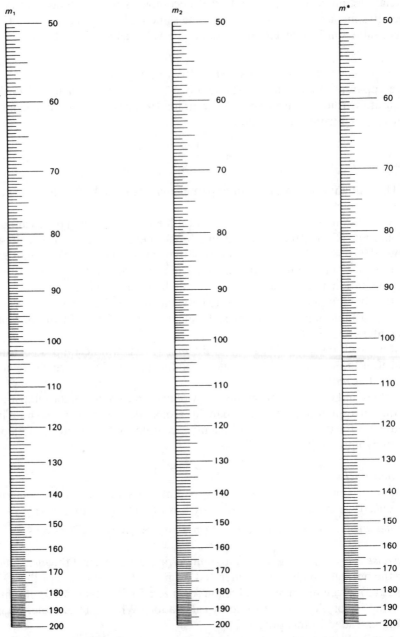

Figure 5.5 *Nomogram to aid identification of parent and normal daughter ions from the position of a related metastable ion. From the mass of m*, a straight line cuts the m_1 and m_2 columns at the masses of the parent and daughter ions, respectively.*

5.5.4 SIGNIFICANCE OF METASTABLE IONS

The presence of a metastable ion in a mass spectrum is taken as very good evidence that the parent ion undergoes decomposition *in one step* to the daughter, so that it is of considerable mechanistic importance to investigate metastable ions.

It follows from the discussion in section 5.5.2 that there may be one-step processes occurring in the mass spectrometer that do not produce metastable-ion peaks, so the absence of such peaks cannot be used to infer the absence of a one-step transition.

5.6 FRAGMENTATION PROCESSES

5.6.1 REPRESENTATION OF FRAGMENTATION PROCESSES

If a molecular ion loses a methyl radical ($CH_3\cdot$), the mass spectrum will show an ion 15 mass units below the molecular ion: we can write this process as

$$M^{+\cdot} \rightarrow CH_3^{\cdot} + (M - 15)^+$$

An alternative is to write the fragment ion as $M - CH_3$. Throughout this chapter we shall use this convention frequently, referring to ions as $M - 18$, $M - 24$, $M - CO$, $M - H_2S$, etc., it being understood that these may in fact be even-electron ions—for example, $(M - 15)^+$—or odd-electron radical ions—for example, $(M - 18)^{+\cdot}$.

The same convention can be used to represent fragment ions as m^+ (or $m^{+\cdot}$), and these may, in turn, fragment to $m - 1$, $m - CH_3$, $m - C_2H_5$, etc.

We saw in section 5.4 that it is, on the whole, not possible to write the structures of molecular ions with any degree of certainty, but in the absence of structural proof it is a legitimate convenience to indicate that ionization takes place from the most easily ionized orbitals in the molecule.

In representing the molecular ion of an alcohol as I, we at least give ourselves a framework for discussing its behavior on fragmentation. For more complex molecules such as II, we may find it expedient to present one decomposition as proceeding from the molecular ion III, and another as proceeding from IV to V. In doing so, we are using an extension of resonance theory.

These are, of course, postulates for ion structures; some techniques which help to verify the postulates are discussed in section 5S.3.

When we come to represent on paper the mechanics of how a particular fragment ion (m^+) is produced from a molecular ion ($M^{+\cdot}$), the difficulties are twice compounded, since we may have no precise knowledge of the molecular ion structure, and we cannot always know with precision the structure of the fragment ion (although we may know its element composition).

We shall proceed to write such mechanisms, largely because most of them are similar in type to the more familiar mechanisms of 'wet' organic chemistry, and are therefore a help in classifying the observable facts concerning the fragmentations associated with functional groups. To do so we must make an initial assumption that ions, during fragmentation, undergo minimum structural change. This is often contrary to observations carried out on the randomization of atoms (particularly H) during fragmentation, and notable examples will be highlighted: isotope labelling is here an enormously important adjunct, but its discussion will be delayed until section 5S.3.

5.6.2 BASIC FRAGMENTATION TYPES AND RULES

Electron-pair processes (such as bond heterolysis) are represented by the conventional curved arrow, and one-electron processes (such as bond homolysis) by the fish-hook arrow: many fragmentations can be represented as occurring either by one-electron or two-electron processes and mechanisms given in this book do not claim to represent actuality. The molecular ion, because of its excess energy, may take part in processes that have no counterpart in test-tube chemistry.

σ-Bond rupture in alkane groups. This can really only be represented by assuming that, at the instant of ionization, sufficient excitation energy is concentrated on the rupturing bond to ionize it.

$$RCH_2-CH_2R' \equiv RCH_2{:}CH_2R' \xrightarrow{\ -e\ } RCH_2{\overset{+}{\underset{\cup}{}}}CH_2R'$$

$$RCH_2{}^+ + {}^{\bullet}CH_2R'$$

$$\text{similarly} \quad R-H \equiv R{:}H \xrightarrow{\ -e\ } R{\overset{+}{\underset{\cup}{}}}H \longrightarrow R^+ + H^{\bullet}$$

σ-Bond rupture near functional groups. This may be facilitated by the easier ionization of that group's orbitals, as in alcohols, where the nonbonding orbitals of oxygen are more easily ionized than the σ orbitals.

Other groups in this category are ethers, carbonyl groups and compounds containing halogen, nitrogen, double bonds, phenyl groups, etc.

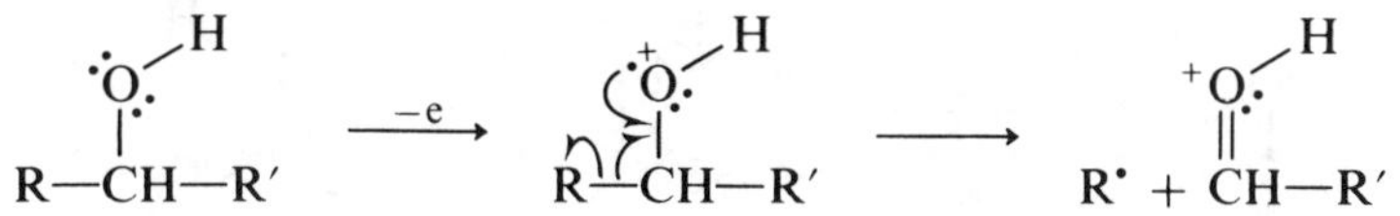

Elimination by multiple σ-bond rupture. Elimination by multiple σ-bond rupture may occur, leading to the extrusion of a neutral molecule such as

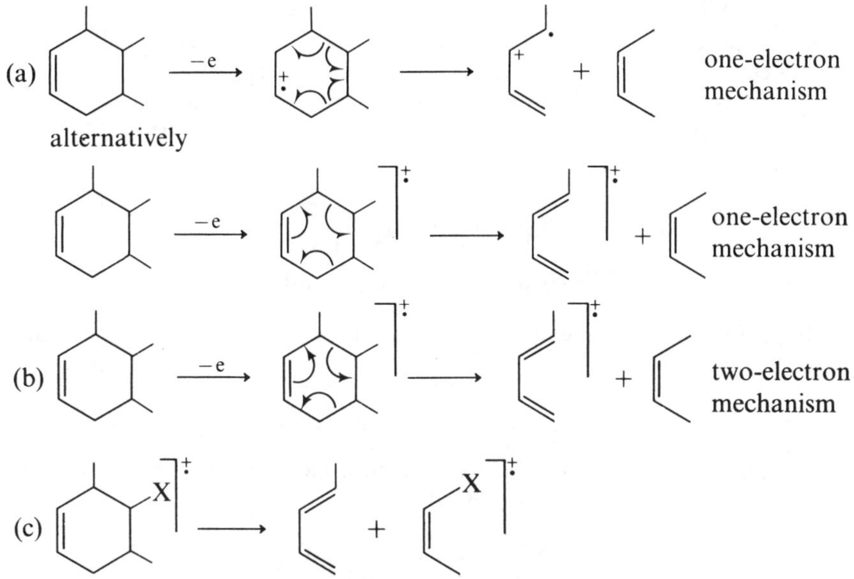

(a)   one-electron
                                                                       mechanism

alternatively

                                                                       one-electron
                                                                       mechanism

(b)   two-electron
                                                                       mechanism

(c)

CO, C_2H_4, C_2H_2, etc. A well-known example is the retro-Diels–Alder reaction of cyclohexenes, which can be represented as in (a) or (b). Highly stabilized ene fragments may cause charge retention to be in part reversed, as in (c).

Rearrangements. These are common, the most frequently encountered having been described by F. W. McLafferty (see McLafferty, 1980, page 123), and named after him. It is exemplified in the case of a carbonyl compound (I) by the extrusion of an alkene, but is also exhibited by ions such as II, III, IV, etc.

The even-electron rule. *The even-electron rule is a rule-of-thumb interpretation of sound thermodynamic principles: in essence it states that an even-electron species (an ion, as opposed to a radical ion) will not normally fragment to two odd-electron species (that is, it will not degrade to a radical and a radical ion), since the total energy of this product mixture would be too high:*

ester
II

carboxylic acid
III

amide
IV

$$A^+ \;\not\longrightarrow\; B^{\ddagger} + C^{\cdot}$$

even   odd   odd

In preference, an ion will degrade to another ion and a neutral molecule:

$$A^+ \longrightarrow B^+ + C$$

even   even even

Radical ions, being odd-electron species, can extrude a neutral molecule, leaving a radical ion as coproduct:

$$A^{\ddagger} \longrightarrow B^{\ddagger} + C$$

odd   odd   even

Radical ions can also degrade to a radical and an ion

$$A^{\ddagger} \longrightarrow B^+ + C^{\cdot} \text{(or } B^{\cdot} + C^+)$$

odd   even odd   odd even

5.6.3   FACTORS INFLUENCING FRAGMENTATIONS

Functional groups. Some functional groups may direct the course of fragmentation profoundly, while other functional groups may have little effect. This is discussed fully in section 5.7.

Thermal decomposition. Thermal decomposition of thermolabile compounds may occur in the ion source, and commonly leads to difficulty in interpreting the mass spectra of alcohols, which may dehydrate *before* ionization. In the case of alcohols, loss of water gives rise to a peak at M − 18 whether the loss occurs before or after ionization, but thermal dehydration may be extensive enough to eliminate entirely the appearance of a molecular ion in the spectrum. If thermal decomposition is suspected, the compound can be ionized in a *cooled* ion source, so that electron bombardment of the whole molecule takes place. An alternative solution is to convert the alcohol to the more volatile trimethylsilyl derivative; this is discussed in section 5S.4.

Bombardment energies. For routine organic spectra these are ≈ 70 eV. It is worth noting that, even with these high energies, molecular ions possess a maximum of ≈ 6 eV in excess of their ionization potentials, and there is little change in the fragmentation *pattern* if this 70 eV is reduced to ≈ 20 eV; however, the *ion yield* (that is, the efficiency of ionization) is reduced, and the spectra are weaker in intensity overall. From ≈ 20 eV down to the ionization potential of the molecule, the spectrum becomes progressively simpler, since only the most favored fragmentations are occurring: recording low-energy spectra is, therefore, a useful tool in bond-energy studies. It follows from these observations that the relative abundances of ions in a spectrum are only reproducible when bombardment energies are constant.

Relative rates of competing fragmentation routes. These also are important in dictating relative abundances. In the simple case of $A^{\ddagger}$ going either to B^+ and $C^{\cdot}$ or to $B^{\ddagger}$ and C, the equilibrium abundances of $A^{\ddagger}$, B^+ and $B^{\ddagger}$ depend on the relative rate constants for the two competing reactions: these rate constants may, in turn, depend on the excitation energy possessed by $A^{\ddagger}$, and will certainly depend on the heats of formation of *all* the products. Calculations involving these and other parameters are the basis of the so-called *quasiequilibrium* theory (QET) of mass spectrometry.

It is misleading to use the intensity of an ion peak as a simple measure of the importance of a particular fragmentation route, unless it is certain that the ion cannot be produced by another route. At low resolution there is the additional complication that the peak may be associated with two ions of equal mass (such as $C_3H_7{}^+$ and $CH_3\overset{+}{C}O$, at m/z 43).

5.7 FRAGMENTATIONS ASSOCIATED WITH FUNCTIONAL GROUPS

In simple monofunctional compounds we can with reasonable certainty predict that the mass spectrum will bear a compound relationship to (a) the nature of the carbon skeleton (whether alkane, alkene, aromatic, etc.) and (b) the nature of the functional group. For difunctional compounds, a more complex interaction can be expected, depending on the relative powers of the two functional groups to direct the fragmentation.

It is always easier to rationalize the mass spectrum of a known structure than to deduce the structure of an unknown compound from its mass spectrum; other spectroscopic evidence for the presence of functional groups should ideally always be available.

5.7.1   ALKANES AND ALKANE GROUPS

The molecular ion will normally be seen in the mass spectra of the lower
n-alkanes, but its intensity falls off with increased size and branching of the
chain.

Figure 5.6 shows the typical mass spectrum for an *n*-alkane—dodecane
in this case. The relative abundances of the ions are also typical, showing
maximum abundance around $C_3H_7^+$ and $C_4H_9^+$, with a weak $M^{\ddagger}$ peak.
Although *n*-dodecane is unbranched, the alkane ions from $C_4H_9^+$ up
rearrange in the mass spectrometer to the branched-chain form: this is
quite analogous to the Wagner–Meerwein shifts which occur in (for
example) S_N1 reactions involving cations, the more stable branched
structures being preferred.

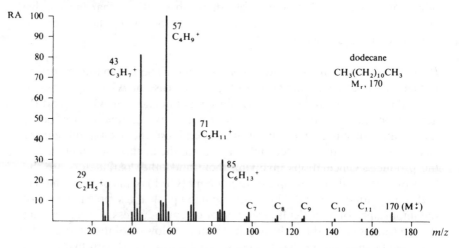

Figure 5.6   *Mass spectrum of dodecane, showing the typical distribution of
fragment ions for unbranched alkanes.*

Associated with each of these C_nH_{2n+1} ions (from C_2 to C_5) are lesser
amounts of the corresponding alkenyl ion (C_nH_{2n-1}) formed by loss of two
hydrogen atoms: they appear at *m/z* 27, 41, 55, 69. Loss of one hydrogen is
also seen. Metastable ions of very low intensity can be detected for the
fragmentations in which the alkyl ions extrude a smaller molecule: thus,
$C_2H_5^+$ and $C_3H_7^+$ extrude H_2; $C_4H_9^+$ extrudes CH_4; $C_5H_{11}^+$ and $C_6H_{13}^+$
extrude C_2H_4; $C_6H_{13}^+$ and $C_7H_{15}^+$ extrude C_3H_6. These are of too low
abundance to be shown in figure 5.6.

Branched-chain alkanes rupture predominantly at the branching points,
and the largest group attached to the branching point is often preferentially
expelled as a radical. The normal rules of chemistry hold, in that the
preference is for the formation of tertiary over secondary over primary
cations.

We might predict, therefore, in the mass spectrum of 2,2-
dimethylpentane (a) no molecular ion peak, (b) substantial peaks for

$C_3H_7^+$, $C_4H_9^+$ (because of their inherent stability), (c) a much higher abundance of $C_5H_{11}^+$ than would be seen for the fragmentation of n-C_7H_{16} (because of the ease of C_2H_5 expulsion) and (d) a reasonable peak for $C_6H_{13}^+$ due to M − CH_3.

5.7.2   CYCLOALKANES

Complex fragmentations usually occur for cycloalkanes, ring size obviously being important in relation to ion stability. Typically, for the simple members the molecular ion peak will be easily seen, its intensity reducing as branching increases.

Common fragmentations are loss of alkenes or alkenyl ions and the splitting off of the side-chains with charge retention by the ring remnant; side-chain loss is simply a special case of fragmentation at a branching point.

Figure 5.7 is the mass spectrum of n-propylcyclohexane, which should be interpreted with these points in mind.

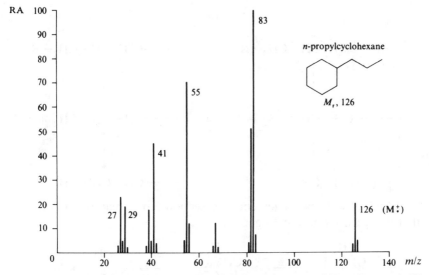

Figure 5.7   *Mass spectrum of n-propylcyclohexane: because of branching, m/z 57 is no longer a prominent fragment.*

Example 5.2

Question. Predict the structure of the base peak in the mass spectrum of (a) n-octane and (b) 2-methylpentane.

Model answer. (a) As an unbranched alkane, we would predict the base peak to be the same as for other long-chain n-alkanes such as n-decane (figure 5.6) and be m/z 57, $C_4H_9^+$. (b) The molecular ion will fragment preferentially at the branch in the chain, and since rupture at the bond between C-2 and C-3 gives two fragments each with m/z 43 ($C_3H_7^+$), this will be the base peak: see figure 5.2.

Exercise 5.2   Write formulae for the fragment ions in the mass spectrum of
n-propylcyclohexane (figure 5.7) with m/z values 83, 55, 41, 29 and 27.

Exercise 5.3   In the mass spectrum in figure 5.7, what reactions give rise to
the peaks at (a) m/z 125 and (b) m/z 127? Which of these fragments is a
cation and which a radical cation?

5.7.3   ALKENES AND ALKENE GROUPS

Molecular ion peaks for simple alkenes are normally distinctly seen. The
commonest fragmentation in alkene groups involves rupture of the allylic
bond (β to the double bond), which gives rise to the resonance-stabilized
allylic cation. Since the allylic radical is also stabilized, the fragmentation
may give rise to peaks corresponding to charge retention by either of the
fragments.

$$-\!\!-\!\!-\text{CH}\!=\!\text{CH}\!-\!\text{CH}_2\!\!+\!\!\!\text{R}$$

$$-e \qquad \text{or} \qquad -e$$

$$-\!\!-\!\!-\text{CH}\overset{+\cdot}{=}\text{CH}\!-\!\text{CH}_2\!-\!\text{R} \qquad\qquad -\!\!-\!\!-\text{CH}\overset{\cdot+}{=}\text{CH}\!-\!\text{CH}_2\!-\!\text{R}$$

$$-\!\!-\!\!-\overset{+}{\text{C}}\text{H}\!-\!\text{CH}\!=\!\text{CH}_2 + \text{R}^{\cdot} \quad \text{or} \qquad -\!\!-\!\!-\overset{\cdot}{\text{C}}\text{H}\!-\!\text{CH}\!=\!\text{CH}_2 + \text{R}^{+}$$

$$-\!\!-\!\!-\text{CH}\!=\!\text{CH}\!-\!\overset{+}{\text{C}}\text{H}_2 \qquad\qquad\qquad -\!\!-\!\!-\text{CH}\!=\!\text{CH}\!-\!\overset{\cdot}{\text{C}}\text{H}_2$$

   A McLafferty rearrangement (see section 5.6.2) may occur, provided
that the γ-carbon has hydrogen on it.

Example 5.3

Question. What are the masses of the two ions produced in the mass
spectrum of 2-hexene by β-fragmentation?

Model answer. The allylic bond, β to the double bond, is between C-4 and
C-5, and this breaks to give two fragments, either of which will retain the
charge, at m/z 29 and 55. The reaction producing m/z 29 ($C_2H_5^+$) also
produces the $C_4H_7^{\cdot}$ radical, which is not detected in the spectrometer.

Conversely, the reaction which produces m/z 55 ($C_4H_7^+$) produces the $C_2H_5^{\cdot}$ radical, also undetected.

Exercise 5.4 Predict two abundant ions in the mass spectrum of 3-hexene. Which fragments are lost in producing these ions, and are they radicals, cations or radical cations?

Exercise 5.5 Predict the masses of the ions produced in the mass spectrum of (a) 1-hexene and (b) 2-heptene by the McLafferty rearrangement. Are these radical cations? What happens to the alkene fragments produced in these rearrangements?

5.7.4 CYCLOALKENES

Fragmentation of cycloalkenes is directed by the double bond and by the nature of any acyclic alkane residues present, so that allylic rupture and McLafferty rearrangements are common.

In addition, cyclohexene derivatives give the important retro-Diels–Alder reaction discussed in section 5.6.2.

Exercise 5.6 What are the masses of the ions produced in the mass spectrum of (a) 2,5-dimethylcyclohexane and (b) 4,5-dimethylcyclohexane by the retro-Diels–Alder process? Are these fragments cations? What happens to the alkene fragments extruded in these processes?

5.7.5 ALKYNES

No simple pattern emerges for the fragmentation of alkynes, which can be applied to complex molecules. Thus, for 1-butyne and 2-butyne the molecular ion peak is the base peak, but the molecular ion peak for higher members is weak. Loss of alkyl radicals gives prominent peaks in many cases (at M − 15, M − 29, etc.) and extrusion of alkenes may give M − 28 and M − 42 peaks, etc.

5.7.6 AROMATIC HYDROCARBON GROUPS

The molecular ions of aromatic hydrocarbons are always abundant, and M^{+} is commonly the base peak; accordingly, M + 1 and occasionally M + 2 peaks are easily seen. Polynuclear aromatic hydrocarbons have particularly stable molecular ions, and doubly or triply charged ions are possible. Doubly charged molecular ions, $m/2z$, will appear at integer m/z values; for example, for naphthalene ($C_{10}H_8$) M^{+} is at m/z 128 and $m/2z$ is at m/z 64. However, the corresponding $(M + 1)/2z$ peak must appear at noninteger m/z values (m/z 64.5 for naphthalene), which makes it easy to pinpoint the presence of doubly charged ions. Triply charged ions must appear at noninteger values: for naphthalene, $m/3z$ is at m/z 42.6.

Alkylbenzenes, I, are the commonest hydrocarbons in this class; here the dominant fragmentation is at the benzylic bond, for reasons analogous to those discussed under allylic fission in alkenes. Peaks corresponding to M − H are also common.

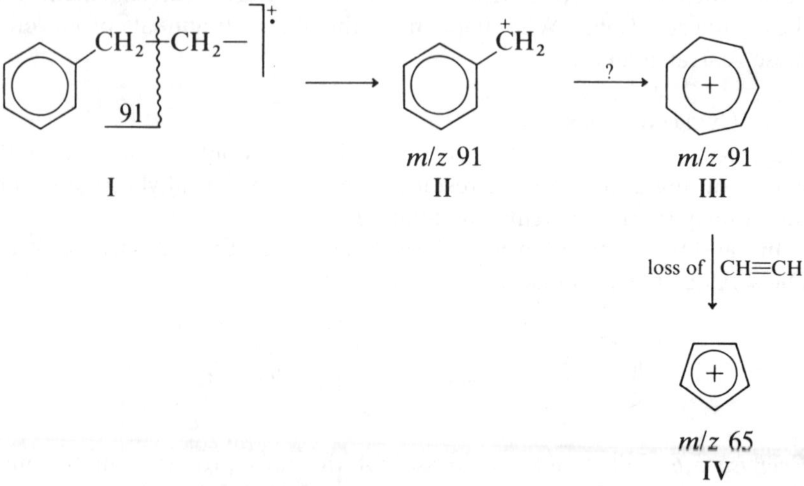

The stable benzyl cation, II, would certainly explain the abundant m/z 91 peak seen in the mass spectra of all compounds of this type, but structure II cannot explain the randomization of C and H atoms shown to occur by isotope substitution studies (see section 5S.3). The best explanation is that the ion m/z 91 is the equally stable tropylium ion, III; the C_7 ring must be formed by rearrangement of the benzyl group *either before or immediately after* expulsion of the m/z 91 fragment.

The m/z 91 ion subsequently expels C_2H_2 (acetylene), giving m/z 65, which may have the stable structure IV, and this fragmentation gives rise to a metastable ion at $65^2/91 = m/z$ 46.4.

McLafferty rearrangements are observed in alkylbenzenes, provided that the side-chain has hydrogen on the γ carbon atom. In monoalkylbenzenes this gives rise to an ion at m/z 92, which may be confused with the m + 1 peak for the tropylium ion; in this latter case the intensity of m/z 92 would be ≈ 7.7 per cent (7 × 1.1 per cent) of the tropylium ion intensity.

A feasible structure for the m/z 92 ion is the methylenecyclohexadienyl radical ion shown below.

The phenyl cation, $C_6H_5^+$, at m/z 77 is produced from many aromatics by rupture of the bond α to the ring, and this ion extrudes C_2H_2 (acetylene),

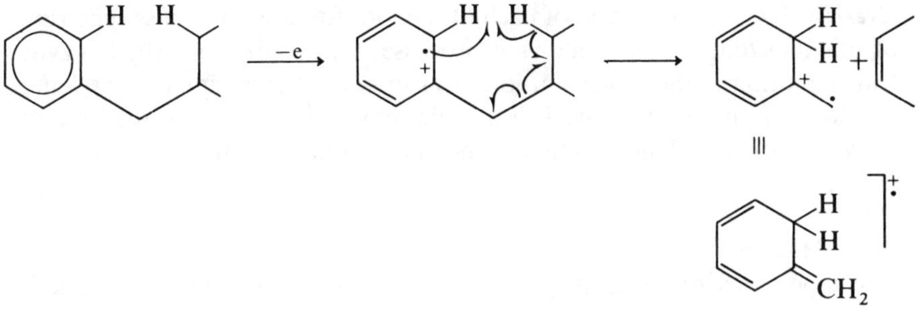

III

m/z 92
methylenecyclohexadienyl
radical ion

to give m/z 51. This fragmentation gives rise to a metastable ion at
$51^2/77 = m/z$ 33.8.

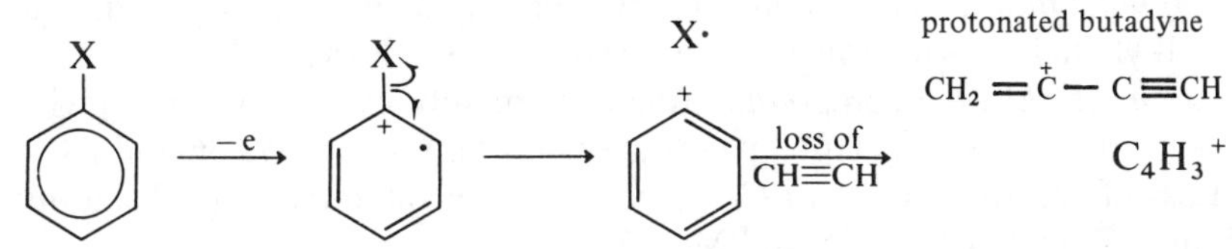

X·

protonated butadyne

$CH_2 = \overset{+}{C} - C \equiv CH$

m/z 77

loss of $CH \equiv CH$

$C_4H_3^+$

m/z 51

Example 5.4

Question. What are the masses of the ions produced in the mass spectrum
of 4-*n*-butyltoluene, $CH_3C_6H_4CH_2CH_2CH_2CH_3$, by (a) benzylic fission and
(b) the McLafferty rearrangement; label the ions as cations or radical
cations.

Model answer. (a) Fission at the benzylic bond expels the C_3H_7 fragment of
mass 43 (as a radical), leaving the methyltropylium ion, $C_8H_9^+$ (m/z 105),
as a cation. (b) In the McLafferty rearrangement one hydrogen is
transferred to the ring from the butyl residue (from the position γ to the
ring), and simultaneously the benzylic bond breaks. The molecular ion
(radical cation) loses the neutral molecule propene, C_3H_6, with mass 42,
leaving the fragment $C_8H_{10}^+$ (also a radical cation) at m/z 106.

Exercise 5.7 Anthracene (molecular formula $C_{14}H_{10}$) is a polycyclic
aromatic hydrocarbon with three linearly fused benzene rings. What is the
mass of (a) its molecular ion and (b) its doubly charged molecular ion?
Where does the latter appear in the mass spectrum of anthracene?

Exercise 5.8 What is the mass of the (M + 1) peak for anthracene?
Where on the mass spectrum will (a) the singly charged and (b) the doubly
charged ions of this species appear?

Exercise 5.9   List the mass of each of the ions found in the mass spectrum of ethylbenzene by the following processes: (a) ionization; (b) benzylic fragmentation of the molecular ion; (c) subsequent extrusion of acetylene; (d) loss of an ethyl radical from the molecular ion; (e) subsequent extrusion of acetylene. State whether each ion is a cation or a radical cation.

5.7.7 HALIDES

The appearance of the mass spectrum of a halogen-containing compound is profoundly affected by the number of halogen atoms present, because of isotopic abundances (as discussed in section 5.3). Fluorine and iodine, being monoisotopic, present no problems in this respect.

The fragmentation of mixed halogen compounds is very complicated, and we shall deal here with compounds containing only one of the halogens.

Aliphatic fluorine compounds, apart from fragmentations appropriate to the alkyl chain, show principally a peak at M − HF (M − 20).

Aliphatic chlorine compounds fragment mainly by loss of HCl (giving M − 36 and M − 38): HCl$^+$ peaks may also be seen at m/z 36 and 38.

Loss of chlorine as Cl$^+$ or Cl˙ gives rise to low-abundance peaks at m/z 35 and 37, and at M − 35 and M − 37.

Aliphatic bromine compounds fragment similarly to chloro compounds, but loss of Br˙ is the preferred fragmentation.

Aliphatic iodine compounds show peaks corresponding to I$^+$ (m/z 127), M − I (M − 127) and M − H$_2$I (M − 129).

Aryl halides, with halogen directly attached to the ring, show abundant molecular ion peaks, but the fragmentation is dominated by the relative stability of the aryl cations. Consequently, halogen is mainly expelled as a radical, and the (M − halogen) ion fragments as discussed in section 5.7.6.

Acyl halides are discussed with other carboxylic acid derivatives: see section 5.7.18.

Example 5.5

Question. Use the formula given in section 5.3 to calculate the relative abundances of the halogen isotope peaks in the molecular ions of 1,3-dichloropropane. Ignore any contribution from carbon-13.

Model answer. The molecular ions will appear at m/z 112, 114 and 116. To calculate the ion abundances, we use the formula $(a + b)^n$, where $a = 3$ (relative abundance of chlorine-35), $b = 1$ (relative abundance of chlorine-37) and $n = 2$ (number of chlorines in the molecule). Expansion gives $a^2 + 2ab + b^2$, indicating that the first ion has two chlorine-35 atoms, the third has two chlorine-37 atoms and the middle ion has one of each. The relative abundance of the first ion is $a^2 = 9$, that of the second ion is

$2ab = 6$ and that of the third ion is $b^2 = 1$: thus, the three ions in the molecular ions cluster appear in the abundance ratio 9:6:1. This is sufficiently accurate for most analytical uses; a more exact prediction would use the data in table 5.1, which shows the ratio of the abundances of chlorine-35 and chlorine-37 to be 75.8:24.2, or 3.13.

Exercise 5.10   Predict the general appearance of the molecular ion peaks for 1-chlorobutane, considering the chlorine and carbon isotopes only.

Exercise 5.11   What would you predict in the cluster of ions around the molecular ions of (a) 1-bromobutane and (b) 1,4-dibromobutane, allowing for the bromine and carbon isotopes only?

5.7.8   ALCOHOLS

Molecular ion peaks for primary and secondary alcohols are weak; for tertiary alcohols the $M^{\ddot{+}}$ peak is usually absent.

A number of fragmentations are open to alcohols, and their relative importance depends on whether the alcohol is primary, secondary, tertiary, aliphatic or aromatic; the most important fragmentation is normally rupture of the bond β to oxygen.

Alcohols of low volatility can be converted into their trimethylsilyl ethers ($ROH \rightarrow ROSiMe_3$), which are much more volatile because of the loss of hydrogen bonding present in the alcohols themselves (see section 5S.4).

Primary aliphatic alcohols (alkanols) show M − 18 peaks corresponding to loss of H_2O.

An accompanying loss of water together with an alkene is shown by alcohols with more than four carbons in the chain, and this simultaneous elimination of an alkene and water can be indicated mechanistically as shown below. Peaks corresponding to this process appear at (M − 18 − C_nH_{2n})—for example, (M − 18 − 28) for loss of water and C_2H_2, etc.:

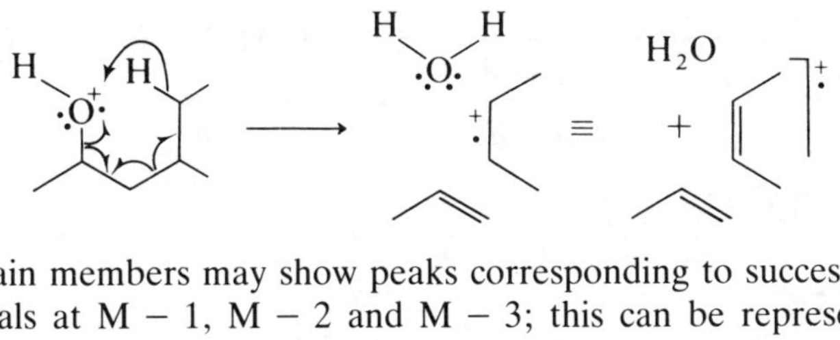

Long-chain members may show peaks corresponding to successive loss of H radicals at M − 1, M − 2 and M − 3; this can be represented as shown below:

$$\overset{\underset{|}{\overset{|}{C}}}{\underset{H}{\overset{H}{\underset{|}{C}}}}\!\!-\!\!\ddot{\underset{\cdot\cdot}{O}}\!-\!H \longrightarrow \overset{H}{\underset{|}{C}}\!=\!\overset{+}{\underset{\cdot\cdot}{O}}\!\!\frown\!\!H \longrightarrow \overset{H}{\underset{}{C}}\!\!=\!\overset{+\cdot}{\underset{\cdot\cdot}{O}} \longrightarrow -C\!\equiv\!\overset{+}{\ddot{O}}.$$

$$M^{\overset{+}{\cdot}} \qquad\qquad M-1 \qquad\qquad M-2 \qquad\qquad M-3$$

Secondary and tertiary aliphatic alcohols preferentially fragment by loss of alkyl radicals, the ease of elimination increasing with increased size and branching in the radical. Thus, the alcohol shown would be expected to give rise to a prominent peak at $M - C_4H_9$, with less abundant peaks at $M - C_3H_7$ and $M - C_2H_5$:

$$C_4H_9\!\!-\!\!\overset{\overset{+}{O}H}{\underset{\underset{C_2H_5}{|}}{\overset{|}{C}}}\!\!-\!\!C_3H_7 \longrightarrow C_4H_9^{\cdot} + \overset{\overset{+}{O}H}{\underset{\underset{C_2H_5}{|}}{\overset{\|}{C}}}\!\!-\!\!C_3H_7$$

Aromatic alcohols with the OH group in the benzylic position fragment so as to favor charge retention by the aryl group; thus, in 1-phenylethanol the base peak corresponds to elimination of $CH_3^{\cdot}$. Peaks corresponding to $ArCO^+$ and Ar^+ are also shown. The peak at m/z 107 given by these alcohols (the base peak in the case of 1-phenylethanol) is best represented as the hydroxytropylium ion. Loss of CO from this ion gives m/z 79, and loss of H leads on to the phenyl cation at m/z 77, which loses C_2H_2 to give m/z 51:

$$\underset{\text{1-phenylethanol}}{\overset{\overset{\displaystyle OH}{\overset{|}{\bigcirc\!\!-\!CHCH_3}}}{}}$$

$$\underset{m/z\ 107}{\overset{OH}{\bigcirc\!\!\!\!+}} \xrightarrow{-CO} \underset{m/z\ 79}{\overset{H\quad H}{\bigcirc\!\!\!\!+}} \longrightarrow \underset{m/z\ 77}{C_6H_5^+} \longrightarrow \underset{m/z\ 51}{C_4H_3^+}$$

hydroxytropylium
ion

Example 5.6
Question. The mass spectrum of 1-(4-methylphenyl)-ethanol (*p*-$CH_3C_6H_4CH(OH)CH_3$) shows an abundant fragment ion at m/z 121, and a less abundant ion at m/z 119. What are feasible structures for these ions?

Model answer. Benzylic alcohols readily fragment at the benzylic carbon, expelling a radical, and leaving an ion which rearranges to the hydroxy-tropylium structure. In this case the molecular ion will appear at *m/z* 136 and therefore the ion at *m/z* 121 corresponds simply to loss of $CH_3\cdot$, and its structure is methylhydroxytropylium, $C_8H_9O^+$. The ion at *m/z* 119 corresponds to loss of CH_5 ($CH_3\cdot$ and two hydrogen atoms) and is the acyl ion $CH_3C_6H_4CO^+$.

Exercise 5.12 Glycerol (1,2,3-trihydroxypropane) has too low a vapor pressure for successful gas chromatography: it can be converted to its tris-trimethylsilyl derivative and this is easily submitted to combined GC–MS. Calculate the mass of the molecular ion, assuming silicon-28 and carbon-12 only.

Exercise 5.13 Calculate the mass of the fragment ion produced in the mass spectrum of 4-methyl-2-hexanol by simultaneous loss of water and alkene. Complete the description of this process by supplying the missing digits in (M − 18 − ?).

5.7.9 PHENOLS

Simple phenols give strong molecular ion peaks.

The commonest fragmentation is loss of CO (M − 28) and CHO (M − 29), which can only be represented as shown below:

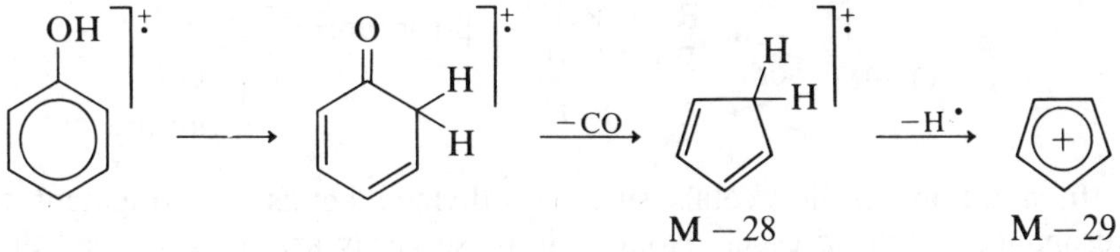

$$M - 28 \qquad\qquad M - 29$$

Phenols with alkyl side-chains also undergo benzylic fission, leaving variants of the hydroxytropylium ion (see section 5.7.8):

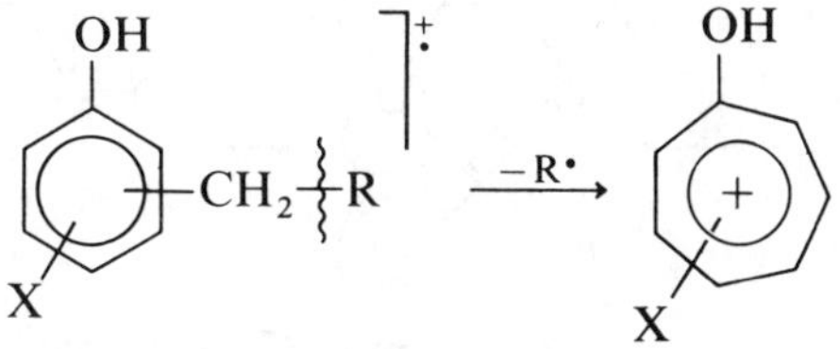

m/z 107 when X = H

Exercise 5.14 Consider the fragmentation of alcohols with benzylic OH (section 5.7.8) and of phenols with alkyl side-chains (section 5.7.9). Would mass spectrometry be able to distinguish 1-phenyl-1-propanol from the isomeric *p*-propylphenol?

5.7.10 ETHERS, ACETALS AND KETALS

On the whole, molecular ion peaks for these classes are weak.

Aliphatic ethers principally fragment at the bond β to oxygen, and the largest group is expelled preferentially as a radical (compare alcohols). Peaks appear therefore at M − CH_3, M − C_2H_5, etc.:

$$-\overset{|}{\underset{|}{C}}\overset{\cdot+}{\underset{\ddot{}}{O}}-\overset{|}{\underset{|}{C}}- \quad \xrightarrow{\;-R^{\cdot}\;} \quad -C=\overset{+}{\underset{\ddot{}}{O}}-\overset{|}{\underset{|}{C}}-$$

Where β hydrogen is present, the oxonium ion may fragment further to eliminate an alkene:

$$-C=\overset{+}{\underset{\ddot{}}{O}}\overset{H\frown C-}{\underset{|}{C}}- \quad \longrightarrow \quad -C=\overset{+}{\underset{\ddot{}}{O}}H \;+\; \overset{|}{\underset{|}{\underset{C-}{\overset{C-}{\|}}}}$$

Fission of the C—O bond (the α bond) occurs to a small extent, charge retention by carbon being favored over retention by oxygen (see below).

Acetals and ketals show simple extensions of the ether fragmentation processes, rupture of the β bond being favored over the α:

$$R-O-R \quad \begin{array}{l} \nearrow \; R^+ + RO^{\cdot} \;\text{(preferred)} \\ \searrow \; R^{\cdot} + RO^+ \end{array}$$

β-rupture — $\underset{\text{α-rupture}}{\overset{OR}{C{<}_{OR}}}$

In addition, cyclic acetals such as ethylene ketals will fragment to resonance-stabilized cyclic oxonium ions: so easily are these formed that they often dominate the fragmentation pattern and give rise to the base peak:

$$:O: \quad :O^+ \quad \xrightarrow{\;-R^{\cdot}\;} \quad :O: \quad \cdot O^+ \quad \longleftrightarrow \quad :O^+ \quad \cdot O.$$

m/z 87, 101, etc.

For cyclic ketals in which the ketone residue is itself cyclic (as in the ethylene ketals of cyclohexanone derivatives) the initial bond rupture is modified by a second rupture so that an alkonium alkene ion is formed:

$$m/z\ 99$$

$$+ C_3H_7^{\bullet}$$

Forming the ethylene ketal constitutes a valuable derivatization technique for directing the fragmentation of ketones (see section 5S.4).

Aromatic ethers other than methyl ethers commonly fragment by variants of the β-hydrogen transfer discussed above for aliphatic ethers: a peak at m/z 94 is formed, which extrudes CO, giving m/z 66:

$$m/z\ 94$$

$$+ \overset{\bullet}{C}H_2$$

$$-CO$$

$$m/z\ 66$$

Methyl phenyl ethers undergo two main fissions at the C—O bonds: loss of HCHO (giving M − 30) and loss of CH_3 (giving M − 15), the latter process giving an ion that further splits out CO, leaving m/z 65:

$$m/z\ 78$$

$$+ HCHO$$

$$-CH_3^{\bullet}$$

$$m/z\ 93 \qquad \xrightarrow{-CO} \qquad m/z\ 65$$

Exercise 5.15 When the molecular ion from *n*-butyl phenyl ether fragments with β-hydrogen transfer, what neutral (uncharged) fragment is extruded, and what is its mass?

5.7.11 CARBONYL COMPOUNDS GENERALLY

Before discussing the idiosyncratic behavior of individual carbonyl classes, we should draw attention to the important similarities in their fragmentation modes.

α-Cleavage. This occurs in all classes; the bonds at the carbonyl group rupture, and the ion abundance can be roughly predicted on the basis of resonance stabilization, etc. In summary,

$$
\begin{array}{c}
\left[\begin{array}{c} X \\ \diagdown CO \\ \diagup \\ Y \end{array} \right]^{\ddagger} \xrightarrow[\text{or}]{} \begin{array}{l} X^+ + YCO^\bullet \\ \\ X^\bullet + YCO^+ \end{array}
\end{array}
$$

Thus, in RCHO aldehydes, where X or Y is H, we find peaks corresponding to R^+, RCO^+ and HCO^+. For aryl aldehydes, the stability of $ArCO^+$ makes this a major contributor.

For carboxylic acids (RCOOH), where X or Y is OH, we find peaks corresponding to R^+, $COOH^+$ (*m/z* 45) and RCO^+. For aryl carboxylic acids, $ArCO^+$ and Ar^+ are particularly stable.

For all aryl carbonyl compounds such as PhCOX, the ion $PhCO^+$ (*m/z* 105) will lose CO, to give Ph^+ (*m/z* 77), which loses C_2H_2, to give *m/z* 51. Metastable ions appear with the sequence $105 \rightarrow 77 \rightarrow 51$, the metastable *m/z* values being 56.5 and 33.8, respectively.

β-Cleavage. This often occurs with expulsion of alkyl ions from aliphatic aldehydes.

β-Cleavage with McLafferty rearrangement. This is very much more common, provided that γ-hydrogen is present (see section 5.6.2):

For the simple unbranched aliphatic aldehydes, X = H and R = H, and we can expect an ion at m/z 44 for this fragmentation. For the corresponding ketones, X must be CH_3, C_2H_5, etc., giving peaks at m/z 58, 72, etc. For carboxylic acids, with X = OH, the peaks appear at m/z 60, 74, etc.

We can now go on to the individual carbonyl classes, noting common ion values and any fragmentation in addition to α- and β-cleavage.

Example 5.7

Question. The mass spectrum of butyrophenone (*n*-butyl phenyl ketone, $C_6H_5COCH_2CH_2CH_2CH_3$) shows peaks at m/z 162, 120, 105 and 85: interpret these.

Model answer. With molecular formula $C_{11}H_{14}O$, the m/z 162 peak is the molecular ion. The m/z 120 fragment therefore corresponds to loss of 40 daltons from $M^{\ddagger}$, which is accounted for by expulsion of propene, C_3H_6, in the McLafferty rearrangement. The ions m/z 105 and m/z 85 have formulae $C_6H_5CO^+$ and $C_4H_9CO^+$, respectively, and both arise from α-fragmentation.

Exercise 5.16 In the mass spectrum of acetophenone (methyl phenyl ketone, MeCOPh) predict a very stable fragment ion containing the aromatic ring, and state how this will, in turn, fragment.

Exercise 5.17 What is the mass of the 'McLafferty ion' from the following aldehydes: butanal, pentanal, hexanal and heptanal?

Exercise 5.18 What is the mass of the 'McLafferty ion' from the following ketones: 2-pentanone, 2-hexanone, 2-heptanone and 2-octanone?

5.7.12 ALDEHYDES

Aliphatic aldehydes give weak molecular ion peaks, whereas aryl aldehydes give strong $M^{\ddagger}$ peaks.

Loss of H˙ from the molecular ion is particularly favored by aryl aldehydes because of the stability of $ArCO^+$; nevertheless, M − 1 peaks are prominent in the mass spectra of all aldehydes, whether aliphatic or atomatic. α-Cleavage gives R^+ peaks and the HCO^+ ion at m/z 29: $C_2H_5^+$ also appears at m/z 29, and for higher aliphatic aldehydes $C_2H_5^+$ is more likely to be the source of m/z 29 than HCO^+.

Other cleavages are discussed in section 5.7.11.

5.7.13 KETONES AND QUINONES

Molecular ion peaks for all ketones are usually strong. Most of the abundant ions in the mass spectra of ketones can be accounted for by α-cleavages and McLafferty rearrangements, and indeed the base peak for methyl ketones and phenyl ketones is frequently CH_3CO^+ (m/z 43) and

$PhCO^+$ (*m/z* 105), respectively: $PhCO^+$ fragments as usual to *m/z* 77 and *m/z* 51.

When RCOR′ undergoes α-cleavage, the larger group is preferentially expelled as the radical, with concomitant charge retention by the smaller version of RCO^+.

Common values for the McLafferty rearrangement ions of aliphatic ketones are *m/z* 58, *m/z* 72, *m/z* 86, etc., but the higher members of the series may undergo a second elimination of an alkene, to generate a further series of ions. For example, *m/z* 86 may lose ethylene, giving an ion at *m/z* 86 − 28, and loss of C_3H_6 will correspond to loss of 42 mass units. For simple aromatic ketones the McLafferty rearrangement ions might arise at *m/z* 120, *m/z* 134, etc. All of the McLafferty rearrangement ions appear at even *m/z* values.

Ethylene ketal derivatives of ketones (mentioned in section 5.7.10) have very strongly directed fragmentations, and are discussed further in section 5S.4.

Quinones mainly undergo α-cleavage, both α bonds being capable of rupture.

      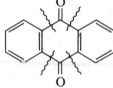

p-benzoquinone   α-naphthoquinone   9,10-anthraquinone

In *p*-benzoquinone and naphthoquinone these fragmentations lead to peaks at *m/z* 54 and at M − 54, whereas α-cleavage in anthraquinone gives the whole range of ions corresponding to M − CO, M − 2CO and *m/z* 76 ($C_6H_4^+$).

5.7.14 CARBOXYLIC ACIDS
Molecular ion peaks are usually observable, but weak. Common peaks arising from α-cleavage and McLafferty rearrangements have already been mentioned in section 5.7.11.

5.7.15 ESTERS
Methyl esters are usually more convenient to study than the free acids, because they are more volatile. The molecular ion peak is weak but discernible in most cases, and the fragmentation is a mixture of α-cleavage and (where appropriate) McLafferty rearrangements. Common ion values for α-cleavage are, therefore, RCO^+, R^+, CH_3O^+ (*m/z* 31) and CH_3OCO^+ (*m/z* 59); the McLafferty ion at *m/z* 74 $(CH_2{=}C(OH)OCH_3)^{+\cdot}$ is the base peak in the saturated straight-chain methyl esters from C_6 to C_{26}.

Higher esters are complicated by the possibility of two fragmentation modes—fragmentations of the acyl group (RCO—) and of the alkyloxy group (ROCO—). The acyl group fragmentation is a simple extension of that discussed under methyl esters, but where the alkyloxy group is C_2H_5O (ethyl esters) or higher, the McLafferty ion (m/z 88 for ethyl esters) undergoes loss of both an alkene and an alkenyl radical, to give ions at m/z 60 and m/z 61 ($CH_3CO_2H^+$ and $CH_3CO_2H_2^+$). The number of possible esters is extremely high, and the relative importance of the different fragmentation modes is almost unique for each member, so that we must be content with these general pointers.

Exercise 5.19 In combined GC–MS analyses (see section 5S.2) fatty acid methyl esters can be identified by the presence of a peak at m/z 74: why is this?

5.7.16 AMIDES

Primary aliphatic amides, $RCONH_2$, undergo α-cleavage to R· and $CONH_2^+$ (m/z 44). Where possible, McLafferty rearrangement occurs to yield homologous variants at $CH_2\!=\!C(OH)NH_2^+$ at m/z 59, m/z 73, etc. Loss of NH_2 gives M − 16 peaks.

Primary aryl amides usually have this as their primary fragmentation, leading to $ArCO^+$ ions.

For secondary (RCONHR′) and tertiary amides (RCONR′R″), the number of individual variations is enormous, as in esters, when we consider that fragmentation must take into account the nature of the three groups R, R′ and R″.

5.7.17 ANHYDRIDES

Molecular ion peaks are weak or absent.

Saturated acyclic anhydrides (acetic anhydride, etc.) fragment mainly to RCO^+ (with m/z 43, 57, etc.), although chain-branching may cause such easy fragmentation that the complete RCO^+ is not detected. Ions at M − 60 and m/z 60 are common ($CH_3CO_2H = 60$), as are m/z 42 ($CH_2\!=\!CO^+$) and McLafferty ions at m/z 44, m/z 58, etc.

Cyclic aliphatic anhydrides such as succinic anhydride show a strong or base peak at M − 72 caused by loss of CO_2 +CO from M^+, and this is also shown by the cyclic aromatic anhydrides (phthalic anhydride, etc.). For these latter, other common ions are $ArCO^+$, $ArCO_2H^+$ and M − CO: loss of H from $ArCO^+$ can also occur.

5.7.18 ACID CHLORIDES

Aliphatic members show fragmentations associated with the Cl and CO groups, so that the following ions are common: HCl^+, M − Cl, $COCl^+$, RCO^+, etc. The effect of isotope abundance (^{35}Cl: $^{37}Cl \approx 3{:}1$) makes it easy to identify the chlorine-containing peaks. For aryl acid chlorides (for example, benzoyl chloride), the stable $ArCO^+$ makes loss of $Cl^{\cdot}$ from M^{+} a dominant process.

5.7.19 NITRILES

Molecular ion peaks are usually weak or absent, although an $M - 1$ ion $(R—CH=C=N^+)$ may be seen.

For the lower aliphatic members, $M - 27$ (corresponding to $M - HCN$) is seen; but from C_4 on, the McLafferty ion is frequently the base peak: these ions appear at m/z 41, or m/z 55, or m/z 69, etc., and can be represented as the homologs of $CH_2=C=NH^{+}$.

Aryl nitriles often show $M - HCN$ peaks, although if alkyl side-chains are present, benzylic rupture will give rise to the main series of ions.

5.7.20 NITRO COMPOUNDS

Molecular ion peaks for aliphatic members are usually absent, while for many aromatic nitro compounds the M^{+} peak is strong.

The mass spectra of aliphatic nitro compounds mainly correspond to fragmentation of the alkyl chain, although peaks for NO^+ (m/z 30) and NO_2^+ (m/z 46) appear also.

For aryl nitro compounds, peaks corresponding to NO^+ (m/z 30), NO_2^+ (m/z 46), $M - NO$ ($M - 30$) and $M - NO_2$ ($M - 46$) all appear commonly, and successive loss of NO and CO ($M - 58$) is also observed. Loss of NO from the molecular ion leaves an ion of structure ArO^+, which can only arise if $ArNO_2^{+}$ rearranges in the spectrometer.

Ortho effects are frequently encountered in aryl nitro compounds where a group *ortho* to NO_2 contains hydrogen: peaks corresponding to $M - OH$ ($M - 17$) appear, indicating loss of oxygen from NO_2 together with H from the *ortho* substituent.

5.7.21 AMINES AND NITROGEN HETEROCYCLES

An odd number of nitrogen atoms in the molecule means an odd relative molecular mass (molecular weight).

For primary aliphatic amines, the base peak is $CH_2=\overset{+}{N}H_2$ (at m/z 30) formed by expulsion of a radical from M^{+}. Higher homologs may also appear at m/z 44, m/z 58, etc., but these are less abundant ions. Loss of an alkene (for example, loss of C_2H_4) may give peaks at $M - 28$, etc.

Secondary and tertiary amines behave analogously, and loss of the substituent alkyl radicals is also observed.

Primary aryl amines principally fragment by loss of H ($M - 1$) and HCN ($M - 27$): thus, aniline gives rise to an ion at m/z 66 ($93 - 27$) whose structure can be represented as the cyclopentadienyl radical cation. For *N*-alkylanilines, α-cleavage of the alkyl group is common.

Aromatic heterocyclic bases give rise to abundant M^+ peaks, and for pyridine and quinoline the principal fragmentation is loss of HCN (M − 27). Alkyl-substituted derivatives fragment at the 'benzylic' bond, as for alkylbenzenes, and loss of HCN commonly follows on this process.

Exercise 5.20 Two amines have respective relative molecular masses of 107 and 108 and in their mass spectra both show M − 1 and M − 27 peaks. What can be deduced from this, in relation to the number of nitrogen atoms in each?

5.7.22 SULFUR COMPOUNDS

Fragmentations of thiols (mercaptans and thiophenols) and sulfides bear close comparison with their oxygen analogs, with the additional complication due to M + 2 and m + 2 peaks because of ^{32}S and ^{34}S isotope abundances. Normally, molecular ion peaks are clearly seen.

Aliphatic thiols may also give rise to the following ions: S^+, HS^+, H_2S^+ and M − H_2S.

The commonest sulfur heterocycles are thiophene derivatives: thiophene itself gives a strong M^+ peak, together with HCS^+ (the thioformyl ion) and M − HCS. As for Ph^+, loss of $CH{\equiv}CH$ produces a peak at M − 26.

For sulfides (RSR') α-cleavage gives RS^+, often followed by loss of CS, to give RS − CS peaks. Ions also arise due to M − HS processes.

SUPPLEMENT 5

5S.1 ALTERNATIVES TO ELECTRON-IMPACT IONIZATION

The ability to measure accurately the relative molecular mass (molecular weight) of organic compounds by mass spectrometry is only possible if a sufficiently stable molecular ion can be formed, and we have seen that many classes of compound do not do so when electron-impact ionization is used. Partly, the reason lies in the large amount of excess energy imparted to the molecular ion by 70 eV bombardment; not only does this lead to rapid decomposition of many molecular ions, but also very complex fragmentation patterns often result. Some useful alternative methods of ionization are worth noting, each of which goes some way toward complementing the data obtained from conventional 70 eV spectra.

5S.1.1 Chemical ionization

This is brought about by mixing the sample at 1.3×10^{-2} N m^{-2} $\equiv 10^{-4}$ Torr) with a reactant gas (at 1.3×10^{2} N $m^{-2} \equiv 1$ Torr) and submitting this mixture to electron bombardment. The reactant gas most commonly used is methane, although other gases such as ammonia or isobutane have also been used: on electron impact, it is the methane which is ionized, and two ensuing ion–molecule reactions are important:

$$CH_4^{+} + CH_4 \rightarrow CH_5^{+} + CH_3^{·}$$
$$CH_3^{+} + CH_4 \rightarrow C_2H_5^{+} + H_2$$

The CH_5^{+} and $C_2H_5^{+}$ ions then react with sample molecules, inducing them to ionize, and these ions are separated magnetically and electrostatically in the normal way. Unfortunately, CH_5^{+} and $C_2H_5^{+}$ do not react with all classes of organic compound in the same way: for n-alkanes the base peak is normally the M − 1 peak at $C_nH_{2n+1}^{+}$, whereas for many basic compounds (amines, alkaloids, amino acids) the base peak is the M + 1 peak. The M + 1 peaks arise by protonation of nitrogen, and for the alkanes the M − 1 peaks can be explained by the following reactions:

$$C_nH_{2n+2} + CH_5^{+} \rightarrow C_nH_{2n+1}^{+} + CH_4 + H_2$$
$$C_nH_{2n+2} + C_2H_5^{+} \rightarrow C_nH_{2n+1}^{+} + C_2H_6$$

The principal advantages of chemical ionization over electron impact are: (a) more abundant peaks related to the molecular ion, whether M^{+} or M + 1 or M − 1; (b) simpler fragmentation patterns, which make it easier in many cases to study the kinetics of reaction of individual ions; (c) easy application of gas chromatography–mass spectrometry interfacing, since methane can be used not only as reactant gas (in the chemical ionization), but also as the carrier gas in the gas chromatograph (see section 5S.2).

5S.1.2 Field ionization and field desorption

An organic compound in the gas phase can be ionized when the molecules pass near a sharp metal anode carrying an electric field of the order of 10^{10} V m^{-1}. Electrons are 'sucked' from the sample molecules into incomplete orbitals in the metal, and the resulting molecular ions are then repelled toward a slit cathode. Primary focusing takes place at the cathode slit before the ions pass through the entrance slit of the mass spectrometer to be focused magnetically and electrostatically, as in electron-impact studies.

As in the case of chemical ionization, the principal advantages of field ionization from an organic chemist's point of view are the increased abundance of molecular ions and the minimization of complex fragmentations and rearrangements. Disadvantages are the lower sensitivity and resolution obtained.

Outstanding advantages can be achieved by a modification of the technique in which the sample is deposited directly onto the anode, and the high field produces not only ionization, but also desorption. Unstable and involatile material can be handled in this way, and molecular ion peaks have thus been produced from complex naturally occurring compounds (notably the carbohydrates) that do not show $M^{\ddot{+}}$ on electron impact.

This method is named *field desorption* (FD).

5S.1.3 Desorption by lasers, plasmas, ions and atoms—LD and LIMA, PD, SIMS and FAB

In the search for soft ionization techniques for measuring the relative molecular masses of large biomolecules, irradiation or bombardment of the sample by several species has been developed; these include lasers, nuclear fission fragments, ions and neutral atoms or molecules. Recordings of the molecular masses of the molecule of insulin (*m/z* 5733), chlorophyll oligomers (*m/z ca* 6000) and a dodecanucleotide dimer (*m/z* 12 637) have been successful examples. Studies have been made on molecules with M_r values in excess of 20 000 daltons.

In some of these techniques the sample is coated on to a metal surface before being bombarded, and the ions produced often include $(M + H)^+$ and $(M - M)^-$ in addition to $M^{\ddot{+}}$, this being dependent *inter alia* on the type of molecule (acidic, basic, etc.) being bombarded.

Laser ionization mass analysis (LIMA) involves irradiation of the sample with a pulsed laser, of output up to 10^5 W cm^{-2}, which vaporizes a minute amount of material from the surface of the sample: this vaporized plume contains ions and neutrals, which are then passed to a mass spectrometer for analysis. Ionization of the vapor plume may be enhanced with a second high-power laser pulse, or by electron impact, etc. Using microscopy, the initial laser pulse can be focused on extremely small areas, one or two micrometers across, which makes LIMA a valuable analytical tool for surface analysis in the polymers and microelectronics industries for detection of impurities in printed-circuit boards, microchips, etc. The layer structure of devices can be investigated (*depth profiling*) by applying a succession of pulses to the same area, each pulse cutting away a few micrometers at a time, enabling mass analysis over an effective cross-section of the material. LIMA is widely used for elemental analysis, but organic materials are amenable to the method.

Plasma desorption (PD) uses as the impacting species a fission fragment from ^{252}Cf, which produces about 3 per cent of various ion pairs; these ion pairs decompose in turn, giving, for example,

^{142}Ba^{18+} and ^{106}Te^{22+}. These enormously energetic ions strike the target molecule and generate about 10^{12} W of power and a localized plasma at a temperature of around 10 000 K; it is the temperature which causes the target molecules to ionize.

Secondary ion mass spectrometry (SIMS) involves generating a beam of ions, such as Ar^{+} (although others may be used, including Xe^{+} and Ne^{+}). The ion beam is directed on to the target molecule and the energy of the ions is transferred to the molecule, causing it, in turn, to ionize (hence 'secondary ion').

Fast atom bombardment is one step beyond SIMS, since Xe ions (for example) are introduced into xenon gas, producing a beam of mixed Xe^{+} and energized Xe atoms; the Xe ions are removed electrostatically and the residual beam of 'fast atoms' is used to bombard the target. In both SIMS and FAB, sample bombardment leads to ions being 'sputtered' from the surface, but some significant differences should be noted. The primary ion beam in SIMS has more energy, and can be more easily focused than can the neutral atom beam in FAB, but in FAB the sample is dispersed in a nonvolatile liquid matrix, commonly glycerol, and this is a critical innovation. Atom kinetic energies in FAB are of the order of 10 keV (10^{6} kJ mol^{-1}) and the inevitable breakdown of some of the sample molecules is overcome by the continuous diffusion of fresh sample to the surface of the matrix.

FAB can generate molecular ions from highly polar nonvolatile compounds such as proteins, but its most striking feature is that the subsequent fragmentation of these ions often allows the amino acid sequence in the protein to be determined. An example is discussed in section 5S.5: see figure 5.10.

5S.2 GAS CHROMATOGRAPHY–MASS SPECTROMETRY (GC–MS) AND HIGH-PERFORMANCE LIQUID CHROMATOGRAPHY–MASS SPECTROMETRY (HPLC–MS)

The coupling of gas chromatography with mass spectrometry has been overviewed in section 1S.3.1, and we also saw in section 5S.1.1 that spectrometers operating with the CI mode can be coupled directly to the eluent of the GC using (for example) methane as both carrier gas and CI reactant gas.

If the gas chromatograph is using packed columns, the flow of carrier gas may be in excess of 30 ml min^{-1}, which would collapse the vacuum of the mass spectrometer. Therefore, the carrier gas must be substantially removed, and various designs of interface have been developed, all of which depend for their success on the higher diffusion rate of the carrier gas (usually helium) compared with the sample molecules. In the *jet separator* (see figure 5.8) effluent from the column passes through a fine orifice into an evacuated chamber,

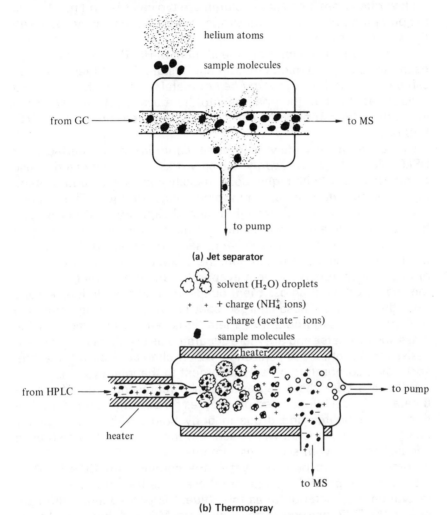

helium atoms

sample molecules

from GC →   → to MS

to pump

(a) Jet separator

solvent (H_2O) droplets

+   +   + charge (NH_4^+ ions)

–   –   – charge (acetate$^-$ ions)

sample molecules

heater

from HPLC →   → to pump

heater

to MS

(b) Thermospray

Figure 5.8   (a) *Jet separator for GC–MS interfacing: the lighter helium atoms diffuse away from the jet stream more than the sample molecules.* (b) *Thermospray for HPLC–MS interfacing: evaporation of solvent from droplets leaves (statistically) some droplets positively and some negatively ionized.*

and in the rapidly expanding jet thus formed, helium carrier gas diffuses more rapidly to the outside of the jet than do the sample molecules. A second orifice (coaxial with the first and at a distance of about 1 mm from it) connects to the mass spectrometer's high-vacuum system: about 90 per cent of the helium and about 40 per cent of the sample miss this orifice, so that the mass spectrometer receives a considerably enriched sample.

Flow rates from a capillary column are typically 1–5 ml min^{-1}, and the pumping system of the mass spectrometer can usually cope with this without any enrichment.

The removal of solvent from the effluent of HPLC columns has also been tackled by ingenious instrument design. The effluent solution can be deposited on a *moving belt* of metal, from which the solvent evaporates; the belt then passes into the mass spectrometer, carrying with it the sample molecules, which are then ionized by SIMS, FAB or LD.

One of the most widely used *direct liquid insertion* interfaces for HPLC–MS is the *thermospray*, which works best in reversed phase chromatography (with highly polar, usually aqueous, eluent, containing an electrolyte such as ammonium acetate buffer). The effluent from the column is led through a heated capillary, and discharged from this into an evacuated heated chamber as an aerosol; see figure 5.8. The droplets consist of solvent, sample and electrolyte, and in this heated discharge environment the droplets carry a slight positive or negative charge. As solvent evaporates and the droplets contract, the charge gradient across them builds up until sample ions escape from the droplets: these charged species (which may be combined with proton or ammonium or metal ions) are accelerated into the mass spectrometer and are there separated as usual. It is common to assist the spontaneous ionization which characterizes the thermospray by incorporating a heated filament in the inlet system.

HPLC–MS with thermospray interface represents one of the ultimate analytical hyphenated procedures for detection and identification of metabolites and drugs in body fluids. Although originally quadrupole instruments were used, the thermospray has also been adopted for use with sector instruments.

Potentially more powerful yet is the combination GC–FTIR–MS. Although it is possible to operate these in series, chromatographic resolution may deteriorate as the effluent is passed along the light pipe in the FTIR instrument *en route* to the MS, and therefore the GC effluent is usually split so that about 2 per cent goes directly to the more sensitive MS instrument, with the remaining 98 per cent going to the FTIR instrument. When the complementary IR and MS information is used, each for its separate library search, the combined ability to identify components from the GC is very great indeed.

5S.3 ISOTOPE SUBSTITUTION IN MASS SPECTROMETRY—ISOTOPE RATIOS

The incorporation of less abundant isotopes into organic molecules makes it relatively easy to follow by mass spectrometry the mechanisms of a number of reactions, both in conventional laboratory or biological sequences and in the fragmentations in the mass spectrometer inself. A few examples will be given to illustrate the scope of the technique, but the variations are too numerous to list.

The isotopes that might be used include ^2H, ^{13}C, ^{18}O, ^{15}N, ^{37}Cl, etc., but the cost of enriched-isotope sources is so high that most work has been done on the cheapest (^2H) and comparatively less on ^{13}C and ^{18}O. To be certain of a mechanistic step in a reaction, we must often be able to show nearly 100 per cent inclusion of the isotope label in the product of the reaction (or 100 per cent exclusion), since values substantially less than 100 per cent could be associated with scrambling of atoms *within* the mass spectrometer itself.

An early successful application of isotope labeling was the investigation of ester hydrolysis, which could conceivably involve acyl–oxygen fission or alkyl–oxygen fission (or a mixture of both). Labeled with ^{18}O as shown, acyl–oxygen fission leads to ^{18}O incorporation in the alcohol, while alkyl–oxygenfission leads to incorporation in the acid. (In general, the former is observed, except for esters or tertiary alcohols.) The distinction is easily made by mass spectrometry, since the acid and alcohol can be isolated and the $M^{\ddagger}$ peak for each measured: ^{18}O incorporation gives the $M^{\ddagger}$ peak two mass units higher than the 'normal' ^{16}O analog.

$$R-\overset{\overset{\displaystyle ^{16}O}{\|}}{C}-^{18}O-R' \left\langle \begin{array}{l} \xrightarrow{H_2O} \quad R-\overset{\overset{\displaystyle ^{16}O}{\|}}{C}-^{16}OH + H-^{18}OR' \quad \text{acyl–oxygen fission} \\ \text{or} \\ \xrightarrow{H_2O} \quad R-\overset{\overset{\displaystyle ^{16}O}{\|}}{C}-^{18}OH + H-^{16}OR' \quad \text{alkyl–oxygen fission} \end{array} \right.$$

In structural organic chemistry it is often desirable to know how many enolizable hydrogens are adjacent to a carbonyl group. The replacement of $-CH_2CO-$ by $-CD_2CO-$ can be executed rapidly by treating $-CH_2CO-$ with D_2O and base, and subsequent measurement of $M^{\ddagger}$ will show the degree of deuterium incorporation. This method has been elegantly extended so that in-column deuterium exchange takes place *during* gas chromatography of the material, which can then be studied by the combined GC–MS method.

The biochemical applications of isotope labeling are legion, and to mention any one would perhaps invidiously imply its relative importance: of outstanding interest to organic chemists have been the acetate-labeling studies of biosynthetic pathways. The modes of incorporation of ^{13}C into steroids, carbohydrates, fatty acids, etc., have been fundamental revelations.

In mass spectrometry itself, solutions to the problems of ion structure and fragmentation mechanisms have been keenly sought. The fragmentations of the toluene molecular ion are archetypal, and illustrate well the complexities involved.

We saw earlier (section 5.7.6) that the toluene molecular ion (m/z 92) gives rise to a series of daughter ions at m/z 91, m/z 65 and m/z 39, and that the ion m/z 91 is best represented as the tropylium ion: the crudest representation of these processes would be as shown below. The evidence which belies this simple interpretation is summarized as follows.

i. ^{2}H labeling shows that loss of H˙ from the toluene molecular ion is almost random, and does not take place exclusively from the benzylic carbon.

ii. ^{13}C labeling of the side-chain shows that this carbon is not exclusively expelled in the transition m/z 91 → m/z 65:

iii. Double ^{13}C labeling (the methyl carbon and its neighbor) shows that although the two ^{13}C atoms are adjacent in the original toluene molecule, complete randomization occurs somewhere between there and the decomposition of ion m/z 91.

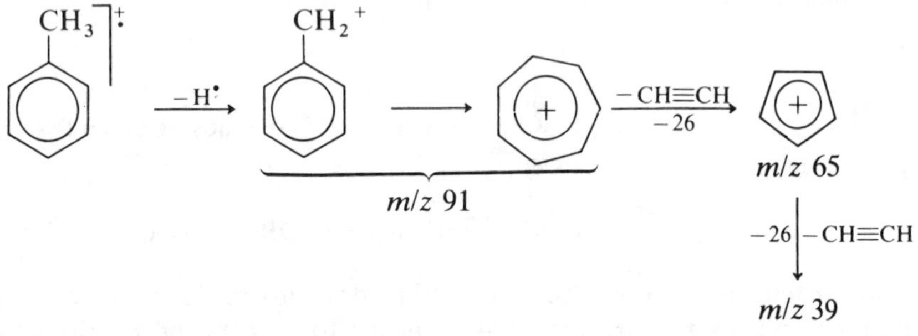

Calculations of the relative energies of benzyl and tropylium ions suggest that both can be formed in the spectrometer from selected precursors and under controlled reaction conditions. In the absence of proof, and in rejecting any simplistic 'mechanism', it is better to write the ion m/z 91 in the form of 'either or' structures.

Radioactive tracer techniques are used to tag the fate of many species in chemical and biochemical reaction pathways, but the use of radioisotopes is not always appropriate or hazard-free. Stable isotopes can be used in a similar way by artificially increasing the relative abundance of one of the isotopes and thereafter using mass spectrometry to measure the isotope ratio in the reaction products.

The isotopic constitution of naturally occurring elements is surprisingly constant, but small variations, often of the order of one part in 10^9, do arise in different   geological or biological circumstances. As an example, petroleum and coal are both fossil materials, but the ratio between carbon-12 and carbon-13 (the stable isotopes of carbon) is different in each; indeed, coals from the northern hemisphere have different carbon isotope ratios from those of the southern hemisphere. When coal and oil are coprocessed, to produce distillate fuels, the differences in stable isotope ratios are sufficient to determine how much of the coprocessed product is coal-derived and how much is petroleum-derived. The method involves oxidizing the material to carbon dioxide, and mass spectrometry is then used to measure the isotope ratios; the difficulty of obtaining accurate measurements of such minute ratio differences is alleviated by comparison with the isotope ratio in international standard substances held in the US Bureau of Standards.

An interesting application of the method is in the detection of adulteration in commercial vanilla extracted from vanilla beans. The major flavor constituent in vanilla is vanillin (4-hydroxy-3-methoxybenzaldehyde), which can also be synthesized more cheaply from wood lignin. The presence of the wood-derived vanillin as an adulterant can only be detected by the fact that the carbon-13/carbon-12 ratio is different in the bean-derived vanillin from that in the adulterating vanillin, because the different biosynthetic routes to the two forms preselect the two isotopes to different degrees.

Other elements whose stable isotope ratios have been used in similar analyses include hydrogen/deuterium, nitrogen-14/nitrogen-15, oxygen-16/oxygen-18 and sulfur-32/sulfur-34.

The mass spectrometer used for stable isotope ratio measurements is usually modified so that the ion beams of the species to be measured (for example, carbon dioxide containing either carbon-12 or carbon-13) are passed through separate slits to separate detectors, and the ratio of ion current is measured by quantitatively attenuating the stronger signal until it matches the weaker signal.

5S.4 DERIVATIZATION OF FUNCTIONAL GROUPS

Two main benefits may accrue from converting the functional group in a molecule to one of its functional derivatives. The derivative may be more volatile than the parent, or the derivative may give a simpler mass spectrum, possibly also with an enhanced molecular ion peak. A few examples will illustrate the value of the method.

Carbohydrates are very difficult to handle in 70 eV mass spectrometry, since they are very involatile and give no molecular ion peaks. They can be converted rapidly and cleanly into their trimethylsilyl ethers by treatment with a mixture of hexamethyldisilazane ($Me_3SiNHSiMe_3$) and trimethylsilyl chloride (Me_3SiCl): the trimethylsilyl ethers are sufficiently volatile to be easily capable of GC–MS analysis.

The following conversions are widely used to effect similar improvement in other functional classes:

$$RCO_2H \rightarrow RCO_2Me$$
$$ROH \rightarrow ROMe \text{ or } ROSiMe_3 \text{ or } ROCOMe$$
$$ArOH \rightarrow ArOMe \text{ or } ArOSiMe_3 \text{ or } ArOCOMe$$
$$RCONH- \rightarrow RCONMe-$$

Functional groups do not possess an equal ability to direct molecular fragmentations, since the activation energies for the formation of the fragment ions will be different (depending on the stabilities of the fragment ions). An approximate ranking order (most strongly directing function last) would be carboxyl, chloride, methyl ester, alcohol, ketone, methyl ether, acetamido, ethylene ketal, amine. We can often make use of this ranking order to simplify the mass spectrum of a molecule, by converting the functional group to a derivative of greater directing propensity. A good example is found in the case of ketones: if a ketone molecular ion gives rise to a large number of

ketone ethylene ketal stabilized oxonium ion *m/z* 99

m/z 99 and *m/z* 113

fragment ions, we can convert the ketone to the ethylene ketal and usually the simple fragmentation of this group will dominate the spectrum. Ethylene ketals fragment to stable oxonium ions, and the low energy of activation for their formation means that alternative reactions cannot compete. The resulting simplification in the fragmentation pattern makes structural deductions much clearer: in the example shown, the position of the two substituents (Me, Et) in the substituted cyclohexanone can be deduced from the two ions' *m/z* values, given the known fragmentation modes of the parent cyclohexanone ethylene ketal.

The phenomenon of influencing fragmentation modes by derivatization is usually termed *directed fragmentation*.

5S.5 ALTERNATIVES TO MAGNETIC/ELECTROSTATIC FOCUSING—TIME-OF-FLIGHT, QUADRUPOLE, ION CYCLOTRON, FTICR AND TANDEM MASS SPECTROMETERS

We have seen that routine organic mass spectrometry is taken to mean 70 eV electron-impact ionization, followed by mass separation on the basis of magnetic (and additionally electrostatic) focusing of positive ions. Three principal alternative means of mass separation have been developed, each of which has advantages and disadvantages, although on balance they have not overtaken magnetic/electrostatic focusing as the method of choice for the greatest number of instruments in current use.

Time-of-flight mass spectrometry differentiates the positive-ion masses by measuring the times that they take to traverse a flight tube of approximately 1 m length. The ions are first of all generated in short pulses and accelerated to uniform kinetic energy; heavier ions travel more slowly to the collector than do light ions, although the very short time differences between ions can be of the order of 10^{-8} s. The ion pulses enter the flight tube at intervals of around 10^{-4} s (a frequency of 10 kHz), and the mass scan must be of the same time scale: usually the spectrometer conditions are optimized by presenting the spectrum on an oscilloscope (also scanning at 10 kHz), and a permanent record is obtained from an analog recording of the oscilloscope trace. The extremely fast scan time and wide-slit dimensions of time-of-flight spectrometers make them particularly useful for dealing with transient species—for example, those produced in shock tubes or flames: by using rapid photographic techniques (drum cameras, etc.), the oscilloscope trace can be recorded over a very small time span.

The ions of any particular species may leave the ion source with a spread of kinetic energies. To obviate the lack of resolution which this would entail (they would arrive at the detector over a spread of times), a sequence of electrostatic 'mirrors', with increasing repelling potential, is arranged as in figure 5.9. Faster ions penetrate the mirrors further, their path lengths are longer, and, therefore, they arrive at the detector together with the slower ions.

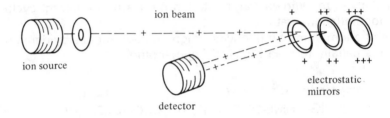

(a) **Time-of-flight MS — schematic**

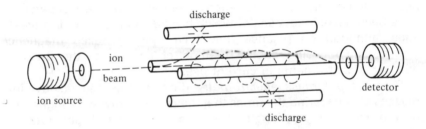

(b) **Quadrupole MS — schematic**

Figure 5.9 (a) *Time-of-flight mass spectrometer, showing the sequence of electrostatic reflectors which improve resolution.* (b) *Quadrupole mass spectrometer: for each m/z one value of the ratio DC:RF will permit that ion alone to pass through the ion filter.*

Quadrupole mass spectrometers have as their core an ion tube (the 'mass filter') containing four accurately aligned metal rods, arranged symmetrically around the long axis of the tube. Opposite pairs of rods are coupled together, and a complex electric field is set up within the four rods by applying direct current voltage (a few hundred volts) across the coupled rods, superimposed by radiofrequency potential. In operation, positive ions are propelled into the mass filter; the ratio between DC voltage and RF voltage is chosen, and these voltages are scanned (at constant ratio) from zero to a maximum value, when they return to zero to repeat the scan. As the DC and RF voltages are built up, a hyperbolic potential field is established within the four rods: at any particular value for this field (that is, at any particular point in the scan), most ions will be deflected toward the rods and discharged there. For ions of the appropriate mass-to-charge ratio, however, there will be a particular value of the DC/RF scan which will induce the ion to describe a modulated wave-like path along the hyperbolic field. These ions will reach the collector (usually an electron multiplier) and be recorded there. See Fig. 5.9.

The entire mass spectrum is obtained by the DC/RF scan through the correct values for each m/z ratio.

Quadrupole mass spectrometers are small and relatively cheap, and are widely used for fast-scan GC–MS work associated with kinetic or pyrolysis experiments in the gas phase. The major limitation in the cheaper versions is low resolution, which restricts their use to the study of relative molecular masses (molecular weights) up to around 500. Within this limit, however, they find wide use in monitoring processes—for example, in the analyses of respiratory gases, of atmospheric gases in tanks and spacecraft, and of hazardous components in working atmospheres.

A modified form of quadrupole, consisting of a circular electrode with two end caps, is available as a detector in GC; in this form it is usually called an *ion trap*, but other, more elaborate, devices also are so named.

Ion cyclotron resonance spectrometry involves generation of positive ions by electron impact, after which they are drawn through a short analzyer tube by a small static electric field. While in the analyzer, a magnetic field (of the order of 0.8 T) is superimposed: this causes the ions to perform a series of cycloidal loop-the-loops along the analyzer, the angular frequency of which lies in the radiofrequency range (around 300 kHz). This frequency (the *cyclotron resonance frequency*, ω_c) is a function of both the electric field strength and m/z value for the ion ($\omega_c = Ez/m$), and the 'mass spectrum' is obtained by scanning the fields until ω_c comes to resonance with a fixed radiofrequency source beamed onto the analyzer tube. As in NMR, the resonance condition is reached when ω_c equals the frequency of the radiofrequency source, and, at resonance, measurable radiofrequency energy passes from the RF circuit to the ion beam.

Ion cyclotron resonance spectrometry is relatively new, and its principal application has been in the study of ion–molecule reactions. Although pressures used in the spectrometer are low ($\approx 1.3 \times 10^{-4}$ N m$^{-2} \equiv 10^{-6}$ Torr), the cycloidal path is long and drift times are long (of the order of milliseconds), so that the probability of ions colliding with molecules is enhanced. A general example of how the technique can be applied to the investigation of ion structure might be a situation where two molecules, A and B, both give rise to a daughter ion at $m/z = x$: has the $m/z = x$ ion from A the same structure as the $m/z = x$ ion from B? If we can demonstrate by isotope substitution that the daughter ion from A reacts in the ion cyclotron resonance spectrometer with an admixed neutral molecule, to produce a new ion at $m/z = y$, then we can repeat the experiment with the daughter ion of B mixed with the same neutral molecule. If the new ion at $m/z = y$ is not produced in this test, then the daughter ions of A and B, even though they have the same m/z value, cannot have the same structure.

Fourier Transform ion cyclotron resonance spectrometry (FTICR) is a variant on the above in which, instead of scanning the field until the cyclotron resonance of each ion comes to resonance with the radiofrequency, a short scanning pulse of RF is applied, spreading over all of these frequencies. This generates a current in the receiver, which consists of an interferogram containing information on *all* of the ion frequencies present; this *time-domain* response is converted to *frequency-domain* by Fourier Transform, to reveal the individual ion cyclotron frequencies (which are then correlated with their masses as before). See sections 1S.2, 2.4, 3.3.2 and 4.3.1 for the application of FT to other methods.

There are several advantages with the FTMS method.

As with other applications of FT (e.g. to IR or NMR), the method is fast (by a factor of about 100 compared with a normal ICR experiment) and lends itself to spectra summation techniques to increase sensitivity.

On the basis of frequency measurement, very high resolution is achieved, and since the ion frequencies are all recorded at (substantially) the same time, many rate-dependent reactions in the gas phase can be studied, including the isomerization of benzyl and tropylium cations (m/z 91).

Tandem mass spectrometry can be described as the recording of a mass spectrum in one mass spectrometer of an ion generated in another mass spectrometer; the technique is thus MS–MS. Its principal application is as a complement to the soft ionization methods (see section 5S.1) used to generate the molecular ions of large molecules; such molecular ions can then be transported to the second mass spectrometer and there fragmented, the fragmentation pattern being used as usual as an aid to structural identification. Virtually all of the ionization methods can be combined one with another in MS–MS, each bringing unique advantages.

Figure 5.10 shows an analysis of a pentapeptide (M$_r$ 565) in which the molecular ion is generated by FAB (see 5S.1.3) from a methanol/ glycerol matrix, subsequent analysis being by FTMS—that is FTICR mass spectrometry; the spectrum also shows considerable contamination from the glycerol fragments. The lower spectrum is the MS (recorded on a FTICR analyzer) of the molecular ion produced by FAB and separated by the first analyzer: this much cleaner MS–MS spectrum allows more certain deductions to be made about the fragmentation of the peptide.

FAB – FTMS OF TYR – PRO(CH2) – PHE – PRO – GLY OH (MW 565)

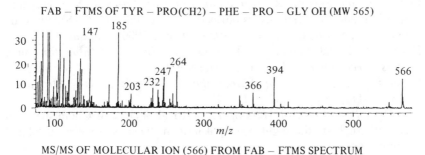

MS/MS OF MOLECULAR ION (566) FROM FAB – FTMS SPECTRUM

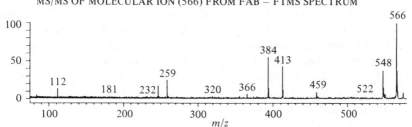

Figure 5.10 *Mass spectrum (above) of a pentapeptide, ionized by FAB, and mass analyzed by FTICR mass spectrometry. The molecular ion generated in the first MS was thereafter selectively analyzed in a second FTICR mass spectrometer: hence MS–MS (below).*

FURTHER READING

MAIN TEXTS

McLafferty, F. W., *Interpretation of Mass Spectra*, University Science Books, Mill Hill, California (1980).

Hill, H. C., *Introduction to Mass Spectrometry*, Heyden, London (2nd edn, 1972).

Biemann, K., *Mass Spectrometry. Organic Chemical Applications*, McGraw-Hill, New York (1962).

Rose, M. E. and Johnstone, R. A. W., *Mass Spectrometry for Organic Chemists*, University Press, Cambridge (2nd edn, 1982).

Howe, I., Williams, D. H. and Bowen, R. D., *Principles of Organic Mass Spectrometry*, McGraw-Hill, London (2nd edn, 1982).

Beynon, J. H. and Brenton, A. G., *Introduction to Mass Spectrometry*, University of Wales Publications (1982).

Davis, R. and Frearson, M., *Mass Spectrometry*, Wiley, Chichester (1987).

Chapman, J. R., *Practical Organic Mass Spectrometry*, Wiley, New York, (1985).

SPECTRA CATALOGS

Cornu, A. and Massot, R., *Compilation of Mass Spectral Data*, Heyden, London, and Presses Universitaires de France, Paris (1966–8). Lists the ten principal ions in the mass spectra of over 5000 compounds.

Beynon, J. H. and Williams, A. E., *Mass and Abundance Tables for Use in Mass Spectrometry*, Elsevier, Amsterdam (1963).

Lederburg, J., *Computation of Molecular Formulas for Mass Spectrometry*, Holden-Day, San Francisco (1964).

The Eight Peak Index of Mass Spectra, Royal Society of Chemistry (3rd edn 1983). Lists the eight principal ions in the mass spectra of 52 332 compounds.

Beynon, J. H., Saunders, R. A. and Williams, A. E., *Tables of Metastable Transitions for Use in Mass Spectrometry*, Elsevier, Amsterdam (1965). These last three compilations form a complete kit for deducing molecular ion and fragmentation formulae from the mass spectrum m/z values.

McLafferty, F. W. and Stauffer, D. B., *Registry of Mass Spectral Data*, Wiley, New York (1989). Contains data on 120 000 spectra, both in book form and on database.

SUPPLEMENTARY TEXTS

Lawson, G. and Todd, J. F. J. Radiofrequency quadrupole mass spectrometers. *Chem. Brit.*, **7**, 373 (1971).

Waller, G. R. (Ed.), *Biochemical Applications of Mass Spectrometry*, Wiley, Chichester (1972).

Milne, G. W. A. (Ed.), *Mass Spectrometry. Techniques and Applications*, Wiley, New York (1972).

McLafferty, F. W. (Ed.), *Tandem Mass Spectrometry*, Wiley, New York (1983).

Gaskell, S. J. (Ed.), *Mass Spectrometry in Biomedical Research*, Wiley, Chichester (1986).

McFadden, W. H., *Techniques of Combined GC–MS*, Wiley, New York (1973).

Chapman, J. R., *Computers in Mass Spectrometry*, Academic Press, London (1978).

Wilkins, C. L., Experimental technique of GC–FTIR–MS, *Anal. Chem.*, **59**, 571A (1987).

Smith, R. D. and Udseth, H. R., *Chemistry in Britain*, **24**, 350 (1988). Good account of GC–MS interfaces.

Spectroscopy Problems

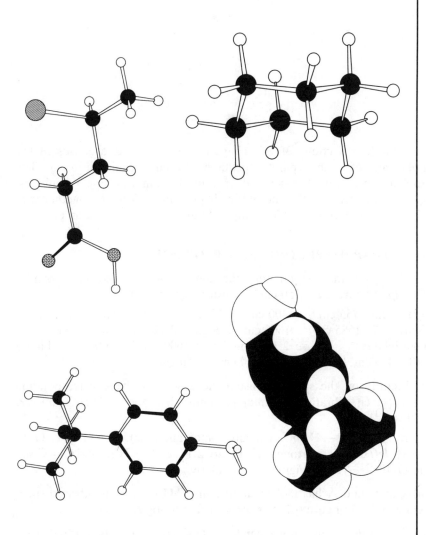

Cyclohexane?   4-Iodopentanoic acid?

4-*Tert*-butylaniline?   (Twice?)

The first four sections of this chapter contain varied exercises in the application of each separate technique to an organic problem. The problems in section 6.5 demand an interplay among these techniques. Solutions are given at the end of the chapter, in section 6.6, together with solutions to the other problems throughout the book, in section 6.7.

6.1 INFRARED SPECTROSCOPY PROBLEMS

(i) What are the assignments for the following absorption bands? (*Example*: figure 2.2, 2150 cm^{-1}. *Answer*: C≡C *str*)

Figure 2.3, 755 cm^{-1}, 3400 cm^{-1}. Figure 2.6, 2900 cm^{-1}, 1060 cm^{-1}. Figure 2.7, 1585 cm^{-1}, 1605 cm^{-1}. Figure 2.11, 700 cm^{-1}, 1601 cm^{-1}, 2000–2600 cm^{-1}. Figure 2.15, doublet at 1600 cm^{-1}, 3100 cm^{-1}. Figure 2.18, 1460 cm^{-1}, 1500 cm^{-1}, 1600 cm^{-1}, 760 cm^{-1}.

(ii) Identify on the spectrum listed the absorption band corresponding to the given vibration. (*Example*: figure 2.2, ≡C—H *str*. *Answer*: 3320 cm^{-1})

Figure 2.15, C—H *str* (aromatic and aliphatic), C═C *str*, OOP C—H *def*, C═O overtone (2v). Figure 2.17, all C—H *str* and C═C *str*, and OOP C—H *def*. Figure 2.18 similarly.

(iii) Compound A has molecular formula C_3H_3N, and its infrared spectrum is given in figure 6.1. Suggest a structure for A.

(iv) Compound B has molecular formula $C_6H_{10}O$, and its infrared spectrum is given in figure 6.2. Suggest a structure for B.

(v) The liquid-film infrared spectrum of 2,4-pentanedione (acetylacetone) shows absorption bands at 1600 cm^{-1} (strong, broad); 1710 cm^{-1} (less strong than the 1600 cm^{-1} band, and more sharp); a very broad band stretching from ≈ 2400 cm^{-1} to 3400 cm^{-1}, which is unchanged on dilution. What are these bands?

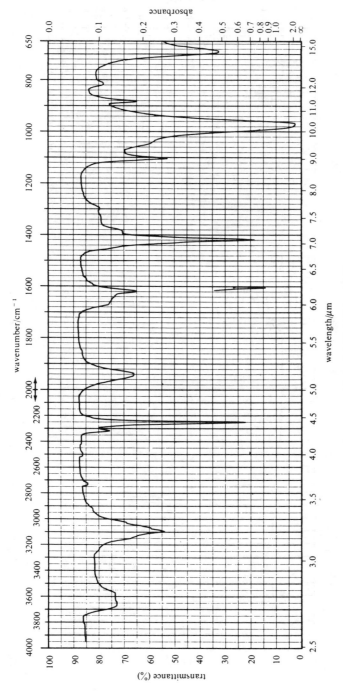

Figure 6.1 *Infrared spectrum for compound A in problem 6.1 (iii). Liquid film.*

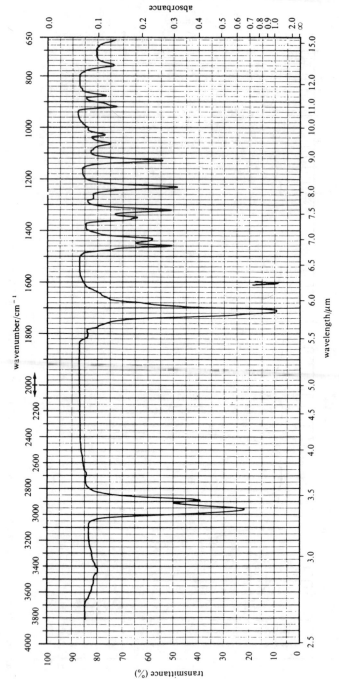

Figure 6.2 *Infrared spectrum for compound B in problem 6.1 (iv). Liquid film.*

(vi) State whether the following pairs of compounds could be distinguished by an examination of their infrared spectra. Give reasons.

> $PhCH_2NH_2$ and $PhCONH_2$
> $H_2N—C_6H_4—CO_2Me$ and $Me—C_6H_4—CONH_2$
> $MeO—C_6H_4—COMe$ and $Me—C_6H_4—CO_2Me$
> cyclohexanone and 3-methylcyclopentanone
> $PhCOCH_2CH_3$ and $PhCH_2COCH_3$

6.2 NMR SPECTROSCOPY PROBLEMS

Examples (i)–(v) deal exclusively with problems in proton NMR spectroscopy; examples (vi)–(x) concentrate mainly on the practice of carbon-13 NMR spectra interpretation, but with cross-references, where significant, to features of the corresponding proton spectra.

Worked solutions to the problems are provided in section 6.6.2. Work through each problem and use the stepwise assistance provided in the solutions only when necessary.

(i) The ^1H NMR spectrum shown (figure 6.3) is that of the dental local anesthetic procaine (at 60 MHz). Assign all signals, accounting for (a) chemical shift values, (b) integrals and (c) coupling constants.

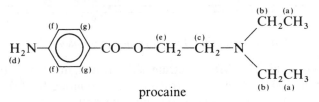

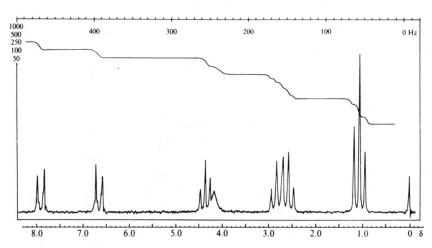

Figure 6.3 ^1H *NMR spectrum for procaine in problem 6.2(i).*

(ii) The ^1H NMR spectrum shown in figure 6.4 is that of 4-vinylpyridine (at 60 MHz). Assign all signals, accounting for (a) chemical shift values, (b) integrals and (c) coupling constants.

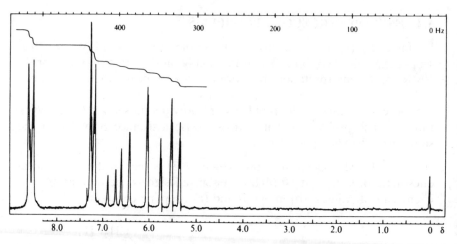

4-vinylpyridine

Figure 6.4 ^1H *NMR spectrum for 4-vinylpyridine in problem 6.2(ii).*

(iii) The 60 MHz ^1H NMR spectrum shown in figure 6.5 is that of a hydrocarbon C_9H_{12}. Deduce its structure by accounting for (a) chemical shift values, (b) integrals and (c) coupling constants.

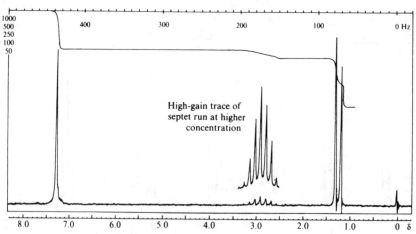

High-gain trace of septet run at higher concentration

Figure 6.5 ^1H *NMR spectrum for C_9H_{12} in problem 6.2(iii).*

(iv) The 60 MHz ^{1}H NMR spectrum in figure 6.6 is of a compound $C_4H_7BrO_2$. Infrared evidence shows it to be a carboxylic acid. Deduce its structure by accounting for (a) chemical shift values, (b) integrals and (c) coupling constants.

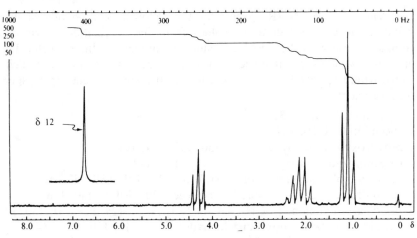

Figure 6.6 ^{1}H NMR spectrum for $C_4H_7BrO_2$ in problem 6.2(iv).

(v) The 60 MHz ^{1}H NMR spectrum in figure 6.7 is of a compound $C_{10}H_{13}NO_2$. Significant features of the infrared spectrum are $C{=}O$ stretch and one $N{-}H$ stretch peak. Deduce the structure of this compound from the chemical shift, integral and coupling data shown on the spectrum.

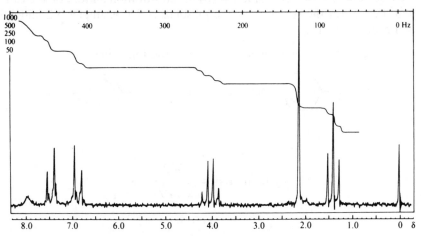

Figure 6.7 ^{1}H NMR spectrum for $C_{10}H_{13}NO_2$ in problem 6.2(v).

(vi) The carbon-13 NMR spectrum of one of the butyl acetate isomers $(C_4H_9OCOCH_3)$ showed signals at δ_C 22, 28, 80 and 170. What is its structure? See example 3.11. Why is the intensity of the peak at δ 28 much more intense than that at δ 22 (by a factor of approximately 8)? How would the multiplicity and signal intensity in the proton NMR spectrum of this compound confirm your deductions?

(vii) The ^{13}C-$\{^1H\}$ NMR spectra for the following compounds show δ values as listed. Predict the δ values from the correlation charts and compare these with the observed values. Comment on the expected line intensities which you might see in these spectra, and state the multiplicity which would appear for each signal in an off-resonance decoupled spectrum of each:

n-butanol, δ 16, 22, 37, 60
succinimide, δ 30, 184
p-hydroxybenzaldehyde, δ 166, 130, 133, 164, 191
catechol (o-dihydroxybenzene), δ 117, 122, 114
paracetamol (p-acetamidophenol), δ 24, 115, 121, 131, 153, 168
terephthalic acid (p-benzenedicarboxylic acid), δ 130, 140, 176
p-nitroacetophenone (p-$CH_3COC_6H_4NO_2$), δ 27, 124, 129, 141, 150, 196
anisaldehyde (p-methoxybenzaldehyde), δ 55, 114, 129, 131, 164

(viii) A plasticizer was shown from its infrared spectrum to be an aromatic ester, and the carbon-13 NMR spectrum indicated that it was a butyl ester, with signals at δ 14, 19, 30, 65 and 167. In addition, there were three signals from a substituted benzene ring, at δ 129, 131 and 132, the last signal being of much lower intensity than the other two. Deduce (a) the nature of the butyl residue (n-, iso-, sec- or tert-); (b) the substitution pattern on the ring; and, hence, (c) the structure of the plasticizer. See examples 3.2 and 3.3 and problem 6.2(vi).

(ix) A compound was known to be one of the following:

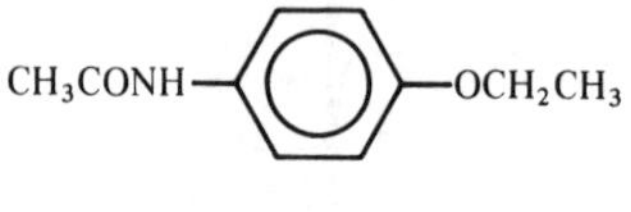

CH_3CONH—⬡—OCH_2CH_3 CH_3CH_2CONH—⬡—OCH_3

I 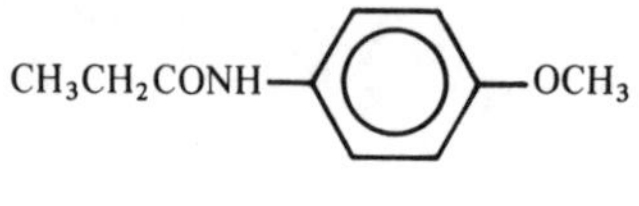   II

$CH_3CH_2CH_2CONH$—⬡—OH

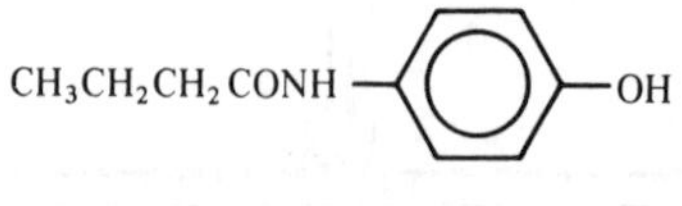

III

Its carbon-13 NMR spectrum consisted of peaks at the δ values listed, and the multiplicities in an off-resonance decoupled spectrum are shown in parentheses: δ 15 (q), 24 (q), 64 (t), 115 (d), 123 (s), 157 (s) and 171 (s). Which isomer is it? See also problem 6.2(v) for the proton NMR spectrum of this compound.

(x) Predict the δ values for the carbons in 2-methylcyclohexanone. Proceed from the δ values for hexane—these are shown on the formula, but should also be predicted from the empirical table of constants in table 3.11. Thereafter consider substitution of a carbonyl group at *both* C-1 and C-5. The observed values for 2-methylcyclohexanone are shown on the formula.

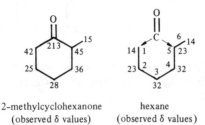

2-methylcyclohexanone   hexane
(observed δ values)   (observed δ values)

6.3   ELECTRONIC SPECTROSCOPY PROBLEMS

(i) Use the Woodward rules in table 4.3 to predict the expected λ_{max} for the following compounds dissolved in ethanol:

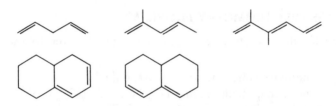

(ii) Use the Woodward rules in table 4.4 to predict the expected λ_{max} for the $\pi \rightarrow \pi^*$ transition in the following compounds (in ethanol):

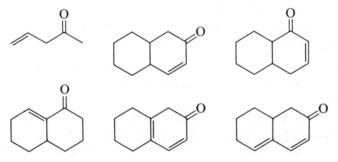

(iii) A ketone was known to have one of the isomeric structures shown below and had λ_{max} (in ethanol) at 224 nm. Which was it?

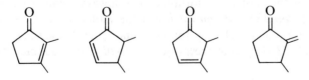

(iv) Could the following isomers be distinguished by their electronic absorption spectra?

(v) For each of the following compounds, write out the structure of an isomer that is likely to have a substantially different electronic absorption spectrum.

6.4 MASS SPECTROMETRY PROBLEMS

(i) Write feasible structures for these ions (found in the following mass spectra):

> 1-methylcyclohexene, m/z 96, 81, 68, 67
> 1-methylnaphthalene, m/z 142, 141, 115, 143, 71, 57.5
> 3-methyl-2-butanol, m/z 45, 43, 55, 73
> trimethylsilylether of this alcohol, m/z 160, 117, 145
> 4-heptanone, m/z 114, 86, 71, 58, 43, 41
> ethylene ketal of this ketone, m/z 113
> benzyl methyl ketone, m/z 134, 119, 92, 91, 65, 51, 43
> phenylacetic acid, m/z 136, 92, 91, 65, 51, 45, 39

(ii) Deduce feasible structures (not necessarily unambiguous) for the compounds whose mass spectra have ions at the following m/z values. Base peak first.

$C_{11}H_{16}$, m/z 91, 119, 148, 41, 27, 39, 92, 77, 51, 29
$C_{11}H_{14}O_2$ (ester), m/z 105, 123, 77, 56, 122, 106, 41, 29
$C_{10}H_{20}O$ (alcohol), m/z 57, 81, 67, 56, 82, 83, 41, 123, 99
$C_{10}H_{12}O_2$ (carboxylic acid), m/z 149, 164, 105, 119, 77, 91, 79, 131, 135, 150

Note: From the molecular formula, it is always useful to calculate the number of *double-bond equivalents* in the molecule. For an alkane, the molecular formula is C_nH_{2n+2}: for an alkene, or cycloalkane, it is C_nH_{2n}, so that loss of 2H (implying the presence of *either* one double bond *or* one ring) is one double-bond equivalent (DBE). Dienes, cycloalkenes or bicycloalkanes have two DBE. Trienes, etc., contain 3 DBE, benzene contains 4 DBE. To find the number of DBE from molecular formula C_nH_mO, use the formula $DBE = \frac{1}{2}[(2n + 2) - m]$. For formulae $C_nH_mN_p$, use $DBE = \frac{1}{2}[(2n + 2) - (m - p)]$. Thus, $C_{11}H_{16}$ contains 4 DBE (probably an aromatic ring?).

(iii) Calculate the m/z value for the parent ions (m_1) that produce the following normal daughter ions (m_2) and metastable daughter ions (m^*):

$m_2 =$	$m^* =$
117	93.8
61	31.8
131	117.5
77	56.5
51	33.8

(iv) How could the following pairs of isomers be differentiated by their respective mass spectra?

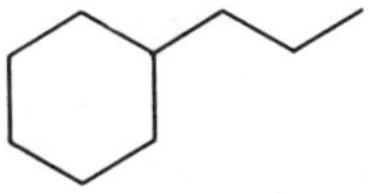

and

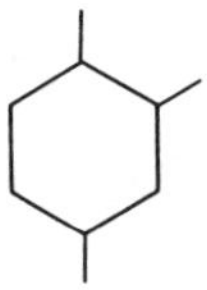

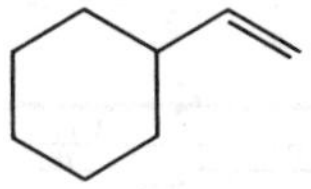

and

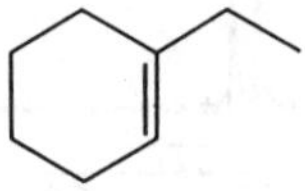

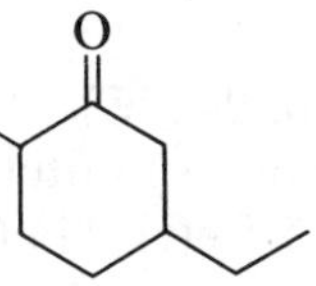

and

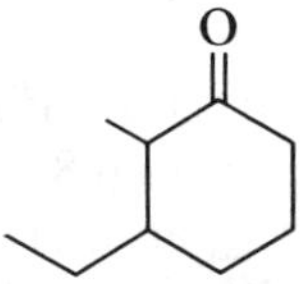

6.5 CONJOINT IR–UV/VIS–NMR–MASS SPECTROMETRY PROBLEMS

General approach. Every spectroscopic problem of this type is unique, but you should begin by quickly perusing all of the spectral data and noting the following. Are there useful and prominent bands in the infrared spectrum ($C=O$, $C\equiv C$, $C\equiv N$, $O-H$, $N-H$, etc.)? Is the compound aromatic, alkene, alkane (NMR and IR spectra)? Is there an ultraviolet chromophore, and can it be tentatively identified? Is there an M^{+} peak in the mass spectrum, from which can be found the molecular formula and the number of DBE?

Thereafter it will usually be found necessary to extract information from the spectra in succession, gradually homing in on an unequivocal solution. When a structure finally emerges, it should be examined scrupulously against all the available data until no shadow of doubt remains.

(i) When acetone is treated with base, a higher-boiling liquid (b.p. 130 °C) can be isolated from the reaction mixture. The spectroscopic properties of this liquid are: infrared, 1620 cm^{-1}(m), 1695 cm^{-1}(s): ^{1}H NMR, δ 1.9 (3H, singlet), δ 2.1 (6H, singlet), δ 6.15 (1H, singlet): UV, λ_{max} 11 700: mass, m/z (RA), 55(100), 83(90), 43(78), 98(49), 29(46), 39(43), 27(42), 53(13), 41(13), 28(8): ^{13}C NMR, δ 20, 27, 31, 124, 154 and 197.

Make sketches of the NMR spectra and construct a mass spectrum bar diagram. Deduce the structure and account for all of the observed data.

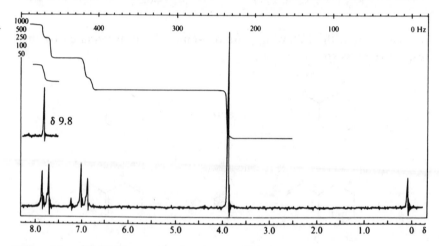

Figure 6.8a ^{1}H NMR spectrum for compound (b.p. 248 °C) in problem 6.5(ii).

(ii) Deduce the structure of the compound (b.p. 248 °C) whose spectral data are given in figure 6.8. (λ_{max} typically that of a substituted benzene, around 250 nm, ϵ_{max} around 12 000. ^{13}C NMR, δ 55 (t), 114 (t), 129 (2C, d), 131 (s), 164 (2C, d) and 190 (s).)

(iii) Deduce the structure of the compound (b.p. 97 °C) whose spectral data are given in figure 6.9. (No UV absorption above 200 nm. ^{13}C NMR, δ 63, 115, 137.)

(iv) 2,2-Dimethylcyclopropanone undergoes ring-opening when attacked by methoxide ion, the product (b.p. 101 °C) having the following spectral properties: IR, 1740 cm^{-1} (s), 1160 cm^{-1} (s), no absorption near 1600 cm^{-1} or 3100 cm^{-1}: ^{1}H NMR, δ 3.6 (3H singlet), δ 1.2 (9H singlet): UV, transparent above 200 nm: mass, m/z 116, 85, 59, 31. Deduce the structure of the product and suggest a mechanism for its formation. From

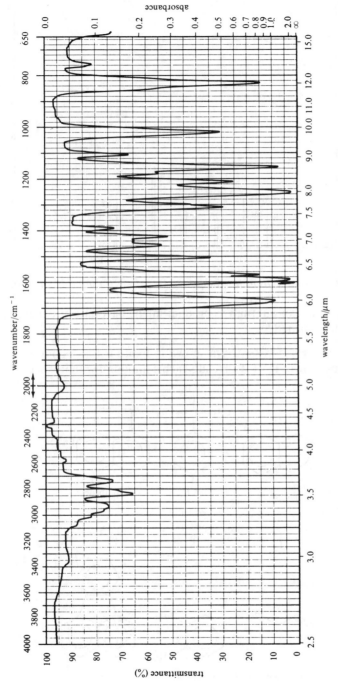

Figure 6.8b *IR spectrum for compound* (b.p. 248°C) *in problem 6.5(ii).*

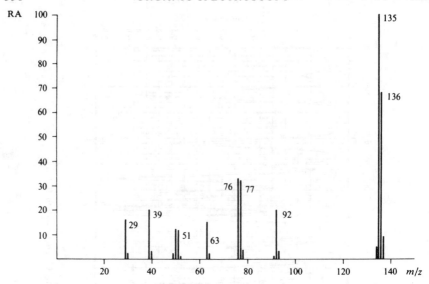

Figure 6.8c *Mass spectrum for compound* (b.p. 248 °C) *in problem 6.5(ii).*

the mechanism, an alternative product might have arisen: what is it and why is it not formed?

(v) The $^{13}C-\{^1H\}$ NMR spectra shown in figure 6.10(a)–(f) are of the six compounds shown, I–VI. Correlate each compound with its spectrum, and

naphthalene

I

$p\text{-}tert\text{-}$butyl-
toluene

II

acenaphthene

III

allyl bromide

IV

o-xylene

V

fluoranthene

VI

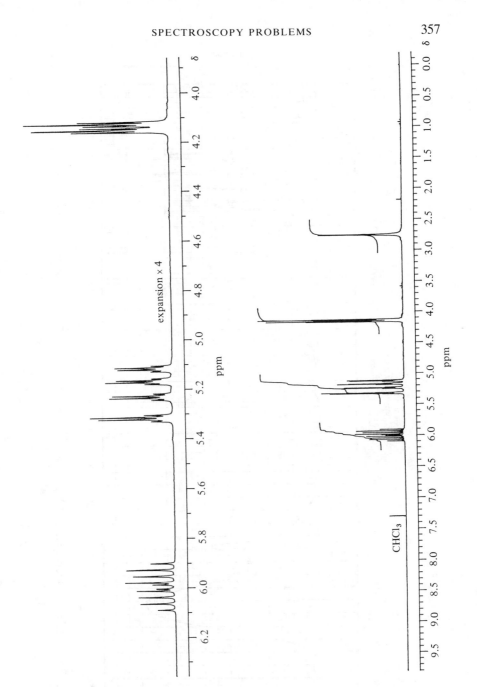

Figure 6.9a ¹H *NMR spectrum for compound* (b.p. 97 °C) *in problem 6.5(iii)* (200 MHz *in* CDCl₃).

358

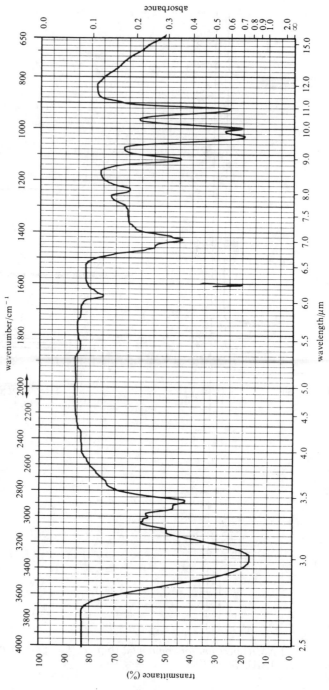

Figure 6.9b *IR spectrum for compound* (b.p. 97°C) *in problem 6.5 (iii).*

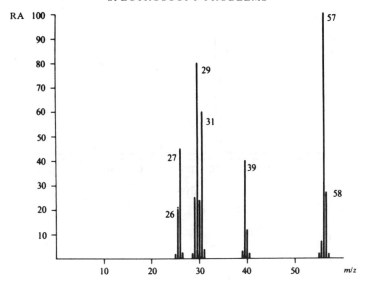

Figure 6.9c *Mass spectrum for compound (b.p. 97 °C) in problem 6.5(iii).*

interpret this with respect to the additional spectroscopic information provided.

(a) and (b) give substantially the same UV spectrum, with absorption around 220 nm and 275 nm (with some fine structure). (c) and (d) also give substantially the same UV spectrum, with absorption around 260 nm. The UV spectrum of (f) has strong absorptions extending to λ_{max} 359 nm.

In the ^{1}H NMR spectrum of (d) there are, among others, two signals in the intensity ratio 3:1, and in the mass spectrum of (e) the molecular ion cluster shows two main peaks, in the intensity ratio 1:1.

(vi) The infrared spectra of six compounds, (a)–(f), are shown in figure 6.11. Deduce their structures, taking into account the following additional spectroscopic information. In each case verify that the observed NMR data fit well with those predicted from the correlation charts. (Note, s = singlet, d = doublet, m = multiplet, etc. In the ^{13}C NMR data these abbreviations apply to off-resonance decoupled spectra or equivalent, and each resonance refers to one carbon atom, unless otherwise stated.)

(a) Proton NMR: δ 2.0 (3H, s), 2.6 (2H, t), 2.8 (2H, t), 7.1 (5H, nearly s).
^{13}C NMR: δ 29 (q), 29 (t), 44 (t), 126 (d), 128 (2C, d), 128 (2C, d), 140 (s), 207 (s).
Mass spectrum: m/z 148, 133, 105, 91, 77, 51, 43.

(b) Proton NMR: δ 1.3 (3H, t, J = 8 Hz); 3.4 (2H, s, lost on deuteriation); 3.9 (2H, q, J = 8 Hz); 6.4–6.8 (4H, AA'BB').

(a)

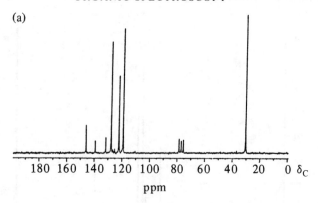

(b)

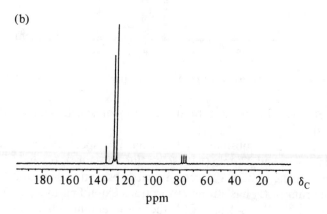

(c)

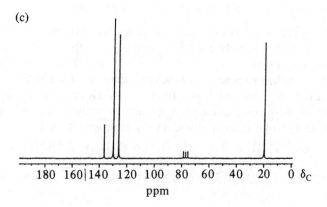

Figure 6.10 ^{13}C-$\{^1H\}$ *NMR spectra for compounds I–VI in problem 6.5 (vi).*

(d)

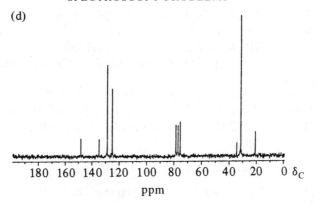

(e)

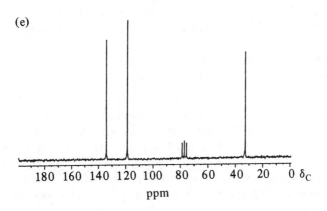

(f)

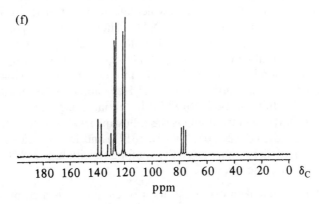

^{13}C NMR: δ 15 (q), 63 (T), 115 (2C, d), 116 (2C, d), 139 (s), 152 (s).
Mass spectrum: m/z 137, 136, 110, 109, 122.

(c) Proton NMR: δ 2.4 (3H, s), 7.0–7.5 (4H, m).
^{13}C NMR: δ 20 (q), 112 (s), 117 (s, low intensity), 126 (d), 130 (d), 131 (d), 132 (d), 141 (s).
Mass spectrum: m/z 117, 116, 90, 89, 63, 51, 39, 38, 27.

(d) Proton NMR: δ 2.4 (3H, s), 7.5 (2H, m), 7.9 (2H, m).
^{13}C NMR: δ 21 (q), 120 (d), 123 (d), 129 (d), 135 (d), 140 (s, low intensity), 148 (s, low intensity).
Mass spectrum: m/z 137, 107, 91, 89, 65, 63, 39, 28.

(e) Proton NMR: δ 3.9 (3H, s), 7.0 (1H, s, lost on deuteriation), 7.1 (1H, apparent d), 7.4 and 7.45 (2 overlapped H, m), 9.8 (1H, s).
^{13}C NMR: δ 56 (q), 109 (d), 114 (d), 127 (d), 130 (s), 147 (s, low intensity), 152 (s, low intensity), 191 (s, low intensity).
Mass spectrum: m/z 152, 151, 109, 81, 53, 52, 51, 29, 18.

(f) Proton NMR: δ 1.3 (6H, d, $J = 8$ Hz), 5.3 (1H, septet, $J = 8$ Hz), 7.4 (3H, m), 8.1 (2H, m).
^{13}C NMR: δ 22 (2C, q), 68 (d), 128 (2C, d), 129 (2C, d), 131 (s, low intensity), 132 (d), 166 (s, low intensity).
Mass spectrum: m/z 164, 106, 105, 123, 122, 77, 59, 51, 43, 41.

(vii) The 'essential oil' (essence oil) from aniseed can be steam-distilled from the crushed seeds, and from its mass spectrum the major constituent (A) has the molecular formula $C_{10}H_{12}O$. Other significant peaks in the mass spectrum appear at m/z 117 and 133. Its UV spectrum had only one band at 250 nm, and its IR spectrum showed evidence of C—O $str.$

Its proton NMR spectrum shows easily analyzable absorptions at δ 1.8 (3H, d, $J = 6$ Hz), 3.8 (3H, s), 6.8 (2H, approximately d, line separation $= 10$ Hz), 7.2 (2H, approximately d, line separation $= 10$ Hz). In addition, there is a set of signals which are non-first-order at 60 MHz but which are nearly first-order at 200 MHz. They can be analyzed as such, as follows: a doublet (integration 1H) with δ $= 6.33$ and coupling constant $J = 16$ Hz; a doublet of quartets (total integration 1H) near δ 6.1, in which the doublet coupling constant is 16 Hz and the quartet coupling constant is 6 Hz. (A further splitting with $J = 1.5$ Hz is apparent in the doublets at δ 1.8 and 6.33; this can be ignored in the initial interpretation.)

The ^{13}C NMR spectrum shows the following signals: δ 18 (q), 55 (q), 113 (2C, d), 123 (d), 129 (2C, d), 130 (d), 131 (s) and 159 (s). Careful oxidation of A with acid dichromate gives the compound with b.p. 248 °C in problem 6.5(ii). Deduce the structure of A.

(viii) The principal flavor constituent of cinnamon is a compound whose mass spectrum shows the molecular ion at m/z 132 (C_9H_8O), with the base peak at m/z 131, and a significant peak at m/z 103. Its IR spectrum has a strong absorption band at 1690 cm^{-1}; the UV spectrum has an intense band at 284 nm and a very weak absorption at 308 nm.

Its proton NMR spectrum consists of the following absorptions: δ 6.7 (1H, double doublet, two coupling constants, $J = 16$ Hz and $J = 8$ Hz), 7.4 (5H, narrow multiplet), 7.45 (1H, d, $J = 16$ Hz, overlapping with the previous signal), 9.75 (1H, d, $J = 8$ Hz). In the ^{13}C NMR spectrum, absorptions appear as follows: δ 128.2 (2C, d), 128.3 (d), 128.8 (2C, d), 131 (d), 134 (s), 152 (d), 193 (d). Deduce the structure of this compound, and comment on its stereochemistry in relation to the 16 Hz coupling constant in the proton NMR spectrum.

(ix) A sample of Dacron (Terylene) and a sample of a nylon were hydrolyzed, and in each case the single carboxylic acidic component was isolated from the reaction mixture. Deduce the structures of these two acidic compounds from their spectral data:
From Dacron (Terylene)
Molecular formula $C_8H_6O_4$. Proton NMR: δ 8.2 (4H, s), 12.5 (1H, broad singlet). ^{13}C NMR: δ 130 (4C, d), 140 (2C, s), 176 (2C, s).
From the nylon
Molecular formula $C_6H_{10}O_4$. Proton NMR: δ 1.5 (4H, distorted triplet), 2.3 (4H, distorted triplet), 11 (1H, broad singlet). ^{13}C NMR: δ 26 (2C, t), 37 (2C, t), 182 (2C, s).

(x) Benzaldehyde, PhCHO, condenses in the presence of base with acetone, CH_3COCH_3, to give a compound (A) whose mass spectrum showed the molecular ion at m/z 146, and two intense peaks at m/z 131 and 103. There was a strong absorption band in the infrared spectrum near 1650 cm^{-1}. The compound showed the following NMR absorptions. Proton NMR spectrum: δ 2.3 (3H, s), 6.7 (1H, d, $J = 16$ Hz), 7.4 (5H, narrow multiplet), 7.5 (1H, d, $J = 16$ Hz, overlapping with the previous signal). ^{13}C NMR spectrum: δ 27 (q), 127 (d), 128 (2C, d), 129 (2C, d), 130 (d), 135 (s), 144 (d), 198 (s). Deduce the structure and stereochemistry of A.

6.6 SOLUTIONS TO PROBLEMS

6.6.1 INFRARED SPECTROSCOPY PROBLEMS
(i) and (ii) See correlation charts.
(iii) Acrylonitrile ($C\equiv N$ *str*, $C=C$ *str*, out-of-plane $C=H$ *def* and its overtone at 2ν).
(iv) Cyclohexanone ($C=O$ *str* for six-ring or acyclic ketone. Alkane C—H *str* and *def*. No alkene C—H *str* or $C=C$ *str*. No aldehyde C—H *str*).
(v) 2,4-pentanedione is 85 per cent enolic; intramolecular hydrogen bonding in the enol gives O—H *str* and low $C=O$ *str* (together with C—O *str*). The keto form has normal $C=O$ *str*.)

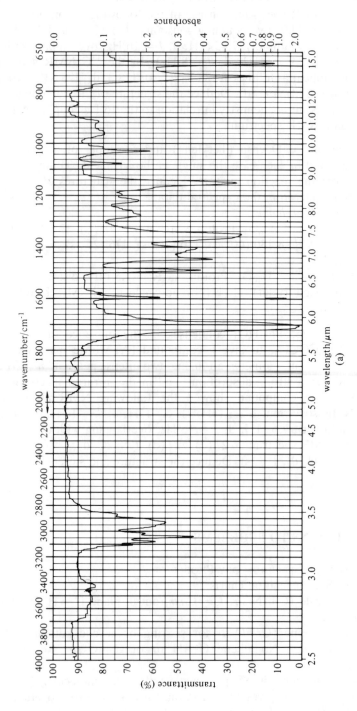

Figure 6.11 *The infrared spectra of compounds (a)–(f) in problem 6.5 (vi).*

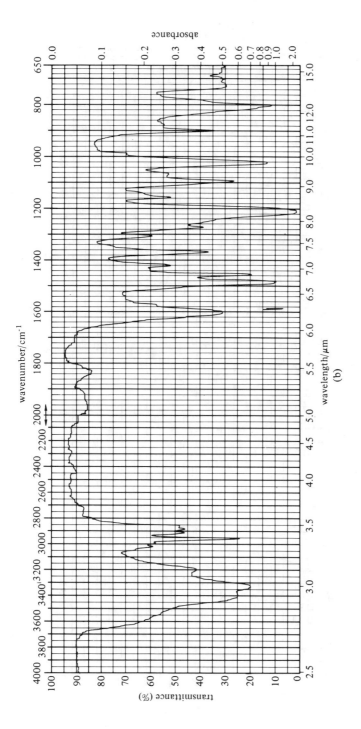

365

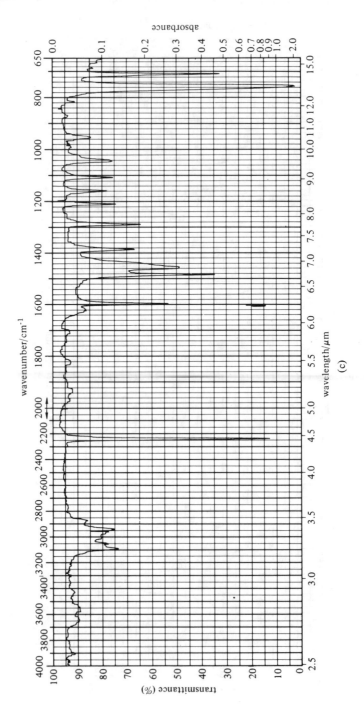

(c)

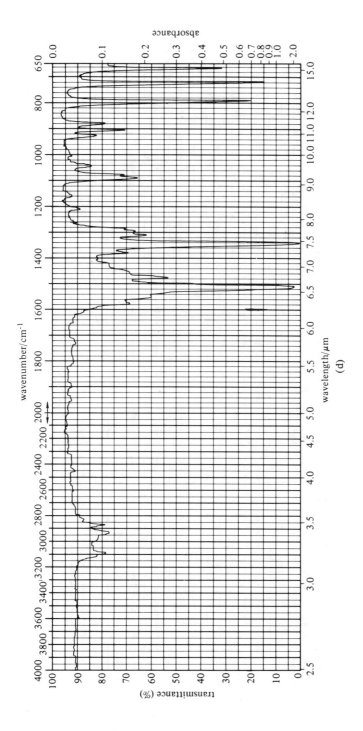

(d)

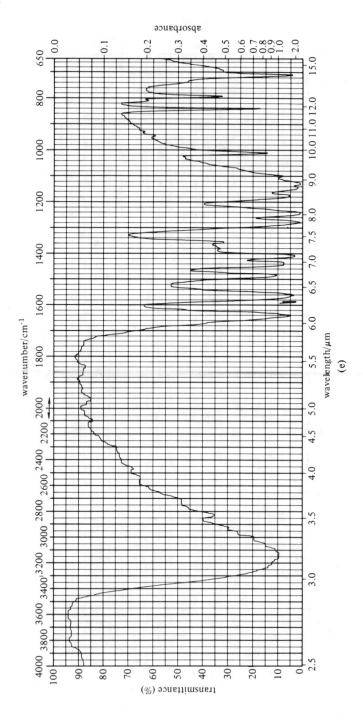

(e)

369

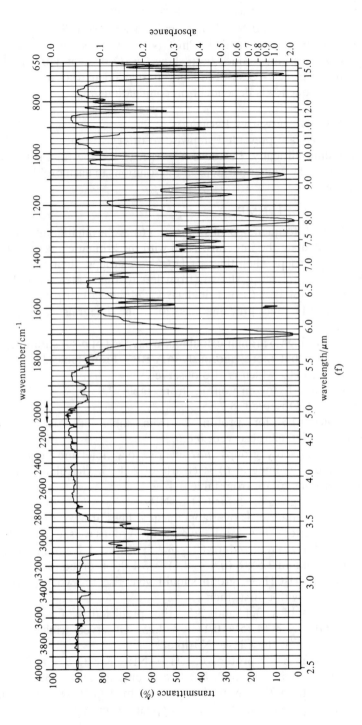

(f)

(vi) Yes (C=O *str* and N—H *def* (amide I and II) in PhCONH$_2$; wide separation of N—H *str* bands in PhCONH$_2$). Yes (as before). No (or at least not with certainty, since both show C=O *str* near 1700 cm^{-1}, and C—O *str*). Yes (C=O *str* frequency is dependent on ring size). Yes (C=O *str* frequency is dependent on conjugation).

6.6.2 NMR Spectroscopy Problems

These solutions are presented stepwise to offer assistance to the student who is finding difficulty, but does not wish to see the final solution too soon. Identify the sequential steps in the solutions to each problem as i(1), i(2), i(3), etc.

Problems (i)–(v): ^{1}H NMR

(i)(1) The high-frequency peaks (above δ 6) are aromatic; all others are aliphatic. Calculate the integrals. Deal with the aromatic peaks first, then the two ethyl groups, then the —CH$_2$CH$_2$— groups, then NH$_2$.

(i)(2) The integral ratios shown (from left to right) are: 1 (doublet): 1 (doublet): 1 (triplet): 1 (broad singlet): 3 (quintet?): 3 (triplet). Since this only totals 10H, and procaine has 20H, the integrals correspond to 2:2:2:2:6:6.

(i)(3) The two high-δ aromatic doublets are typical of *p*-substitution, being similar to AX, but more rigorously AA'BB'. From table 3.10, J_{ortho} is predicted as 10 Hz, which is also observed.

(i)(4) Predicted δ for proton g, from table 3.9, is δ (7.27 + 0.8 − 0.15) = δ 7.29. Observed, δ 7.85.

(i)(5) Predicted δ for proton f, from table 3.9, is δ (7.27 − 0.8 + 0.15) = δ 6.62: observed, δ 6.6.

(i)(6) The two *N*-ethyl groups *must* give rise to a triplet and a quartet. The triplet (protons a, integral 6) is clearly at δ 1.05. Calculate where the quartets (protons b) are from table 3.4

(i)(7) Table 3.4 gives δ 2.5 for CH$_2$—NR$_2$.

(i)(8) The ethyl quartet is overlapped; the 4 lines at lowest δ, integral 4, contain the quartet. Thus, for the two *N*-ethyl groups we see the triplet and quartet, J = 8 Hz. Predicted for CH$_2$, δ 2.5; observed, δ 2.6. Predicted for CH$_3$ (tables 3.4 and 3.5), δ (0.9 + 0.1) = δ 1.0; observed, δ 1.05.

(i)(9) The —CH$_2$CH$_2$— group should give rise to two triplets, each integral 2; deal with e first, then c.

(i)(10) Using tables 3.4 and 3.5, the predicted δ for protons e, CH$_2$—OCOR, is δ (4.15 + 0.1) = δ 4.25. Observed triplet at δ 4.3, J = 8 Hz. Where is the CH$_2$—N triplet?

(i)(11) Predicted δ for protons c, CH$_2$—NR$_2$, is δ (2.5 + 0.3) = δ 2.8 (tables 3.4 and 3.5). The high δ 3 lines (centered on δ 2.8, integral 2; J = 8 Hz) are the CH$_2$—N triplet; that is, this triplet overlaps with the ethyl quartet (same J value), giving an apparent quintet.

(i)(12) The broad singlet at δ 4.15 is $ArNH_2$ (protons d); predicted from table 3.8, between δ 4.0 and δ 3.5.

(ii)(1) Table 3.7 gives the δ values for all these protons. The integral ratios can only be significantly interpreted if you appreciate the multiplicity in the spectrum.

(ii)(2) The pyridine ring protons are predicted (table 3.7) at δ 8.5 (α to N), and δ 7.0 (β to N); the observed values are δ 8.5 (approximate doublet), and δ 7.2 (approximate doublet).

(ii)(3) The 'doublets' (δ 7.2 and δ 8.5, J = 9 Hz) are AA'BB' in type, and the coupling constant is appropriate for *ortho* protons. Integrals correspond to 2H for each 'doublet'.

(ii)(4) The best analogy for the vinyl protons is to consider pyridine ≡ Ph in table 3.7.

(ii)(5) The predicted values for the vinylic protons are similar to

With three different environments, what is the multiplicity of the vinylic system?

(ii)(6) The vinyl protons are coupling AMX, so that each proton gives rise to a double doublet, integral 1. From the approximate chemical shift values in 5, we can make a preliminary allocation of the protons observed as

Although the predicted δ values show only moderate agreement with the observed values, the observed *relative* positions follow the predicted order. Final confirmation of the allocation rests with the AMX coupling analysis: measure the three J values, and note that each multiplet has two J values. Check these against the predicted values in table 3.10.

(ii)(7) Predicted J_{AM}, 11–19 Hz; observed, 19 Hz. Predicted J_{AX}, 5–14 Hz; observed, 12 Hz. Predicted J_{MX}, 3–7 Hz; observed, ≈ 1.5 Hz.

(iii)(1) The peak at δ 7.1 is aromatic H. The others are aliphatic. Are the two peaks near δ 1.3 separate singlets, or do they together constitute a doublet?

(iii)(2) The molecule is aromatic and clearly benzenoid. Integrals show that the aromatic peak corresponds to 5H, and therefore the ring is monosubstituted. This leaves 7H in the other peaks.

(iii)(3) C_6 is accounted for in the benzene ring, leaving C_3 in *one* side-chain, which also contains 7H. Side-chain is C_3H_7.

(iii)(4) Side-chain must be isopropyl. Predicted δ for methine, CH—Ph (table 3.4), is δ 2.87. Coupling with 2Me (6H) gives septet at δ 2.87, $J = 8$ Hz as in table 3.10.

(iii)(5) Predicted δ for CH_3 (tables 3.4 and 3.5) is δ $(0.9 + 0.3) = $ δ 1.2. Coupling with methine (1H) gives a doublet at δ 1.2, $J = 8$ Hz.

(iii)(6) Compound is isopropylbenzene (cumene).

(iv)(1) The presence of carboxyl is confirmed by the δ 12 peak. Taking COOH from the molecular formula leaves C_3H_6Br.

(iv)(2) The C_3 residue can be *n*-propyl or isopropyl, and the Br can be attached to several points within these possibilities.

(iv)(3) The possibilities are

$$\begin{array}{ll}
\text{BrCH}_2\diagdown & \text{CH}_3\diagdown \\
\quad\text{CHCO}_2\text{H} & \text{Br}-\text{CCO}_2\text{H} \\
\text{H}_3\text{C}\diagup & \text{H}_3\text{C}\diagup \\
\qquad\text{I} & \qquad\text{II}
\end{array}$$

$$\begin{array}{lll}
\text{CH}_2\text{CH}_2\text{CH}_2\text{CO}_2\text{H} & \text{CH}_3\text{CHCH}_2\text{CO}_2\text{H} & \text{CH}_3\text{CH}_2\text{CHCO}_2\text{H} \\
| & | & | \\
\text{Br} & \text{Br} & \text{Br} \\
\qquad\text{III} & \qquad\text{IV} & \qquad\text{V}
\end{array}$$

(iv)(4) Consider the predicted multiplicities and integrals for each of these.

(iv)(5) Only III and V agree in multiplicity, but not in integrals, with the spectrum. (J, 8 Hz).

(iv)(6) Predict the δ values for CH_2Br and CH_2COOH in III. (δ 3.3, 2.3).

(iv)(7) Predict the δ values for V. Predicted δ for CH_3 (table 3.4) is $(1.25 + 0.6) = $ δ 1.85; observed, δ 2.1. Predicted δ for CH (table 3.6) is δ $(1.5 + 0.7 + 1.9) = $ δ 4.1; observed, δ 4.25.

(iv)(8) Compound is 2-bromobutyric acid (V).

(v)(1) The compound is aromatic (peaks above δ 6), and broad singlet at δ 7.9 is probably H attached to nitrogen (infrared evidence confirms; see also table 3.8).

(v)(2) It is *p*-substituted (AA′BB′ system, $J_{ortho} = 11$ Hz).

(v)(3) If 4H are on the ring and 1H is attached to N, this leaves 8H remaining.

(v)(4) There is a clear ethyl system present (triplet at δ 1.3, quartet at δ 4.0; $J = 8$ Hz). This accounts for 5H, leaving 3H unaccounted for.

(v)(5) The singlet at δ 2.1 is methyl.

(v)(6) Check all integrals; ratio is 1:2:2:2:3:3.

(v)(7) Fragments of the structure are

—NH   —CO—   —C$_6$H$_4$—(ring)—   CH$_3$CH$_2$—   CH$_3$—   [and —O—]

(One O atom has to be added to bring the total to the molecular formula.)

(v)(8) Which of these functions (or combinations of these functions), attached to CH$_3$, will cause the CH$_3$ to appear at δ 2.1? (See table 3.4.)

(v)(9) Use table 3.4 to decide which the possible neighbors for the ethyl CH$_2$ are (appearing at δ 4.0).

(v)(10) Use table 3.9 to infer which are the likely substituents on the ring.

(v)(11) CH$_3$CH$_2$O— is one aromatic substituent. CH$_3$CO— and CH$_3$NHCO— cannot be aromatic substituents.

(v)(12) CH$_3$CONH is the second aromatic substituent.

(v)(13) The compound is aceto-p-phenetidide (phenacetin, acetamino-phen):

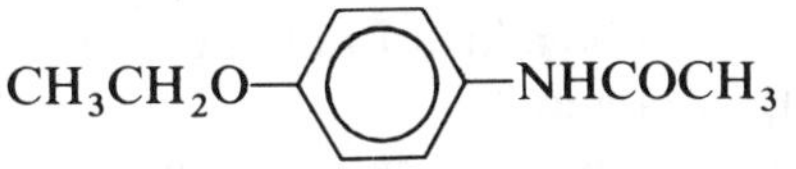

$$CH_3CH_2O—C_6H_4—NHCOCH_3$$

Problems (vi)–(x):^{13}*C NMR*

(vi)(1) How many signals appear in the ^{13}C NMR spectra of the groups n-butyl, isobutyl, sec-butyl and tert-butyl?

(vi)(2) The compound is tert-butyl acetate. The difference in intensity relates primarily to a 3:1 ratio in the number of carbons in the two environments; the factor of 8 is explained in terms of both the NOE and differences in relaxation rates. See section 3.15.5.

(vi)(3) tert-Butyl acetate gives a very simple proton NMR spectrum, consisting of two singlets in the intensity ratio 3:1. All other butyl isomers show proton coupling.

(vii)(1) The *approximate* relative intensities for all signals in a typical spectrum are found to be as follows: recall that these do *not* all necessarily relate to integer ratios of the number of nuclei in each environment, and will vary somewhat if the spectrum is recorded under different conditions:

   n-butanol, 1:1:1:1
   succinimide, 5:1 (use ethane as base, and substitute CON< twice)
   p-hydroxybenzaldehyde, 8:1:7:1:2
   catechol, 4:4:1
   paracetamol (p-HO-C$_6$H$_4$NHCOCH$_3$), 1:2:2:1:1:1
   terephthalic acid, 6:1:1
   p-nitroacetophenone, 6:14:16:2:1:2
   anisaldehyde, 2:6:1:6:1

(vii)(2) The multiplicities for each signal (derived from off-resonance decoupling) are, respectively:

n-butanol, q, t, t, t
succinimide, t, s
p-hydroxybenzaldehyde, d, s, d, s, s
catechol, d, d, s
paracetamol, q, d, d, s, s, s
terephthalic acid, d, s, s
p-nitroacetophenone, d, d, d, s, s, s
anisaldehyde, q, d, s, d, s

(viii)(1)The butyl group is *n*-butyl.
(viii)(2) The plasticizer is di-*n*-butyl phthalate.
(ix)(1) Isomer III would show wrong multiplicity in both the ^{13}C and proton NMR spectra, and is eliminated.
(ix)(2) The predicted alkane chemical shifts for isomer I are δ 16, 24 and 56. Those for isomer II are δ 8, 28 and 56.
(ix)(3) The compound is isomer I, aceto-*p*-phenetidide (phenacetin, acetaminophen).
(x)(1) The increments to be added to the hexane δ values are, from C-1 on, 30, 2, (−3 −3), 2, 24 and 2.
(x)(2) The predicted δ values in 2-methylhexanone are δ 44, 25, 26, 34, 47 and 16, in good agreement with observed values.

6.6.3 ELECTRONIC SPECTROSCOPY PROBLEMS
   (i) < 200 nm, 225 nm, 255 nm, 235 nm, 273 nm.
   (ii) < 200 nm, 227 nm, 227 nm, 242 nm, 338 nm, 286 nm.
   (iii) The first structure.
   (iv) The middle structure would have a biphenyl-like spectrum, while the other two would be indistinguishably naphthalene-like.
   (v) Examples might be

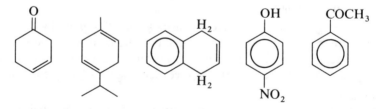

6.6.4 MASS SPECTROMETRY PROBLEMS
(i) See cycloalkanes and cycloalkenes, sections 5.7.2 and 5.7.4.
See aromatic hydrocarbons (remembering tropylium ions and doubly charged ions), section 5.7.6.
See alcohols, section 5.7.8.
See section 5S.4, A_r for Si is 28.
See ketones, sections 5.7.11 and 5.7.13.
See section 5S.4.
See carboxylic acids, section 5.7.14.

(ii) 3-Phenylpentane.
Butyl benzoate.
4-*tert*-Butylcyclohexanol.
4-Isopropylbenzoic acid.
(iii) *m/z* 146, 117, 146, 105, 77.
(iv) Prominent peak at M — C_3H_7.
Retro-Diels–Alder products.
Mass spectrum of ethylene ketals.

6.6.5 CONJOINT SPECTROSCOPIC PROBLEMS
(i) $(Me)_2C$═CHCOMe (mesityl oxide).
(ii) 4-Methoxybenzaldehyde (anisaldehyde).
(iii) CH_2═$CHCH_2OH$ (allyl alcohol).
(iv) Me_3CCO_2Me. Methoxide attacks the carbonyl group, with ring opening to give either a primary or a tertiary carbanion transition state: the latter is disfavored, so that $Me_2CHCH_2CO_2Me$ is not formed.
(v) a = III, b = I, c = V, d = II, e = IV, f = VI.
(vi) (a) 4-Phenyl-2-butanone, benzylacetone; (b) *p*-phenetidine, 4-ethoxyaniline; (c) *o*-tolunitrile; (d) *m*-nitrotoluene; (e) vanillin, 4-hydroxy-3-methoxybenzaldehyde; (f) isopropyl benzoate.
(vii) A is anethole or 3-(4-methoxyphenyl)-1-propene, $4\text{-}MeOC_6H_4CH$═$CHCH_3$.
(viii) Cinnamaldehyde or 3-phenylpropenal, C_6H_5CH═CHCHO; *trans* double bond.
(ix) Terephthalic acid from Dacron: terephthalic acid is benzene-1,4-dicarboxylic acid, and Dacron is poly(ethylene glycol terephthalate). Adipic acid from Nylon 66: adipic acid is butane-1,4-dicarboxylic acid and Nylon 66 is the polyamide formed with hexamethylenediamine.
(x) Benzylideneacetone or *trans*-4-phenyl-3-buten-2-one, C_6H_5CH═$CHCOCH_3$.

6.7 ANSWERS TO SELF-ASSESSMENT EXERCISES DISTRIBUTED THROUGHOUT THE BOOK

1.1 (a) 6.6×10^{13} s^{-1} and 4.55 μm; (b) 9×10^{13} s^{-1} and 3.3 μm.
1.2 (a) 11.11 m; (b) 137 m and 1.91 m; (c) 2.47 m and 1.23 m.
1.3 (a) 1.0×10^{15} Hz (s^{-1}); (b) approximately 300 nm, in the ultraviolet region.
1.4 (a) 6.6×10^{-15} J; (b) 4×10^6 kJ mol^{-1}.
1.5 (a) 1.12×10^{-17} J; (b) 6.74×10^6 J mol^{-1} or 6740 kJ mol^{-1}.

2.1 (a) 0.0418; (b) 0.0224; (c) 1.0000; (d) 1.6990. Check this by examining typical IR spectra (figure 2.9, for example).
2.2 (a) Polyethylene is a long-chain alkane: see figure 2.1 for the IR spectrum of Nujol (a mixture of long-chain alkanes and therefore with a

near-identical spectrum). See also figure 2.9 for the IR spectrum of polystyrene. (b) Nitrile absorption, $C\equiv N$ *str*, is at 2200 cm^{-1}; ester $C=0$ *str* for an aryl ester is near 1710 cm^{-1}.

2.3 Follow the disappearance of the ester $C=O$ *str* band near 1720 cm^{-1}; in aqueous alkali the reaction product shows the carboxylate absorptions near 1600 cm^{-1} and 1350 cm^{-1}. To avoid damaging salt flats with the aqueous medium, use calcium fluoride flats instead.

2.4 Cooking oil. Lubricating oil (mineral oil) is mainly long-chain alkane, similar to Nujol (see figure 2.1), whereas cooking oil is a mixture of triglyceride esters, with $C=O$ *str* near 1720 cm^{-1}.

2.5 Same as in exercise 2.4: polypropylene is a hydrocarbon, and cellulose acetate is an ester. (For interest, they can easily be distinguished by crinkle and tear characteristics, since polypropylene has a very crinkly feel, but is very difficult to tear.)

2.6 The antisymmetric and symmetric $N-H$ *str* bands for primary amide near 3200 cm^{-1} and 3400 cm^{-1}: use calcium fluoride flats, which are resistant to water.

2.7 $O-H$ *str* near 3300 cm^{-1}, $C-H$ *str* near 2900 cm^{-1}, $C-H$ *def* near 1450 cm^{-1} and 1400 cm^{-1}, $C-O$ *str* near 1300 cm^{-1} and 1100 cm^{-1}.

2.8 'Light petroleum', 'petroleum fraction' or 'petrol', but avoiding the name 'ether', since it consists of a mixture of alkane hydrocarbons.

2.9 (a) 10; (b) 1600–2100 cm^{-1}.

2.10 0.0508 mm if you counted 13 fringes from 2000 cm^{-1} to 2080 cm^{-1}; 0.0547 mm for 14 fringes.

2.11 Prepare a working curve (a graph of concentration against transmittance or absorbance) for a series of solutions of a standard (for example, diesel oil) in carbon tetrachloride, using one of the strong $C-H$ *str* absorptions near 2900 cm^{-1}. Extract each aliquot of water with a fixed small volume of carbon tetrachloride and measure the $C-H$ transmittance or absorbance of these solutions: quantify the diesel present in the carbon tetrachloride by comparison with the working curve and thence calculate the diesel concentration in the original water sample. The method would be accurate enough for most diesel fuels but would underestimate gasoline contamination (since gasoline contains more aromatic hydrocarbons, with weak $C-H$ *str* absorptions just above 3000 cm^{-1}).

3.1 1380 Hz.

3.2 (a) 438 Hz; (b) 2190 Hz.

3.3 δ 4.6 (460 Hz) and δ 7.3 (730 Hz).

3.4 (a) 100 Hz; (b) 250 Hz; (c) 500 Hz.

3.5 (a) δ 2.3, 2.9; (b) δ 2.1; (c) δ 3.3; (d) δ 2.1.

3.6 (a) δ 2.1, 2.4, 1.2 (0.9 + 0.3); (b) δ 1.1 (0.9 + 0.2), 2.1, 4.1, 1.3 (0.9 + 0.4); (c) δ 1.1 (0.9 + 0.2), 2.1, 4.6, 4.9, 7.3; (d) δ 2.0, 4.1, 1.6 (1.3 + 0.3), 0.9.

3.7 (a) δ 7.67 (2H, *ortho*), 7.07 (2H, *meta*), 6.97 (1H, *para*), 2.0 (3H); (b) δ 7.47 (*para*), 7.42 (*meta*), 8.07 (*ortho*), 3.9 (Me ester); (c) δ 8.22, 7.62, 3.9; (d) δ 1.3 (0.9 + 0.4, 3H), 3.9 (2H), 6.87 (2H, *ortho* to EtO), 7.47 (2H, *ortho* to NHCOMe), 2.0 (3H).

3.8 (a) δ 2.6, 4.1, 1.3; (b) δ 3.1, 3.6; (c) δ 7.17 (the main influence is the —CH$_2$— group (—R) directly attached to the ring, although oxygen moves the signal slightly to higher frequency), 4.2, 3.7; (d) δ 7.5; observed, δ 7.3.

3.9 (a) δ 2.5 (2.3 + 0.2), near δ 11; (b) ring proton, δ 6.27 (*ortho* to OH), 7.27. OH predicted near δ 4.5, but is observed at higher frequency, owing to H-bonding with C=O. NH around δ 5.0–8.5, broad signal.

3.10 (a) Two doublets; (b) triplet and doublet; (c) quartet and doublet; (d) see figure 3.17.

3.11 Singlet, triplet and quartet in all three.

3.12 Singlet, doublet and septet in all four.

3.13 0.7 Hz, 1.7 Hz and 3.7 Hz, and they do not change with increased field strength.

3.14 Approximately 18 Hz. Yes. (*cis* coupling constants are smaller than *trans*, commonly around 8 Hz.)

3.15 J_{AM} = 1.6 Hz; J_{AX} = 6.3 Hz; J_{MX} = 14.0 Hz. Yes, $J_{MX} > J_{AX}$.

3.16 (a) Three in the *o*-isomer, four in the *m*- and two in the *p*-. (Remember the substituent-bearing carbons.) (b) Four in the *o*-isomer, five in the *m*- and three in the *p*-. (In all isomers the two methoxy carbons are equivalent.) (c) Four, eight and eight, respectively. (Structure IV is correct.)

3.17 (a) Eight in the *o*-isomer, eight in the *m*- and six in the *p*-. (Remember to count the substituent-bearing carbons and the carbonyl carbon. In the *p*-isomer there are only four non-equivalent ring carbons.) (b) Ten in the 2-hydroxy, ten in the 3-hydroxy and eight in the 4-hydroxy.

3.18 Diethyl ether—δ 15.7 and 55.7. Methyl propyl ether—δ 9, 26, 59 and 65. Methyl isopropyl ether—δ 25, 40 and 59.

3.19 They are (a) pentanal, (b) 2-pentanone and (c) 3-pentanone. (a) Pentanal—(start from butane) δ 13, 22, 27, 43 and 200–205 (table 3.17). (b) 2-Pentanone—(start from propane) δ 12, 18, 26–31 (table 3.18), 45 and 205–218. (c) 3-Pentanone—(start from ethane) δ 7.7, 35.7 and 205–218.

3.20 (a) δ 51–52 (table 3.18, methyl group of the ester), 164–169 (carbonyl group of conjugated ester), 130, 130 (2C), 128 (2C) and 133 (ring carbons, beginning with the ipso carbon). (b) δ 66 and 22 (2C) for the isopropyl group; ring carbons as in (a).

3.21 (a) δ 143 (2C), 124 (2C), 135 (2C). (b) All isomers, δ 164–169 (ester carbonyl) and 51–52 (Me of ester). Ring carbons, numbered from ester-substituted carbon; *o*-isomer, δ 130, 136, 128, 134, 131; *m*-isomer, δ 131, 130, 134, 133, 129, 128; *p*-isomer, δ 128, 131 (2C), 128 (2C), 139. (c) For all isomers, consider the benzene ring without OH to have only one C$_6$H$_5$ substituent; the δ values are predicted at 141, 127 (2C), 129 (2C), 127. The second ring has two substituents, OH and C_6H_5. Thus: *o*-isomer, δ 128, 154, 116, 128, 121, 128; *m*-isomer, δ 142, 114, 156, 114, 130, 120; *p*-isomer, δ 134, 128 (2C), 116 (2C), 154.

3.22 The observed chemical shift positions are given in table 3.11; the calculated values show good agreement.

3.23 Both alkene carbons are equivalent. δ value found from $123 + 10 + 7 + 7 - 8 - 2 - 2 = δ\ 135$.

3.24 C-3 value found from $123 + 10 + 7 + 7 - 8 - 2 + 2 = δ\ 139$. C-4 value found from $123 + 10 + 7 - 2 - 8 - 2 - 2 = δ\ 126$.

3.25 4-Hydroxybenzaldehyde. The predicted chemical shifts for the ring carbons are δ 116, 130, 130, 161.

3.26 Isopropyl benzoate. Compare the observed chemical shift values with those predicted in exercise 3.20(b). The *o*- and *m*-carbons cannot be unambiguously assigned.

3.27 Mesityl oxide (2-methyl-2-penten-4-one). (First calculate the alkene shifts for isobutene from table 3.12, then compute the influence of C=O from table 3.16. Deduce the methyl shifts from table 3.18, taking account of the note at the foot.)

3.28 (a) Assign according to whether each carbon atom is attached to three, two or one proton. (b) They appear centered on each individual chemical shift value. The three signals between δ 21 and 23 are difficult to separate.

4.1 (a) First; (b) second; (c) first; (d) first.

4.2 Both materials absorb the ultraviolet light. In the case of glyceryl *p*-aminobenzoate, this is also nontoxic (since both glycerol and *p*-aminobenzoic acid occur naturally in our metabolism).

4.3 All except (c) and (d).

4.4 (a), (b) and (d) are equal, and larger than (c). The longest wavelength—lowest energy—is associated with (c).

4.5 (a) 1500; (b) 3.0.

4.6 16 368 and 5580.

4.7 (a) 6.863 mmol/l or 0.5353 g/l; (b) 2.8.

4.8 The concentrations are, respectively, 3.0 and 1.5×10^{-4} molar, and with molar mass 84 these are equivalent to 0.0252 and 0.0126 g/l.

4.9 III, 323 nm (253, homocyclic diene + 30, conjugation + 15, exocyclic bonds + 25, substituents). IV, 235 nm (215 acyclic diene + 20, substituents).

4.10 III, 237 nm (215, acyclic ketone + 10 + 15, substituents). IV, 296 nm (215, 6-ring cyclic ketone + 30, conjugation + 5, exocyclic bond + 12 + 17 + 17, substituents).

4.11 In water the chromophore is the benzene ring with the phenolic oxygen nonbonding electrons: in alkali a full negative charge is on the oxygen—greater electron 'availability'; therefore, longer wavelength absorption.

4.12 Nitrogen nonbonding electrons are less 'available' as a chromophore after protonation by acid.

4.13 (a) Pyrene; (b) fluoranthene; (c) acenaphthylene.

4.14 (a) 0.036 g/l; (b) 0.014 g/l.

4.15 Excitation 600 kJ mol^{-1}, emission 262 kJ mol^{-1}; therefore 'loss' of 338 kJ mol^{-1}.

4.16 Around 500 nm and 700 nm, respectively.

5.1 (a) m/z 77. CO (28 daltons). (b) m/z 64.2. CH$\equiv$CH (26 daltons).

5.2 $C_6H_{11}^+$ (cyclohexyl), $C_4H_7^+$ (butenyl), $C_3H_5^+$ (propenyl), $C_2H_5^+$ (ethyl), $C_2H_3^+$ (ethenyl, vinyl).

5.3 (a) Loss of one H atom (a radical), leaving a cation. (b) This is the carbon-13 isotope peak, at M + 1; it is a radical cation (see section 5.3).

5.4 m/z 15 is a cation formed by loss of C_5H_9 as an allylic radical; m/z 69 is a cation formed by the loss of a methyl radical.

5.5 (a) m/z 42; (b) m/z 58. Both are radical cations. The alkenes drift down the mass spectrometer, but are not detected, since they are uncharged.

5.6 (a) m/z 54; (b) m/z 68. Both are radical cations. The neutral alkene fragments drift undetected down the flight tube.

5.7 (a) m/z 178; (b) this also has a mass of 178 daltons, but it appears at m/z 89 because it is deflected more than the singly charged ion.

5.8 (a) m/z 179; (b) m/z 89.5.

5.9 (a) m/z 106, the molecular ion, radical cation. (b) m/z 91, benzyl/ tropylium ion, cation. (c) m/z 65 with a metastable ion at m/z 46.4, both cations. (d) m/z 77, phenyl, cation. (e) m/z 51 with a metastable ion at 33.8, both cations.

5.10 Two peaks in the abundance ratio 3:1 at m/z 92 and 94 (chlorine-35 and -37 isotopes), with two much smaller peaks at m/z 93 and 95 (carbon-13 isotope peaks).

5.11 (a) Two equally abundant ions at m/z 136 and 138, with less abundant ions (roughly 4.5 per cent of the abundance of the larger peaks) at m/z 137 and 139. (b) Three peaks with abundance ratio 1:2:1 at m/z 214, 216 and 218; the carbon isotope peaks will appear at m/z 215, 217 and 219.

5.12 m/z 308 (see section 5S.4).

5.13 m/z 56 from (M − 18 − propene, mass 42).

5.14 Both would give rise to the hydroxytropylium ion at m/z 107, but the alcohol spectrum will have a peak at M − 18 (loss of water) and the phenol spectrum peaks at M − 28 (loss of CO) and M − 29 (loss of CHO).

5.15 Butene, mass 56.

5.16 $PhCO^+$, m/z 105, going to m/z 77, then m/z 51, with metastable ions at m/z 56.5 and 33.8.

5.17 m/z 44 for all of them.

5.18 m/z 58 for all of them.

5.19 m/z 74 is the McLafferty ion formed from the methyl esters of longer-chain fatty acids; it is highly characteristic.

5.20 The amine with odd molecular mass must contain 1, 3, 5, etc., nitrogen atoms; the other amine must contain 2, 4, 6, etc., nitrogens. (They could be a toluidine, $CH_3C_6H_4NH_2$, and a phenylenediamine, $C_6H_4(NH_2)_2$.)

Appendix 1
Useful Data: Correlation Tables and Charts

Infared Spectroscopy

NMR Spectroscopy

Ultraviolet and Visible Spectroscopy

Mass Spectroscopy

Appendix 2
Acronyms in Spectroscopy

ACAC   Acetylacetonate (NMR)
ATR   Attenuated total reflectance (IR)

BB   Broad band (decoupling) (NMR)

CAT   Computer averaging of transients (NMR)
CD   Circular dichroism (UV/VIS)
CI   Chemical ionization (MS)
CIDNP Chemically induced dynamic nuclear
       polarization (NMR)
CIR   Cylindrical internal reflectance (IR)
COSY   Correlation spectroscopy (NMR)
CW   Continuous wave (NMR)

DPM   Dipivaloylmethane (NMR)
DR   Diffuse reflectance (IR)
DRIFTS DR infrared FT spectroscopy (IR)
DTGS   Deuteriated triglycine sulfate (IR)

EI   Electron impact or electron ionization (MS)
EPR   Electron paramagnetic resonance
ESCA   Electron spectroscopy for chemical analysis
ESR   Electron spin resonance

FAB   Fast atom bombardment (MS)
FD   Field desorption (MS)
FI   Field ionization
FID   Free induction decay (NMR)
FOD   Fluoro–octanedionato
FT   Fourier transform (IR, NMR, UV/VIS, MS)
FTIR   Fourier transform infrared
FTIR   Frustrated total internal reflectance
FTMS   Fourier transform mass spectrometry
FTNMR Fourier transform NMR

GC   Gas chromatography
GC–IR Gas chromatography infrared
GC–MS Gas chromatography mass spectrometry

HPLC	High performance liquid chromatography (formerly High pressure liquid chromatography)
ICR	Ion cyclotron resonance (MS)
INDOR	Internuclear double resonance (NMR)
INEPT	Insensitive nuclei enhanced by polarization transfer (NMR): now superseded by DEPT
IR	Infrared
IR	Internal reflectance
LASER	Light amplification by stimulated emission of radiation
LC	Liquid chromatography
LC–IR	Liquid chromatography infrared
LC–MS	Liquid chromatography mass spectrometry
LD	Laser desorption (MS)
LDA	Linear diode array (UV/VIS)
LIMA	Laser ionization mass analysis
LIMS	Laboratory information management system
LIMS	Laser ionization mass spectrometry
LRS	Laser Raman spectroscopy (IR)
MCT	Mercury cadmium telluride (IR)
MIR	Multiple internal reflectance (IR)
MRI	Magnetic resonance imaging (NMR)
MS	Mass spectrometry
NMR	Nuclear magnetic resonance
NOE	Nuclear Overhauser effect (NMR)
ORD	Optical rotatory dispersion (UV/VIS)
PDA	Photodiode array (UV/VIS)
RF	Radiofrequency
S/N	Signal–to–noise
SFC	Supercritical fluid chromatography
SIMS	Secondary ion mass spectrometry
TLC	Thin layer chromatography
TMS	Tetramethyl–4–silane (NMR)
TOF	Time–of–flight (MS)
UV/VIS	Ultraviolet/visible spectroscopy
XPS	X–ray photoelectron spectroscopy

Index